Microwave Radiometry

Construction, Measurement and Analysis

Motivation to write this book began after retiring from NOAA in 2005 where I studied satellite microwave radiometry for 30 years. Upon retirement I constructed ground - based Dicke radiometers at 4, 12 and near the 22 GHz water vapor line. This book begins by describing the construction, measurement and analysis of the radiometers as well as their surface and atmospheric observations. It concludes by discussing satellite radiometers.

Since around 1970, when NASA first launched satellite microwave radiometers to view earth, they have evolved from single channel to multi-channel high resolution sensors. Due to the cloud penetration of microwaves, they have become the primary instruments to globally measure surface and atmospheric parameters by different organizations. As described in this book, these parameters include temperature, rain and cloud liquid water, water vapor, snow cover and sea ice concentration.

Microwave Radiometry

Construction, Measurement and Analysis

Norman C. Grody

2nd Edition, June 2023

Preface

Beginning in the 1960's the National Aeronautic and Space Administration (NASA) solicited scientists and engineers to develop satellite infrared and microwave radiometers to remotely measure the atmospheric and surface properties of planets within our solar system. In 1971, after completing my PhD in Electrophysics where I studied plasmas I began my government career working at NASA at the time when they first began launching experimental microwave radiometers aboard satellites to view Earth from space. I then joined the National Oceanic and Atmospheric Administration (NOAA) in 1972 working on more advanced operational microwave radiometers. Subsequently, they were also developed for operational use by the Air Force and Navy. Throughout the years, these sensitive instruments measured the extremely low level thermally emitted natural radiation (~10^{-14} watts) emanating from the Earth's surface and atmosphere at frequencies as low as 1.4 GHz and up to 183 GHz. To detect such weak signals in the presence of instrumental noise, radiometers require very high gain receivers with large noise reduction. Such capability is analogous to being able to hear a person speak next to a roaring jet engine. This feat is accomplished using the lock-in amplifier or synchronous detection approach developed around 1944 by Dr. Robert Dicke [1] and described here at different levels of detail. Due to its simple design, this analog approach is used, which besides its historical significance is still used for ground-based and satellite radiometers.

The development of satellite-borne radiometers initiated a new field of study, Satellite Microwave Radiometry which is highly interdisciplinary, drawing on results from *electrical engineering, oceanography, geophysics, atmospheric and earth sciences*. At NOAA I was mainly involved in evaluating the performance of radiometers and developing algorithms to derive surface and atmospheric parameters from their measurements. These parameters such as temperature, water vapor, rain rate, snow cover and sea ice concentration are used by various organizations to monitor, analyze and forecast the global weather and climate. Upon retiring from NOAA in 2005, I first considered developing improved algorithms for these products. However, I found it more challenging to construct ground-based microwave radiometers using components available through the Internet such as from EBay. This interest was spurred by a 2003 article I read from the internet (http://www.qsl.net/oh2aue/dicke) by Michael Fletcher who describes a homebuilt 11 GHz Dicke radiometer and gave references. One reference, the September 1978 article in Sky and Telescope by Swenson and Yang [2] was most helpful since it gave detailed circuit diagrams. Another reference is the book "Microwave Radiometer Systems: Design & Analysis" by Neils Skou [3], which describes a 5, 17 and 34 GHz radiometer, each operating as a total power, Dicke and noise injection instrument. These, as well as all other references are listed in Chapter 15.

Although much of my work at NOAA was devoted to analyzing satellite data, my main interest following retirement was to construct microwave radiometers using components available from the Internet. This became a reality when I realized how inexpensive the radiometer components could be. In fact the most expensive part was the test equipment needed to measure the radiometer performance. Such test equipment included a Tektronix oscilloscope and a Hewlett Packard spectrum analyzer, sweep generator, power meter and a slotted line to measure the high emissivity calibration target. The test equipment was of high quality from the 1970's purchased from eBay. While some equipment was relatively expensive (*i.e.,* 0.01 to 22 GHz spectrum analyzer cost $400), I was able to construct Dicke radiometers operating at 4, 12, 20.5 and 22.2 GHz from parts costing less than $200, with the highest cost being at the two higher frequencies. Most costly were the front end components consisting of a waveguide to SMA (SubMiniature version-A) transition or adapter, followed by a Pin Diode Switch, Isolator and Low Noise Block (LNB) amplifier, which I could not construct. However, I constructed the lower frequency components which consist of a multiplexer, square law detector, AC amplifier and synchronous demodulator. These were built using standard circuits containing operational amplifiers whose components were connected together on a printed circuit board. Figure A5-2 in Appendix A5 shows an example of the construction used. Also, the antenna and radiometer cabinet were constructed using sheet metal.

I began this project by first building a total power radiometer that only required an LNB, square law detector and DC amplifier. However, as discussed here, I found that very small gain change in the LNB and DC amplifier due to incremental temperature variations made it impractical to build a drift free radiometer without very frequent calibration. I therefore decided to forego this approach in favor of the more stable Dicke radiometer design which uses the above mentioned lock-in amplifier approach of signal modulation followed by AC amplification and synchronous demodulation to reduce the effects due to electronic noise and gain change. In fact, because of this design, much concerns involving signal to noise ratio (SNR) or noise figure, stability and sensitivity proved to be unfounded by the instruments unattended long time performance.

The book describes the construction, measurement and analysis of radiometers operating at 4, 12, 20.5 and 22.2 GHz. Although all of the radiometers have similar design, the 4 GHz unit required a narrow band filter to suppress intermittent Radio Frequency Interference (*RFI*) from WiFi, radar and aircraft altimeters as primary examples. In fact, the 4 GHz radiometer without the filter routinely detected approaching aircraft. No such *RFI* was seen for the 12 GHz radiometer. However, unlike the 4 GHz radiometer, the 20.5 GHz radiometer detects intermittent *RFI* in the form of DC offsets that could not be filtered, while the 22.2 GHz radiometer shows no interference. Of greater importance is that these highest frequency radiometers have a peak response approaching the 22.235 GHz water vapor absorption line. As such, they have a higher sensitivity to water vapor as well as clouds and rain than the lower frequency units. Calibration of these radiometers therefore requires water vapor corrections when using clear sky measurements. In addition to calibration, algorithms are applied to derive water vapor and cloud liquid water from the combined 20.5 and 12 GHz radiometer measurements. Because of issues such as *RFI* the results are compared with that obtained using the 22.2 and 12 GHz radiometers as well.

Originally, I wrote this document to keep track of my progress in constructing these radiometers. As the project evolved I added separate chapters and appendices describing the calibration and analysis of the measurements. The appendices were added to provide more detailed analysis of the radiometer components and lock-in amplifier approach. It also includes some less understood issues pertaining to radiation transfer and emissivity modeling. To share my experience with others having similar interest I decided to write this material as a book. Since about half of this book involves microwave engineering, it contains numerous references to the book "Microwave Engineering" by David Pozar [4]. However, to be of interest to a wider audience I added three extensive chapters on satellite instruments and their measurements, much of which is based on my first-hand knowledge. These chapters outline the history of satellite radiometers and show unique examples of measurements that played a large part in establishing their role in earth remote sensing. *Examples also describe some unresolved observations that need to be better analyzed and understood even after many years of launching new satellite instruments.*

Content

Chapter **Page**

1. Introduction 5
2. Radiometers at 4 and 12 GHz 7
3. Radiometer Block Diagrams 9
4. Radiometer Calibration 14
   - 4.1 Near - Field Calibration Measurements 16
   - 4.2 LNB Gain Change with Temperature 19
   - 4.3 Sky Brightness Temperature 20
   - 4.4 Clear Sky Calibration Measurements 25
5. Gain Variation, Noise and *RFI* Mitigation 26
6. Detector Response 29
   - 6.1 Temperature Compensation 33
7. Radiometer Applications 36
   - 7.1 Surface Viewing Measurements 37
   - 7.2 Glass Door Reflectance 42
   - 7.3 Sky Viewing Measurements 43
8. Water Vapor Radiometers 48
   - 8.1 Radiometer at 20 GHz 48
   - 8.2 Radiometer at 22 GHz 53
   - 8.3 Cloud and Rain Measurements 56
   - 8.4 Tipping Curve Calibration Measurements 59
   - 8.5 Water Vapor and Cloud Water Simulations 68
   - 8.6 Water Vapor and Cloud Water Measurements 72
   - 8.7 Water Vapor Retrieval Accuracy 75
9. Transmission Measurements of Sand 78
   - 9.1 Emissivity Measurement of Quartz 82
10. Satellite - Microwave Radiometers 83
11. Satellite - Atmospheric Measurements 89
12. Satellite - Surface and Rainfall Measurements 99
    - 12.1 Desert Measurements 102
      - 12.1.1 Diurnal Effects 106
      - 12.1.2 Scattering Effects 108
    - 12.2 Snow Cover Measurements 112
    - 12.3 Sea Ice Concentration 116
13. Concluding Remarks 126
14. Acknowledgments 129
15. References 129

Content - Continued

Chapter **Page**

Appendices 133

A1. Total Power Radiometer..133

A2. Pyramidal Horn Antenna 137

A3. Temperature Controlled Fan...138

A4. 12 GHz Radiometer 139

A5. Synchronous Demodulator...139

A6. Temperature Compensated Detector...142

A7. AC Amplifier...144

A8. 4 GHz Radiometer Isolator..145

A9. 4 GHz Radiometer *RFI* Filter...146

A10. Glass Door Reflection and Transmission......................................146

A11. Glass Door Insertion Loss Measurements.....................................150

A12. Severe Storm Measurements...151

A13. Tipping Curve Analysis..152

A14. Dual Frequency 21 – 22 GHz Radiometer.......................................157

A15. Spectral Analysis of Dicke Radiometers...161

A16. Modeling Limitations of Random Media..163

A17. Near-Field Emissivity Measurements..166

A18. Nonlinear Calibration Equation...174

A19. Atmospheric Temperature Retrievals...180

A20. Temperature Weighting Function...190

A21. Atmospheric Wind Retrievals...192

List of Acronyms..199
List of Variables..200
Subject Index..201

1. Introduction

After working as a microwave engineer and completing my PhD in Electrophysics I began my government career at NASA and then at NOAA during the evolution of a new interdisciplinary field Satellite Microwave Radiometry, from its onset in 1971 until I retired in 2005. As with radar, radiometers are placed on the ground and aboard satellites to measure atmospheric and surface properties. However, unlike radar which is an active sensor, radiometers passively measures the natural thermal radiation emitted by the earth. At NOAA I spent much of my career evaluating the performance of satellite radiometers and applying them to measure atmospheric and surface parameters. Following my retirement I focused on ground-based radiometry, which is contained in the first nine chapters, ending with three chapters and appendices on satellite radiometry. Theses appendices comprise half of the book, and provide more detail on subjects in the main chapters.

The first eight chapters summarize my experience in constructing ground-based Dicke radiometers at 4, 12 and near the 22.235 GHz water vapor absorption line. It describes the calibration, measurements and analysis of the radiometers. In many ways these chapters are like a diary or notebook since much is written in the chronological order of findings. I must therefore apologize for some repetition and unevenness due to the chronology of the presentation. Furthermore, the book has been updated as more information became available through analysis and measurement. *The revision date is listed on top of the Preface.* This second edition expands upon the first edition by describing the algorithms and analyzing the satellite measurements of atmospheric temperature, winds and sea ice concentration. Such global measurements by satellites are used by different organizations for analysis and prediction of weather events at different space-time scales.

Unlike satellites, ground-based radiometers provide localized and even controlled experiments. As an example, Chapter 8 describes the use of the 12 and 22 GHz radiometers to measure cloud liquid water, precipitation and water vapor. These applications are best obtained outdoors, but for convenience they are also viewed through a glass patio door. Therefore, Appendix A10 analyzes the radiation transfer through the glass door. As mentioned above, these appendices comprise half of the book so as to not interrupt the main flow of the text. Appendices also include subjects requiring more detail. For example, Appendix A16 describes the problems pertaining to modeling random media while Appendix A18 derives the nonlinear calibration equation due to imperfect square law detectors. However, to keep the book small, I omitted derivations available in other books. Therefore, although not a textbook, it should be of interest to those wanting a more in-depth description of ground-based radiometers. Also discussed in detail are satellite radiometers, including their measurements and algorithms used to derive surface and atmospheric parameters.

Although microwave radiometry is less familiar than radar, it is well recognized in the Atmospheric and Earth Sciences due to its unique application in remote sensing of temperature, atmospheric absorption and surface emissivity. Also, while both sensors have their advantage, radiometers are simpler to construct than radar whose phase measurement or echo-return is best detected and analyzed digitally. As mentioned in Sections 7.3 and 12.2 such phase information is only available from radar and used to observe the ice stratification in snow and measure the fall velocity of rain and its vertical distribution. However, although it's easy to find Doppler radar modules from the internet for under $10 to measure speed, this is not true of radiometers. In fact, it took 10 years working on and off until arriving at the final design described in Chapter 3. While it was a great learning experience, the reason it took so long was that I initially tried building total power radiometers. It however became obvious after many failed attempts that the best way to minimize the slow drift and spurious instrumental noise was to use Dicke's method of synchronous modulation and demodulation, which is called synchronous detection. As described in Appendix A15, both of these effects are mitigated using modulation to shift the low frequency noise spectrum away from the signal. The following lists some of the issues I came across while constructing these radiometers.

The first issue involved the testing and evaluation of the radiometer performance. As described in Chapter 4, this ultimately required the construction of a high emissivity calibration target using high quality ferrite materials called Eccosorb manufactured by Emerson & Cuming. Calibration of the radiometer using the target revealed a number of issues. In particular, it became obvious that an isolator [4, pg 465] was needed to suppress the LNB local oscillator signal from being transmitted out and then reflected back into the radiometer due to a slight impedance mismatch of the target. The isolator also reduces reflected radiation due to impedance mismatch between other front end components such as the antenna, pin diode switch and LNB.

The second problem area had to do with the detector. This critical radiometer component is discussed extensively throughout this document. As explained in Chapter 6 and Appendix A6, rather than purchase a square law detector, I constructed one using a Schottky diode [4, pg 509]. However, after considering different diodes and reviewing the literature, I saw the advantage of using a matched pair of diodes in a temperature compensated circuit. Chapter 6 demonstrates the improved radiometer performance upon using such a temperature compensated detector.

The third problem involved the synchronous demodulator which was constructed from information acquired from the internet. As described in Appendix A5, this unit uses operational amplifiers configured as a difference amplifier, an integrator and DC amplifier to process the input signal in the analog domain. Most important was the judicious choice of a J177 p-channel Metal-Oxide Semiconductor Field Effect Transistor (MOSFET) whose gate input is energized by a clock generator to switch the polarity of the difference amplifier output during half the clock cycle. This same clock also energizes a pin diode switch [4, pg 514] to switch the radiometer input from antenna to reference load. I tried different transistors until arriving at the J177 which could handle the large drain signal when viewing space without affecting the switching action.

The fourth issue pertains to *RFI* observed with the 4 GHz radiometer. Besides obvious sources (*e.g.*, aircraft altimeters, radar and WiFi), which I tracked down using my spectrum analyzer, there were many other frequencies that turned up intermittently. I tried using different filters placed after the LNB intermediate frequency (*IF*) output to suppress the *RFI* until arriving at the final filter whose frequency response is shown in Appendix A9. However, as discussed in Chapter 8, *RFI* is not always narrow band and intermittent. This is shown for radiometers having frequencies of 20.5 and 21.1 GHz, whose interference could not be removed using filters. In contrast, my highest frequency radiometer centered at 22.2 GHz displays no significant *RFI* since it resides in a protected region. This as well as other comparisons are discussed in Chapter 8.

Many of the most important applications of microwave radiometers are obtained by placing them aboard satellites to view Earth globally. As discussed in Chapter 10, while satellite and ground-based instruments are similar, depending on the application different antennas are used to scan Earth and calibrate satellite instruments. I therefore included three extensive chapters at the end, including appendices, on satellite radiometers. In summary, Chapter 10 provides a history of satellite radiometers while Chapters 11 and 12 describe their atmospheric and surface applications. Atmospheric applications include water vapor and temperature measurements which serve as input to weather prediction models. Surface applications comprise of snow cover, surface wetness and rainfall for use in hydrology, and measurements of sea ice concentration and winds in oceanographic models. Besides synoptic scale measurements, Chapter 11 also highlights the use of satellite microwave radiometers to measure the globally averaged temperature trend for climate studies, an application requiring exceptionally high accuracy, independent of cloud cover.

2. Radiometers at 4 and 12 GHz

I began this project by first building a total power radiometer, which only requires the minimal configuration, *i.e.*, using DC amplification of the detected signal. A brief description of the instrument and its measurement is described in Appendix A1. It is shown that the slow drift in output due to very small temperature induced gain change makes it impractical to use this type of radiometer without frequent calibration (*i.e.*, less than a minute). I therefore decided to forego this simple approach in favor of the more stable Dicke radiometer which uses AC amplification and synchronous modulation-demodulation to eliminate drift and low frequency electronic noise [4, pg 669]. As such, this book only describes the construction of this unique type sensor, which provides accurate measurements using analog techniques. Although not used, the application of a digital or Software Designed Radio (SDR) approach is briefly mentioned in Section 6.1. Also not considered is any digital processing of the pre-detection signal for *RFI* mitigation and thermal noise detection as discussed in Chapter 5. Due to its complexity only the simpler analog design developed by Dicke in the 1940's is used here, which besides its historical significance is still used today for ground-based and satellite-borne radiometers. Unlike radar which requires digital processing of phase information to analyze the return echo, detection of the incoherent thermal radiation by radiometers can be obtained using analog techniques.

The first radiometers constructed operate at 4 and 12 GHz. In addition to being least costly, the measurements are least affected by atmospheric absorption. In comparison, my highest frequency radiometers operate at 20.5 and 22.2 GHz, which is near the 22.235 GHz water vapor absorption line. As discussed in Chapter 8, their cost is higher and requires larger atmospheric corrections. Figure 1 shows the 4 GHz radiometer where the right-most picture has the lid opened to show the labeled components. A larger picture of the components is shown in Figure 18. The bottom-right shows the radiometer output connector (Rad Out) as well as other outputs used for diagnostics. Also shown is the 15 dB gain horn antenna whose design equations are given in Appendix A2. The appendix shows that in order to increase the gain or beamwidth one needs to greatly increase the axial length of the horn antenna. The bottom-left of Figure 1 shows the power input connectors, the DC fine offset control and two additional diagnostic outputs. A digital voltmeter shown in the top-left picture displays the radiometer output and LNB temperature. For more reliable measurements (see Sections 4.2 and 6.1), a temperature regulated exhaust fan on the right side of the cabinet is used to convectively cool the LNB as well as other components using ambient temperature. The circuit is shown in Appendix A3 although I found that the fan is unnecessary after waiting about an hour for the instrument to reach thermal equilibrium. Similarly, Figure 2 shows the 12 GHz radiometer. It has a 19 dB gain horn antenna and the same input and output connections as the 4 GHz radiometer. The bottom two figures show the lid opened to view the components, which is enlarged in Figure A4 of Appendix A4. Most components are labeled so they can be compared with the block diagrams discussed in the next chapter.

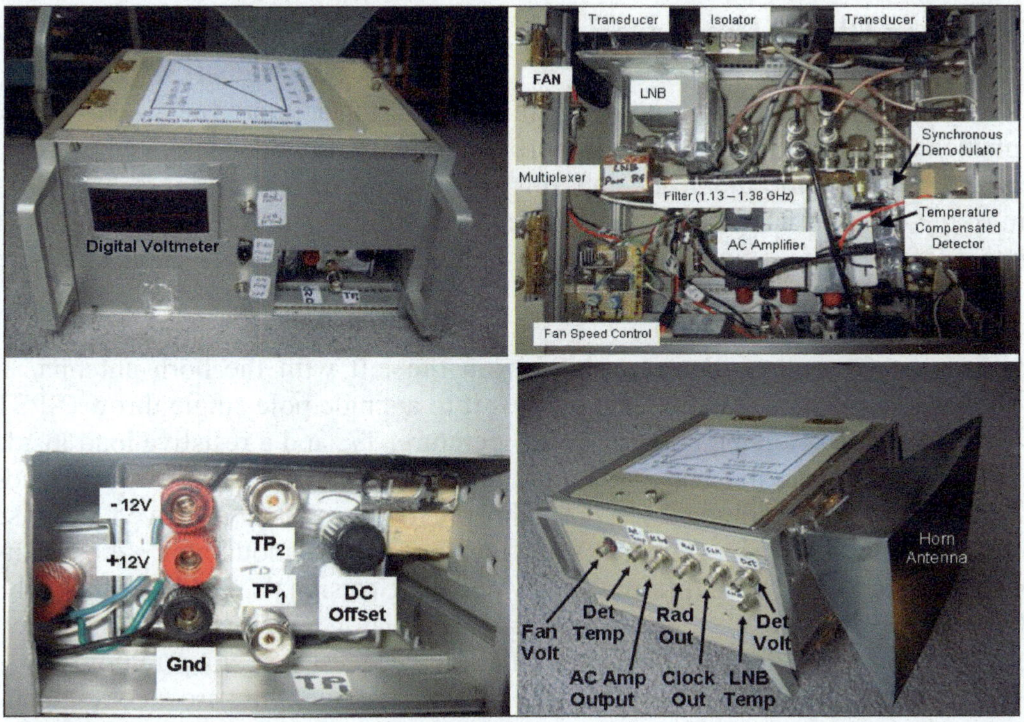

Figure 1 - Different views of the 4 GHz radiometer. The open lid seen on the top-right shows the components while Figure 18 shows a larger image. The bottom-left shows the 12 V power input, DC offset control and diagnostic outputs of the synchronous demodulator, TP_1, TP_2. The bottom-right shows the radiometer output (Rad Out) and other outputs. This view also shows the 15 dB gain horn antenna.

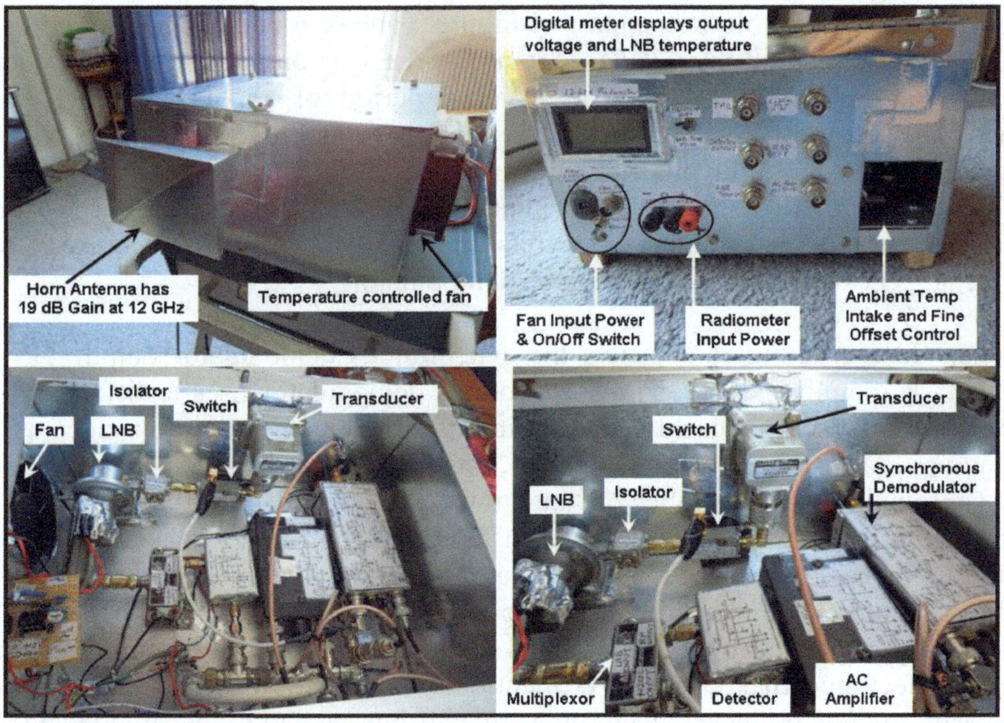

Figure 2 - Different views of the 12 GHz radiometer. The top-left shows the 19 dB gain pyramidal horn antenna. The back view on the top-right image shows the power input and access to the fine offset control. The top lid is opened in the two bottom figures to show the various components. Figure A4 in Appendix A4 shows a larger picture of the components. The top-right image also shows the temperature controlled exhaust fan used to mainly cool the LNB. Its controller circuit is described in Appendix A3.

3. Radiometer Block Diagrams

The radiometers are constructed using separate modules to facilitate the assembly and testing as well as enhancing the isolation. It is also important to use quality connectors when constructing the instrument, particularly for the front end microwave components. Figure 1 and 2 shows the placement of the different modules in the cabinets while Figures 3 and 4 show the block diagrams. A description of the block diagrams is given below, with additional details presented in later chapters and appendices.

The flow diagram for both radiometers begins on the left with the horn antenna. The antenna receives microwave thermal radiation and outputs it to a single pole single throw (SPST) pin diode switch that switches between the input noise temperature, T_A, and a resistive load inside the switch at temperature T_R. As such, the switched output is a modulated microwave signal whose envelope is a square wave with amplitude $T_R - T_A$. The pin diode switch is driven by a square wave generator (clock) that also drives the last stage, synchronous demodulator, whose output is a DC level proportional to the inverted difference between the two signals, *i.e.,* $T_A - T_R$. Although shown separately, the clock is part of the synchronous demodulator which is discussed below and more fully described in Appendix A5. This modulation reduces the slow radiometer drift, while Appendix A15 shows how synchronous demodulation removes low frequency electronic flicker noise by first shifting it to higher frequencies. A low-pass filter then blocks this frequency component while extracting the lower frequency desired synchronized signal. The filter or integrator stage also significantly reduces the third component of fluctuations that is due to broadband electronic thermal noise [4, pg 488]. In summary, *Dicke radiometers are designed so that thermal noise is the major noise source.*

At each output stage of the radiometer, different electronic components are used to process the waveforms illustrated in the block diagrams of Figure 3 and 4. As an example, Figure 5 shows the measured waveforms seen on my oscilloscope at the detector output and after the detector stage when the 4 GHz radiometer views space over a period of about two clock cycles or 10 milliseconds. The left-most Figure displays the very small amplitude modulated signal from the detector, which is about 2 millivolts (mv) with a 10 mv DC bias resulting from the LNB noise which is also detected. Also shown is the output from the next stage, high gain AC amplifier, which increases the modulated signal to 2 volts and removes the constant noise level. This AC amplified signal is next sent to the synchronous demodulator.

The right-most image in Figure 5 shows the synchronous demodulator output TP_2 prior to its integrator and DC amplifier stage. The synchronous demodulator uses a difference amplifier to switch the polarity of the modulated signal during half the clock cycle to produce a 2 volt signal with small jumps. As discussed in Appendix A5, this demodulated waveform shown in Figure 5 is then passed through a low pass filter or "integrator" having time constants of 0.1, 1.0 and 5.0 seconds to smooth the signal and produce a constant voltage. The next stages in the synchronous demodulator are then used to provide low level amplification and set the DC offset. As a result of the switching operation the synchronous demodulator output shown in the block diagrams of Figures 3 and 4 reduces the slow drift T_{Drift} and gain fluctuations of the amplifiers which contain low frequency electronic flicker noise. As mentioned above, the use of *modulation reduces the radiometer drift and low frequency fluctuations* by shifting them to higher frequencies while using a low-pass filter to attenuate them while passing the synchronized signal. The low-pass filter of the synchronous demodulator also reduces the Gaussian-distributed wideband instrumental noise T_N by temporal averaging. This noise reduction due to temporal averaging is contained in equation (7b) of Chapter 5 which is the fluctuation of radiometer output or Noise Equivalent Differential Temperature (NEΔT).

Before reaching the synchronous demodulator, the modulated microwave signal shown in the block diagrams of Figures 3 and 4 is first amplified by the LNB. The LNB has very high gain (60 dB or 10^6 power gain) with a very low noise figure of 0.2 dB or 14 *K* noise temperature for the C and Ku bands. This figure of merit is the noise introduced by the LNB to the signal. It is the difference in dB, or equivalent temperature, of the output noise power to that of a noise-free receiver having the same input noise signal. These microwave amplifiers were developed commercially beginning in the 1970's for direct broadcast satellite TV reception, and is the key radiometer component. Due to mass production, these amplifiers have become relatively inexpensive, particularly in these bands. Also, unlike more recent direct detect amplifiers using Monolithic Microwave Integrated Circuit (MMIC) technology, LNB's use the more traditional heterodyne principle [4, pg 650] to down convert the input microwave signal to a lower intermediate frequency (*IF*) between 1 and 2 GHz. These operations are shown in the block diagram of a generic LNB in Figure 6. To obtain the very low noise figure, the LNB uses very low noise Field Effect Transistors (FET's) configured as amplifiers, a local oscillator (*LO*) and mixer. The stable *LO* frequency is generated using a crystal Dielectric Resonator Oscillator (DRO) driven by an FET which has an *LO* stability within 1 MHz, which is more than adequate. As such, LNB's greatly simplify the radiometer construction by combining the amplifiers, *LO* and mixer into a single unit. Unfortunately, these commercially produced LNB's are only available at specific frequencies in C, Ku and Ka Band. While LNB's at other frequency bands can be obtained by custom order from the manufacturer, they are very expensive to acquire.

The bottom of Figure 6 shows the commercially available LNB's used in my 4 and 12 GHz radiometers, which have waveguide input and coax output for the *IF* signal. As discussed in Section 4.1, these radiometers have a NEΔT of about 0.3 *K* for a 0.1 second integration time. Figure 6 also indicates the availability of Ka band LNB's that generally cover the 18.3 to 20.2 GHz region while some go as high as 22.2 GHz. While I began this project by developing a 4 and 12 GHz radiometer, I later constructed a Dicke radiometer using the Norsat 9000C Ka band LNB shown in Figure 6, whose construction is shown in Figure 28. Its radiometer is measured in Chapter 8 to have a peak response at 20.5 GHz. As such, I also discuss this radiometer as well. The 9000C LNB has an *LO* at 19.25 GHz with a 1.3 dB noise figure or a noise temperature of 100 *K*. The radiometer therefore requires a larger integration time than 0.1 second to obtain the same 0.3 *K* NEΔT of the lower frequency units.

Chapter 8 describes the 20.5 GHz radiometer, its calibration and measurements of water vapor, cloud liquid water and precipitation. Furthermore, I also acquired a still higher frequency LNB from Norsat, the model 9000D. This highest frequency amplifier has a 20.25 GHz *LO* so its input is between 21.2 to 22.2 GHz, which is 1 GHz higher than the 9000C unit. Its upper frequency approaches the 22.235 GHz water vapor absorption line so it has the highest sensitivity to water vapor. Its measurements are compared with the 20.5 GHz radiometer measurements in Section 8.2. Also, as discussed in Section 8.6, the 22 and 12 GHz radiometers are mounted on tripods to view and scan space. Their outdoor measurements are combined to determine the water vapor and cloud liquid water variation. Lastly, Chapter 9 describes a simple outdoor radiometer experiment to measure the absorption of low loss materials such as desert sand.

Referring back to the block diagrams, the square wave envelope of the *IF* signal from the LNB is detected using matched pair Schottky diodes that are temperature compensated using the difference amplifier circuit in Appendix A6. As mentioned above, this very small envelope is amplified using an AC amplifier that also removes the detectors 10 mv DC bias signal resulting from the LNB noise. The AC amplifier circuit is described in Appendix A7. The radiometers also use a wide band isolator between the switch output and LNB input, where Appendix A8 shows the 4 GHz isolator response. Note that it blocks reflections due to an impedance mismatch between the switch and LNB at the input frequencies of 3.76 to 4.01 GHz. It also blocks the 5.15 GHz *LO* radiation generated within the C-Band LNB which is leaked out through the antenna. The Ku Band LNB has its *LO* at 10.75 GHz and therefore requires a different isolator. Without the isolator, this leaked radiation can produce errors using the near field calibration approach in Chapter 4 which places the target over the antenna. This leakage is then reflected back into the radiometer due to impedance mismatch between antenna and target. The isolator also prevents the *LO* from being reflected back into the radiometer by an impedance mismatch between the switch and LNB which produces an additional error.

While *RFI* is found to be insignificant at Ku-Band, there are many sources at C-Band (*e.g.,* WiFi, radar, aircraft altimeters). Chapter 5 describes the latest digital technique for mitigating interference, although analog filtering is used here and found to be sufficient. As such the 4 GHz block diagram contains an optimally chosen narrow band coaxial bandpass filter at the LNB output to suppress most radio frequency interference. Figure A9 of the Appendix shows the filter response to be centered at 1.26 GHz with a 250 MHz bandwidth and very high out of band rejection. As a result of the filter, the frequency band detected by the radiometer is reduced from the full LNB band of 3.30 - 4.30 GHz to 3.76 - 4.01 GHz. Since the antenna input power is given by the Nyquist equation $4kTB$, the four times reduction in bandwidth, B, requires a four times increase in gain. Since the 12 GHz radiometer does not require such a filter, it can use its full LNB bandwidth of 500 MHz so its gain would be half that of the 4 GHz radiometer. It turns out that differences between the LNB's and other components even further reduces the 12 GHz radiometer gain compared to that needed for the 4 GHz radiometer.

The synchronous demodulator shown in the block diagrams, and described in Appendix A5, contain an integrator having different time constants to reduce the random fluctuations due to instrumental noise T_N. More will be said about this noise in Chapter 5. The demodulator also contains a low gain DC amplifier and DC offset control set during calibration. Schematics of the synchronous demodulator, temperature compensated detector and AC amplifier are given in the appendices. Not shown in the block diagram is the transition whose waveguide input connects to the antenna while its SMA output connects to the pin diode switch. A second transition is placed between the LNB waveguide input and SMA isolator output. These components are shown in Figures 1 and 2 along with a multiplexer that is connected to the LNB *IF* output. The multiplexer or bias tee shown in Figure 7 (Left) uses filters to pass DC power to the LNB while extracting its *IF* signal. To improve the insertion loss and bandpass, a commercially built multiplexer shown in Figure 7 (Right) uses Surface Mount Technology (SMT) components on a Printed Circuit Board (PCB) for impedance matching and reducing parasitic capacitance.

Many of the radiometer components have adjustable parameters although they are fixed after calibration. These parameters include the gain of the AC amplifier, G_1, detector, G_d, and DC amplifier, G_2. As shown in the block diagrams Dicke radiometers measure the difference between the input radiation seen by the antenna and that of the warm reference temperature. As such, the smallest signal occurs when viewing earth while the largest signal, which is negative, occurs when viewing cold space as shown in Figure 5. To prevent saturation of the AC amplifier output, its gain must be set so that the radiometer output is less than -10 volts when viewing space. Also, as discussed in Section 6.1, the detector bias and input signal level must be such that the detector output volts varies linearly with input power for both small and large signals. Finally, for convenience, the DC offset in the synchronous demodulator is set so the radiometer output is near zero when viewing a high emissivity target at ambient temperature.

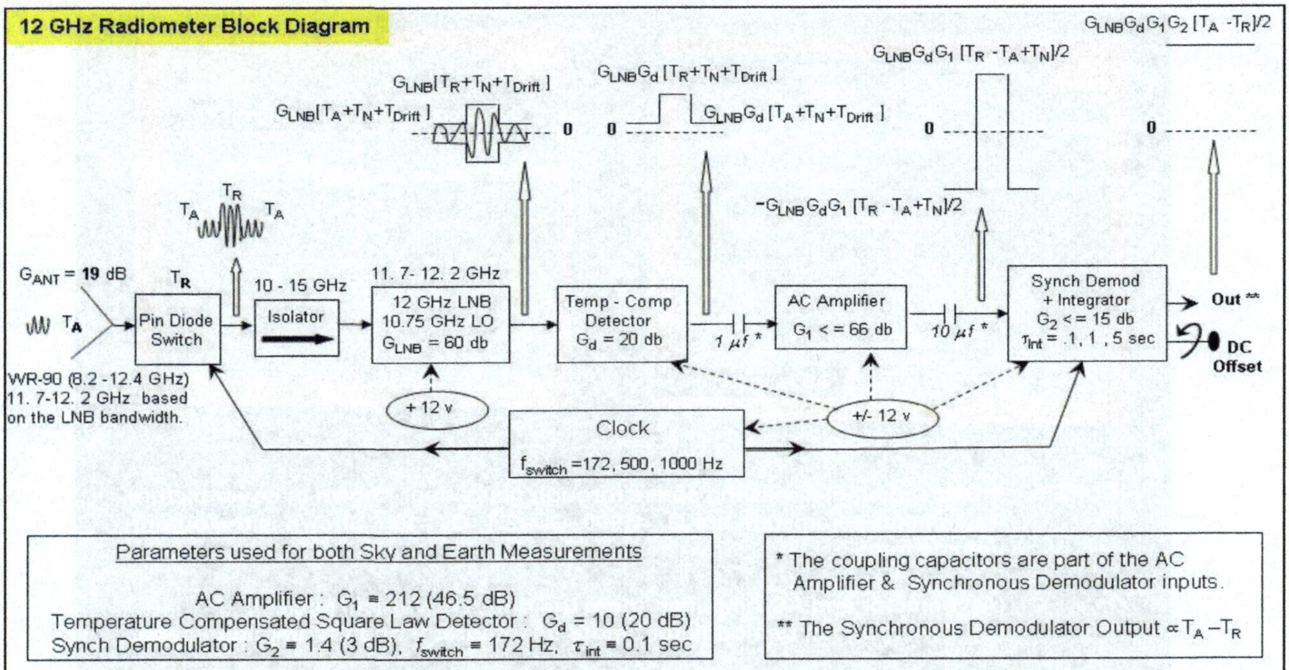

Figure 3 – Block diagram of 12 GHz radiometer showing different modules and waveforms. The detector uses a dual Schottky diode (HSMS-2825 biased at 25 µamp) connected to a fixed gain difference amplifier (G_d=10) with one diode used as reference to compensate for temperature change. The detector's input is 75 Ω to reduce the input to less than -25 dBm when viewing space. It's output then follows a square law response when viewing Earth and Space. The AC amplifier input capacitor removes the 150 mv detector bias. It's gain G1 is set to 212 (47 dB) so its output doesn't saturate when viewing space. Also, the synchronous demodulator's DC amplifier gain G2 is set to 1.4 (3 dB) so its output is less than -10 V when viewing space. The total radiometer gain is 60(LNB)+ 20(Det)+ 47(AC Amp)+ 3(Synch Demod) = 130 dB. *The synchronous demodulator contains the DC offset control.*

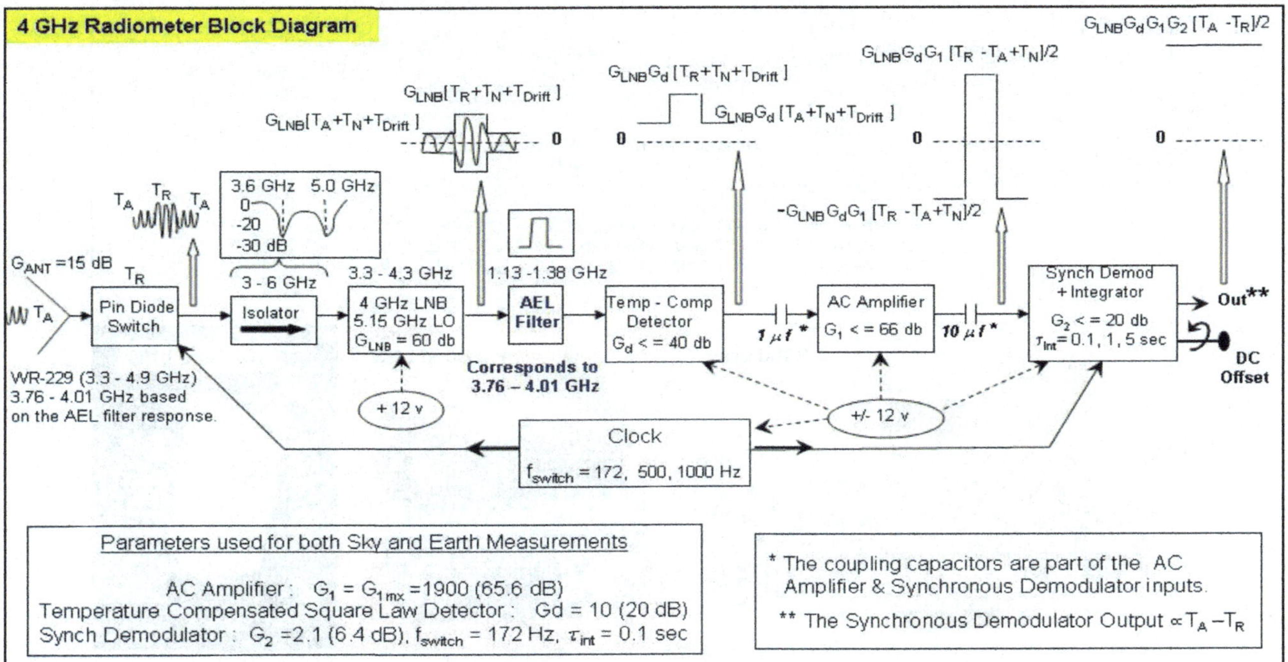

Figure 4 - Block diagram of 4 GHz radiometer showing different modules and waveforms. A narrow band (1.13–1.38 GHz) AEL filter reduces *RFI* by moving the input to a better protected band (3.76 - 4.01 GHz). The detector uses a dual Schottky diode (HSMS-282P biased at 25 microamp) connected to a difference amplifier (G_d=10) with one diode used as reference to compensate for temperature change. The detector's input is 47 Ω to reduce the input to less than -25 dBm when viewing space. It's output then follows a square law response when viewing Earth and Space. The AC amplifier input capacitor removes the 15 mv detector bias. It's gain G_1 is set to 1900 (66 dB) so its output doesn't saturate when viewing space. Also, the synchronous demodulator's DC amplifier gain is set to 2.1 so it's output is less than -10 volts when viewing space. The total radiometer gain is 66(LNB)+ 20(Det)+ 66(AC Amp) + 6.4(Synch Demod) = 158.4 dB. *The synchronous demodulator contains the DC offset control.*

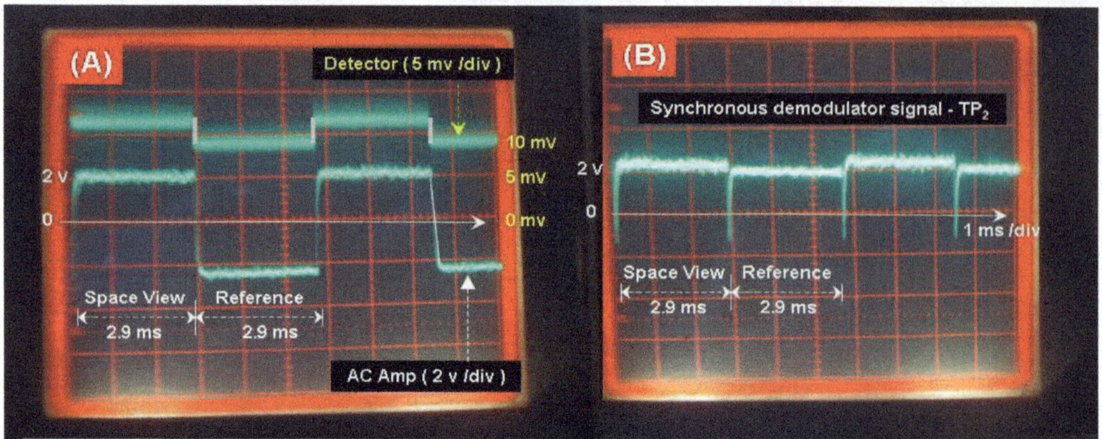

FIGURE 5 - 4 GHz radiometer viewing space through glass door shown in cover picture. Radiometer parameters are G_1 =1900, G_2 =2.1, G_d =10. Shown is the output voltage from

(A) Detector and AC Amplifier (note the different vertical scales).
(B) Synchronous Demodulator before its integrator and DC amplifier stage.

Radiometer Output

The integrator and amplifier stages of the synchronous demodulator smoothes, amplifies and inverts the signal TP_2 so the radiometer output becomes –4.9 volts.

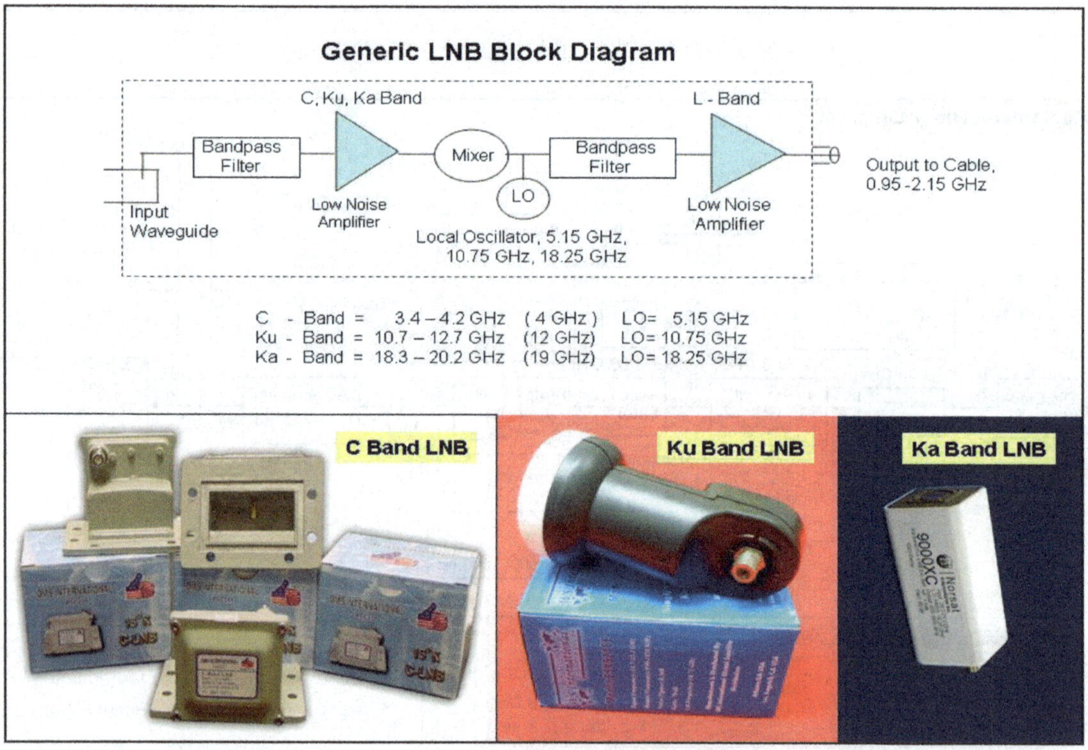

Figure 6 – The key radiometer component is the Low Noise Block (LNB) down converter whose generic block diagram is shown in the top. It uses the heterodyne principle to receive microwave radiation and down convert it to a lower Intermediate Frequency (*IF*) between 1 and 2 GHz using a Local Oscillator (*LO*) and mixer. The LNB has high gain (~ 60 dB) with a very low noise figure and was originally used at the feed point of satellite TV dish antennas. The bottom shows the LNB's used in my 4 GHz (C-Band), 12 GHz (Ku-Band) and 20 GHz (Ka-Band) radiometers which have waveguide input and coax output.

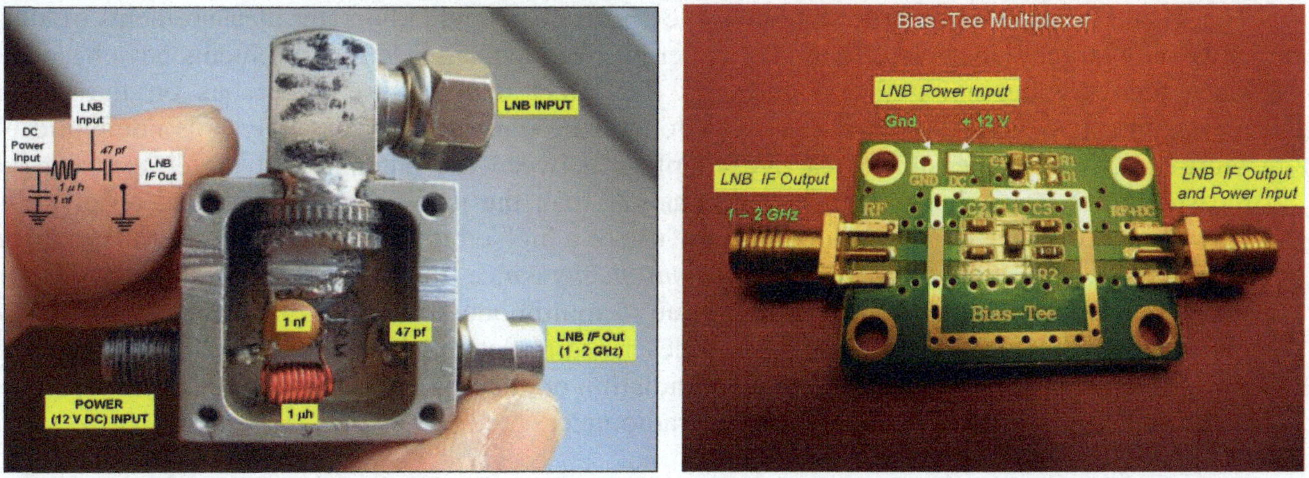

Figure 7 – (Left) Homemade multiplexer or bias-tee used in the 4 GHz radiometer shown in Figure 1. It uses lump circuit elements with the circuit drawn in the insert. The LNB Input connection provides the 12 volt DC power path to the LNB in addition to inputting the LNB's *IF* output signal between 1 and 2 GHz. The inductor and 1 nf capacitor block the *IF* signal from leaking to the power supply while the 47 pf capacitor blocks the DC power from exiting the LNB. (Right) Commercially made multiplexers use an optimally designed layout on a printed circuit board to provide wideband (0.01 to 6 GHz) operation with less than 1.2 dB insertion loss. This one was placed in a metal enclosure and used in the 22 GHz radiometer shown in Figure 32 on page 53.

4. Radiometer Calibration

Much time was initially spent on testing different radiometer configurations as well as different components before coming up with the final designs shown in Figures 3 and 4. I next focused on determining the radiometer's performance and its calibration. Table 1 lists the four different calibration procedures used most frequently by others as well as here. It also indicates the primary advantage and limitation of each technique. In addition to calibration, the radiometer frequency response is obtained using a signal generator connected to the antenna input of the pin diode switch. This procedure is described in Section 8.1 for the 20 GHz radiometer, whose frequency response is shown in Figure 29.

Table 1 - Comparison of Different Calibration Procedures.

Calibration Procedure	Advantage	Limitation
1- Near Field (Section 4.1)	Laboratory Measurement	Moderate Temperatures
2- LN_2 Target (Section 4.3)	Coldest Laboratory Procedure	Cryogenic Facility Needed
3- Clear Sky (Section 4.4)	Coldest Temperature	Need Atmospheric Correction
4-Tipping Curve (Sect. 8.4)	Most Accurate	Weak Absorbing Frequencies

The 1st and 2nd calibration procedure in Table 1 uses near-field radiometer measurements of a high emissivity target placed close to the horn aperture. While near-field measurements have been used since the very beginning (see Figure 41) the effect that a close target has on the antenna measurements is least discussed and understood since standard antenna analysis is not applicable. More specifically, the strong reactive coupling between the antenna and target for near-field measurements is much more difficult to analyze than the more common far-field observations. However, as explained in Appendix A17, the use of a high emissivity *low reflectivity target greatly simplifies the analysis of near-field measurements* since the object radiates at its physical temperature with minimal field distortion due to multimode effects. As such, the target acts as a black body, where its use for calibration is described in Section 4.1. Of additional importance, the low reflectivity target reduces any thermal radiation emitted by the LNB, as well as it's leaked *LO* radiation from being reflected back into the radiometer.

The coherent *LO* radiation generated within the LNB is leaked out of its waveguide port at about a -25 to -30 dBm level. This signal must be properly attenuated so it is not reflected back into the radiometer during near-field calibration measurements. In addition to more than 10 dB attenuation by the low reflectivity target, the *LO* signal is also reduced by the 30 dB isolator in Figures 3 and 4. Note that from Figure A8 in Appendix A8 the 4 GHz radiometer isolator is wideband to not only suppress the 5.15 GHz *LO* signal but also any reflections due to impedance mismatch at the LNB input frequencies between 3.76 and 4.01 GHz. *Without an isolator, the radiometer output due to the LO displays a rising and falling interference pattern whose amplitude increases as the target is brought closer to the antenna.* This test assures the quality of the high emissivity target as well as the isolator. Also, as a more sensitive test of the isolator, a high reflectivity low emissivity metal plate is brought closer to the antenna in a manner similar to the high emissivity calibration target. Incidentally, prior to the development of isolator's in the 1940's, Dicke used a simple innovative approach to minimize this interference problem (see Section 8.4 on page 62).

After considering different materials the best calibration target at 4 GHz is constructed using three sheets of ferrite material, called Eccosorb MCS by Emerson and Cuming, glued together and backed by an aluminum plate. Similarly, at 12 GHz only one sheet was needed to provide the same absorption. According to the manufacturer this highly absorbing target works by having its relative permittivity $\varepsilon/\varepsilon_0$ and relative permeability μ/μ_0 equal over wide frequencies so that its interface impedance $Z=\sqrt{\mu/\varepsilon}$ is that of free space $Z_0=\sqrt{\mu_0/\varepsilon_0}=377\ \Omega$. As such, there is little reflectivity at normal incidence since its reflectance is $|\Gamma|=|(Z-Z_0)/(Z+Z_0)|$ The reflectance is also related to the voltage standing wave ratio (VSWR) and emissivity ε_S by the equations,

$$|\Gamma|=\frac{\text{VSWR}-1}{\text{VSWR}+1} \quad \text{where} \quad \varepsilon_S=1-|\Gamma|^2 . \qquad (1)$$

Figure 8 shows the target in the upper right picture. As an example, the upper left photo shows the layout used to measure the targets VSWR [4, pg 59], which is the ratio of maximum to minimum voltage along a transmission line terminated by the target. The VSWR is measured over the 4 GHz radiometer frequency band of 3.76 to 4.01 GHz as defined by the bandpass filter shown in Figure 4. The setup consists of a coaxial slotted line [4, pg 69] terminated on one side by a coax to waveguide transition that the target rests on. The other side of the slotted line goes to an isolator, which is connected to a signal generator. Also shown is the VSWR meter connected to a crystal detector, which slides along the slotted line to measure the ratio of maximum to minimum voltage. A plot of the VSWR as a function of frequency is shown in the bottom-left image. Within the 4

GHz frequency band the VSWR varies between 1.10 and 1.05 so the calculated emissivity is better than 0.99. Using the same target, the emissivity for the 12 GHz radiometer is expected to be lower due to its higher frequency and wider bandwidth, which extends between 11.7 and 12.2 GHz due to the LNB (see Figure 3). Within these frequencies the VSWR is found to vary between 1.8 and 1.2 so the calculated emissivity is between 0.92 and 0.99, respectively. However, even though the 12 GHz radiometer does not contain a bandpass filter, the radiometer bandwidth is likely similar to the 4 GHz radiometer since the detectors bandwidth is 230 MHz as shown in Figure 29. As such, the emissivity should be similar to the 4 GHz radiometer. In addition to emissivity, its temperature must also be known to calibrate the radiometer. As shown in Figures 8 and 9, the targets temperature is measured by placing a LM34 thermocouple sensor within the Eccosorb.

4.1 Near - Field Calibration Measurements

Although the calibration target can be set at fixed temperatures using thermally controlled sources, I found it easier to use variable temperatures. As part of the near-field calibration procedure, the Eccosorb target was first cooled to 260 K by placing it in a freezer. The target is then placed over the antenna as shown in Figure 9, and the 12 GHz radiometer output voltage and target temperature are recorded as the target warms quickly and then slowly for an hour to reach room temperature. The target is next heated to 330 K using a hair dryer, and cools down to room temperature in an hour after again being placed over the antenna. Figure 10 shows the target temperature (top-right) and 12 GHz radiometer voltage (top-left) plotted as a function of time. To best measure the initially fast temperature change, the radiometer integration time is set to its minimum of 0.1 seconds. However, this does not reduce any errors due to thermal gradients within the target that is expected to be largest initially. For this reason, the measurements are analyzed after waiting about a minute after initially heating or cooling the Eccosorb target.

Least squares regression analysis of the target temperature and radiometer voltage is used to derive a calibration equation relating temperature to voltage. The target temperature and calibration equation is plotted in the top-left of Figure 10 as a function of time. For data storage and analysis, the radiometer output voltage as well as the target and LNB temperature is recorded by connecting a 12 bit analog to digital converter (DI-158U) by DATAQ Instruments to a laptop computer through its USB port. Software supplied by DATAQ enables one to monitor the time variation of up to four signals at once. For comparison, the bottom-left of Figure 10 shows the target temperature and derived calibration equation plotted against the radiometer voltage. The resulting calibration equation obtained from statistical regression analysis of the data is given by

$$T_b(K) = 297.25 + 32.25 \text{ V} . \qquad (2)$$

The equation is shown to have a standard error (SE) of 0.12 *K* and converts the radiometer measured voltage to temperature, which is referred to as brightness temperature, T_b. This linear relationship between brightness temperature and thermally emitted radiation, or voltage, results from the Rayleigh-Jeans law which is valid at microwave frequencies up to about 300 GHz. The calibration constant of 297.25 K in (2) is called the offset which depends on the synchronous demodulator DC amplifier offset control setting. Also, the voltage proportionality constant of 32.25 K/Volt is called the radiometric gain and depends on the radiometers total amplifier gain. In addition to the equation and standard error, Figure 10 (bottom-left) also lists the radiometer amplifier gains G_1, G_2 and G_d. A very similar calibration equation is obtained for the 4 GHz radiometer by appropriately setting its amplifier gains. In this way the 4 and 12 GHz radiometer measurements can be compared directly with one another without having to account for different calibration parameters (*i.e.*, offset and radiometric gain). Alternatively, once the final radiometer design is arrived at and fully calibrated, one can use the calibrated brightness temperatures instead of voltages when comparing radiometer measurements. However, since the radiometer amplifier gains have been changed slightly during the course of this investigation, I often reference the voltage measurements together with the calibration equations.

As mentioned above, the calibration equation has a standard error or NEΔT of only 0.12 *K* for the 0.1 second integration time. This low NEΔT is attributed to its linearity over the 70 *K* temperature range in addition to the LNB's low noise figure of 0.2 dB. While this temperature range is adequate when viewing earth, colder temperatures are preferred for calibration when viewing space since then the clear sky atmospheric radiation can approach the 2.7 K cosmic background. However, assuming the radiometer operates linearly, one could use equation (2) to extrapolate measurements to colder or warmer temperatures. For example, based on the calibration equation the voltage increases from -9.3 volts to 0 volts as T_b increases from 2.7 *K* to 297.2 *K*. Although these voltages are within the +/-10 volt saturation limit, calibration using colder temperatures are preferred to assure the linearity over the large dynamic range observed when viewing both earth and space. As listed in Table 1, for non routine operation one can obtain colder temperatures using a large target immersed in liquid nitrogen whose temperature is 77 *K*. However, the clear sky calibration method or tipping curve procedures listed in Table 1 provide the coldest temperatures. Of these two procedures the tipping curve approach is considered more accurate since it provides the most precise atmospheric corrections due to water vapor and oxygen absorption. Interestingly, as evidence of the radiometers linearity, Section 8.4 obtains nearly the same calibration equation for the 20 GHz radiometer using the outdoor tipping curve method as when applying this near-field laboratory procedure using the Eccosorb target.

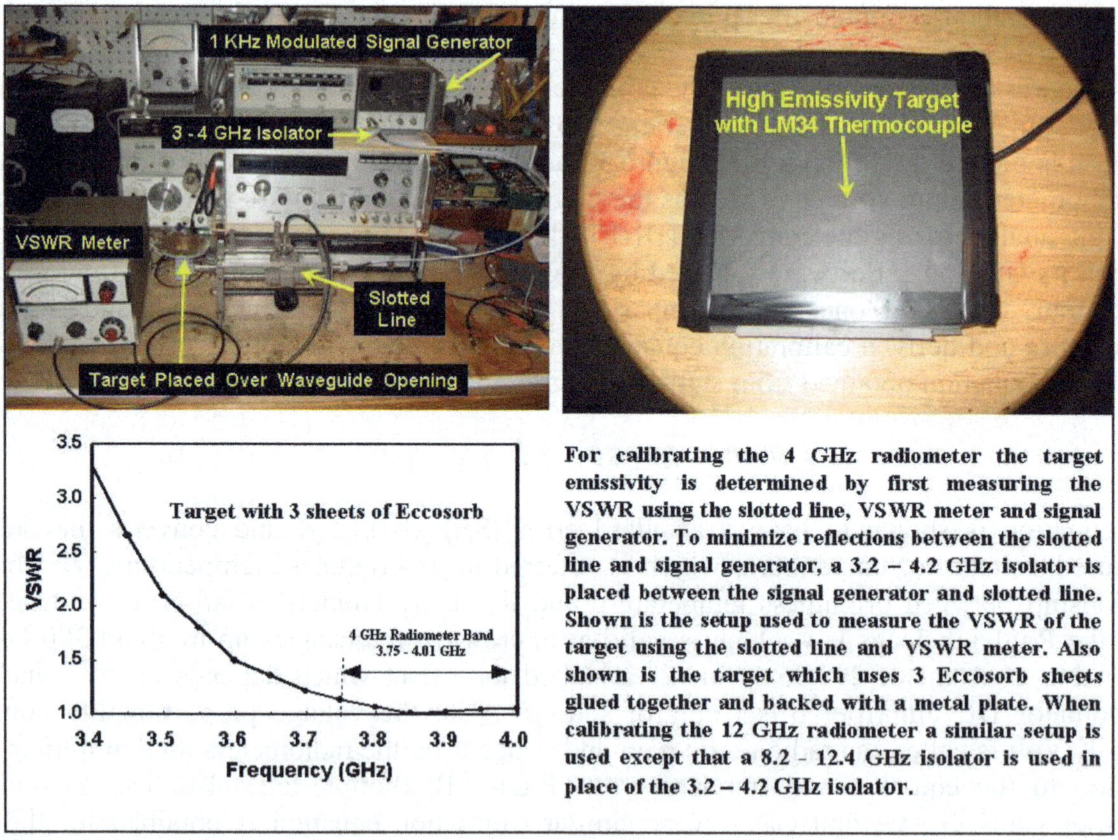

Figure 8 - The target shown in the top-right contains three Eccosorb sheets backed by an aluminum plate. Its temperature is measured using the LM34 thermocouple imbedded in the Eccosorb. The targets VSWR is measured using the slotted line setup in the top-left image. The bottom-left shows the VSWR plotted as a function of frequency, which is used to calculate emissivity based on equation (1).

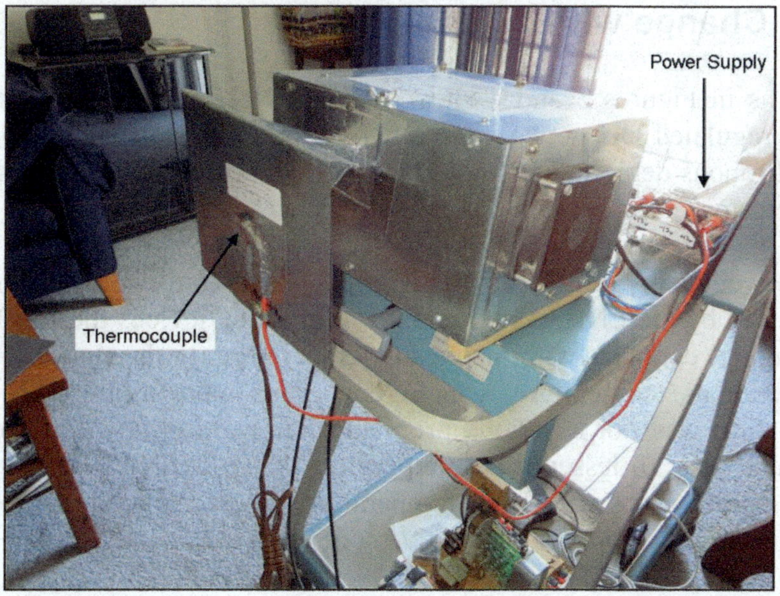

Figure 9 -Calibration of the 12 GHz radiometer by placing a high emissivity target over the antenna and measuring its temperature using the attached thermocouple.

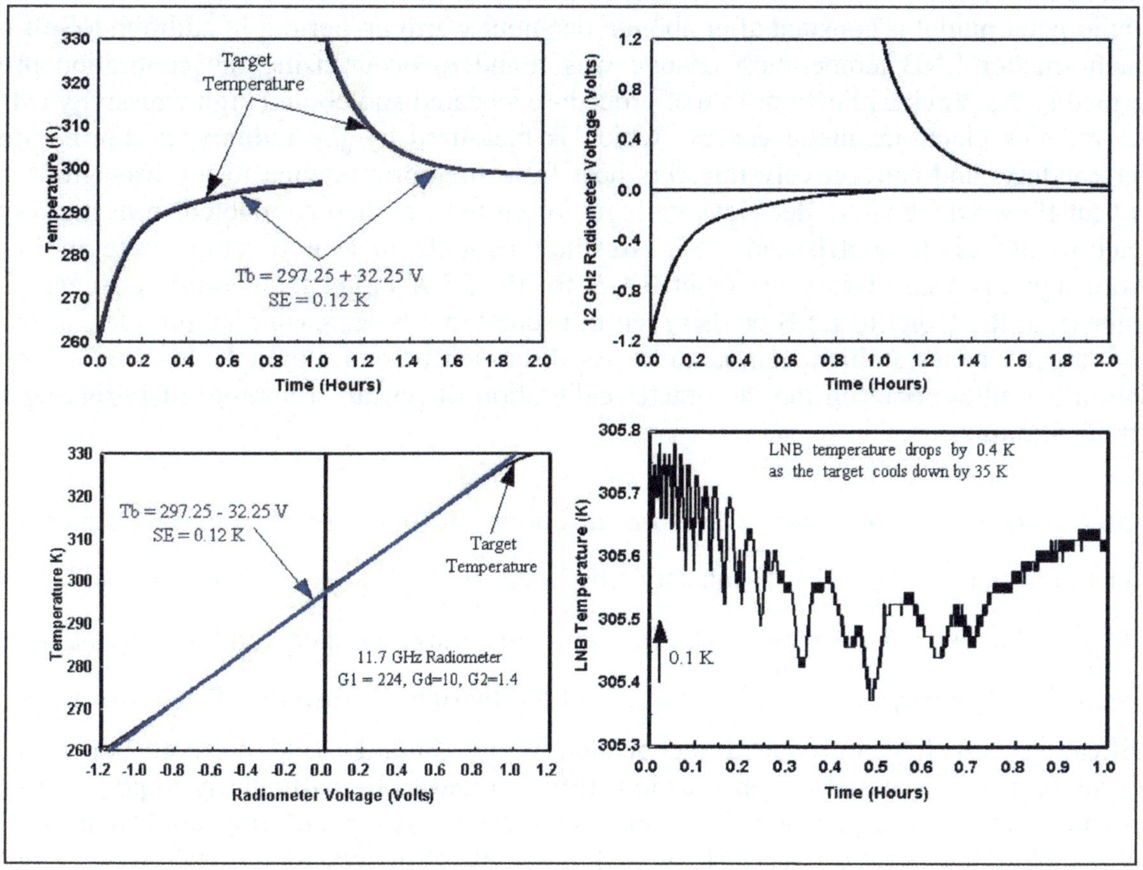

Figure 10 - Calibration of 12 GHz radiometer using a high emissivity target of varying temperature (see Figure 9). The top-right shows the output voltage due to changes in target temperature. Similarly, the top-left plots the target temperature and resulting T_b equation relating temperature to radiometer voltage on the same scale. These two temperatures are plotted against radiometer voltage in the bottom-left, which also shows the best-fit straight line using regression analysis. On a different topic, the bottom-right shows small oscillating temperature variations of the LNB due to thermal conduction during the 1st hour of calibration when the target was initially cooled. Similar features are found when heating the target during the 2nd hour of calibration

4.2 LNB Gain Change with Temperature

The block diagrams in Figures 3 and 4 shows the LNB being powered using a low noise non-switching 12 volt regulated DC power source, while the AC amplifier, temperature compensated detector and synchronous demodulator require +/- 12 volts. Actually, the voltage is +/- 12.5 volts since the LNB, detector, amplifier and demodulator power inputs contain 0.5 volt forward voltage diodes to prevent damage due to accidental polarity reversal. All voltages are obtained from a single regulated power supply which provides more than 200 milliwatts to power each radiometer, with most power used by the LNB. As with many solid state devices the LNB's performance (*i.e.*, Gain and Noise Figure) degrades with increasing temperature. In the case of the LNB, a portion of the DC power raises its temperature, which reduces its gain slightly. However, I found that temperature regulation was not a serious problem after waiting an hour until the LNB reaches thermal equilibrium. Also, to lessen the time period a 12 volt exhaust fan can be used as seen in Figure 2. The fan speed varies with LNB temperature, where Appendix A3 describes the control circuit using ambient temperature for convective cooling.

The largest self heating effect is found for the 20.5 GHz radiometer whose LNB draws 200 ma compared to the lower frequency units which operate at 100 ma. For this radiometer its LNB gain decreases with temperature, resulting in a radiometer voltage decrease of 60 mv/*K*. For comparison, the LNB gain is found to decrease by only 15 mv/*K* for the 12 GHz radiometer. As discussed in Section 4.3, this parameter can be used to adjust the radiometer calibration. However, the radiometer output is constant after about a one hour warm up period. In addition to self heating, a much smaller LNB temperature change was found to occur using the calibration procedure described in the previous Section. In that procedure a heated and cooled high emissivity calibration target radiates electromagnetic energy, which is measured by the radiometer. Furthermore, the target conducts and convectively transfers heat from the horn antenna to the waveguide adapter. This heat flow, which is an electrical analogue of current, is then conducted from the waveguide adapter to pin diode switch and LNB. Its thermocouple measured temperature in Figure 10 (bottom-right) is seen to vary by about 0.4 *K* for the 35 *K* target temperature change. This heat transfer from the target to LNB is also seen to oscillate with decreasing amplitude and frequency as the target reaches ambient temperature. As discussed below, although this small temperature variation has little effect on the radiometer calibration, it reveals a number of intriguing features worth mentioning.

In order to model the oscillatory waveform in Figure 10 (bottom-right), Fourier's parabolic heat conduction equation $\frac{\partial T}{\partial t} = \hat{\mu} \nabla^2 T$ must be modified as $\hat{\tau} \frac{\partial^2 T}{\partial t^2} + \frac{\partial T}{\partial t} = \hat{\mu} \nabla^2 T$. The additional term $\hat{\tau}\, \partial^2 T / \partial t^2$ is due to a lag $\hat{\tau}$ between heat flux and temperature gradient, and produces a *finite speed of thermal energy propagation* $\sqrt{\frac{\hat{\mu}}{\hat{\tau}}}$ where $\hat{\mu}$ is the thermal diffusivity of the horn is. Since this modification was first introduced independently by Cattaneo and Vernotte in 1956, many publications have discussed its ramification. It was found to be particularly important to use this "hyperbolic heat conduction equation" for situations involving suddenly applied heat flux. For example, Nordlund and Kassab [5] applied the hyperbolic equation to show how an abrupt temperature change propagates as a shock front that is reflected in an enclosure as oscillating-traveling waves with increasing wavelength downstream. As discussed next, this feature is similar to the LNB temperature waveform shown in Figure 10.

The LNB temperature waveform in the bottom-right of Figure 10 is due to thermal conduction and convection of the cooled (also heated) high emissivity calibration target to the radiometer components. Both the frequency and amplitude of LNB temperature is shown to diminish as the target temperature approaches thermal equilibrium. When modeled using the hyperbolic heat conduction equation, the temperature variation is equivalent to the voltage (*i.e.*, temperature) and current (*i.e.*, heat flow) propagating along a transmission line. In fact, the hyperbolic equation for one dimensional flow has the same form as the Telegrapher's equation [4, pg 50] whose distributed components of length Δz consist of a series resistor $R\Delta z$ and inductor $L\Delta z$ connected to ground through capacitance $C\Delta z$ and parallel conductance $G\Delta z$. The equivalent transmission line elements then become $RC = 1/\hat{\mu}$, $LC = \hat{\tau}/\hat{\mu}$ and $G = 0$. In the case of unbounded media the voltage propagates as traveling waves with phase velocity $\frac{1}{\sqrt{LC}} = \sqrt{\frac{\hat{\mu}}{\hat{\tau}}}$ that decays exponentially with the time constant $2L/R = 2\hat{\tau}$. Also, in the case of the horn antenna these waves are reflected at the front and back interface so the wavelength λ depends on the horn dimensions with frequency $\frac{1}{\lambda}\sqrt{\frac{\hat{\mu}}{\hat{\tau}}}$. However, as evident by the LNB temperature measurement in Figure 10, its frequency decreases as thermal equilibrium is established. As mentioned above this can result from an abrupt temperature change which propagates as a shock wave with increasing wavelength. Consequently the frequency $\frac{1}{\lambda}\sqrt{\frac{\hat{\mu}}{\hat{\tau}}}$ decreases in time. However, I must mention that while the capacitance is the product of specific heat and mass and physically represents heat storage or thermal inertia, the inductance has no such physical significance.

I found it interesting that the hyperbolic heat equation and supporting measurements in Figure 10 (bottom-right) *have not been reported or described elsewhere even though it was readily observed here*. It is also notable that none of the 0.4 K variations in LNB temperature is seen in the 12 GHz radiometer measurements on the top-right of the Figure. This is due to the fact that the 0.4 K temperature change only produces a maximum voltage variation of 6 mv (15 mv/*K* x 0.4*K*), which is negligible compared to the larger and more rapid change in radiometer voltage as the target warms up. Furthermore, this 0.4 K change in LNB temperature was reduced by more than half by insulating the target by covering the horn aperture with Styrofoam. Lastly, an even smaller LNB temperature variation was measured when replacing the calibration target with an aluminum plate. As discussed in Appendix A17, this is due to the smaller thermal radiation emitted by metal surfaces compared to the high emissivity Eccosorb target.

4.3 Sky Brightness Temperature

As indicated by (2), the linear brightness temperature equation has the form

$$T_b = I + S\,\mathrm{V} \qquad (3)$$

where I is the intercept or offset, S is the slope or radiometric gain and V is the output voltage. The slope is predominately due to the gain from the LNB and detector sensitivity. Both components are relatively constant although as mentioned in Section 4.2 the LNB gain decreases slightly with temperature T_{LNB}. Also, the detector sensitivity is also found to change with its temperature T_{DET} so the equation becomes $T_b = I + S\,\mathrm{V} + [c_1 T_{LNB} + c_2 T_{DET}]$, where the constants c_1 and c_2 are determined from measurements similar to that discussed in Section 4.1, by also including T_{LNB} and T_{DET} as linear predictors in the regression analysis of temperature data. Chapter 6 also explains that detectors can exhibit a small nonlinearity in its power law response. Appendix A18 shows the calibration equation (3) then becomes $T_b = I + S\,\mathrm{V} + \mu S^2(\mathrm{V} - \mathrm{V_C})(\mathrm{V_W} - \mathrm{V})$, where the nonlinear parameter μ is obtained using the laboratory procedures listed in Table 1. In general, however, the modifications of (3) due to LNB gain change and detector nonlinearity are small enough to be negligible for most ground-based and satellite radiometer measurements.

In the **1st** calibration procedure listed in Table 1 a high emissivity target having variable temperature was used to determine the offset and gain parameters and check the linearity over a small temperature range. A more detailed discussion of this near-field calibration method is given in Appendix A17. However, in general, the offset and gain parameters are determined using a two point calibration procedure whereby the radiometer views a high emissivity target having warm and very cold temperatures, T_W and T_C. As such, the calibration parameters become

$$S = \frac{T_W - T_C}{V_W - V_C} \quad \text{and} \quad I = \frac{T_C V_W - T_W V_C}{V_W - V_C} \tag{4}$$

where V_W and V_C are the measured radiometer voltages obtained when viewing the warm and cold targets. To obtain a large dynamic range, the **2nd** calibration approach uses ambient temperature for T_W while the coldest temperature T_C is obtained by immersing the target in liquid nitrogen. Liquid nitrogen (LN_2) is nearly transparent to microwaves and has a boiling point temperature of 77 *K*. As noted in Table 1 this requires a cryogenic facility which is its major limitation, and was therefore not used here. As a **3rd** calibration approach, the more traditional cold sky method is used where the warm target measurement defines the high voltage calibration point while cloud free sky measurements, $T_C = T_{SKY}$, is used to obtain the lowest voltage, $V_C = V_{SKY}$, calibration point. For these measurements the radiometers were placed on my upper patio deck with the antenna viewing the sky as shown in Figure 11 for the 4 GHz radiometer. At such low frequencies or at very high altitudes the atmospheric absorption is small so that T_{SKY} approximates the cosmic background radiation T_{CB} of 2.7 *K*. However, in general $T_{SKY} > T_{CB}$ so that one must accurately determine T_{SKY} based on measurements or model calculations. This is noted to be its major limitation in Table 1. As discussed below, this is rectified using the **4th** and last calibration procedure listed in the Table.

A lengthy analysis of Maxwell's equations is required to derive T_{SKY} for vertically stratified diffuse random media [6]. However, the same solution is obtained more directly using a less rigorous phenomenological theory of radiation transfer[1] which is based on heuristic arguments or experience. As described in Appendix A16, this theory is applicable for sparse discrete random media such as the Earth's atmosphere. In this application of the theory the solution of the radiation transfer equation for the downwelling sky brightness temperature is given by the two terms illustrated on the left of Figure 12. The resulting equation for T_{SKY} then becomes.

$$T_{SKY}(\theta) = \tau^{Sec\theta}\, T_{CB} + [1 - \tau^{Sec\theta}]\, T_M \tag{5}$$

where the 1st term is the attenuated cosmic radiation $\tau^{Sec\theta} T_{CB}$ and the 2nd is the atmospheric thermal emission $[1 - \tau^{Sec\theta}]\, T_M$. Both terms contain the atmospheric transmittance τ which is the fraction of power transmitted vertically through the atmosphere to the receiving antenna. The $Sec\theta$ exponent in τ results from the increased path length when viewed at zenith angle, θ, through a vertically stratified atmosphere. Ignored is any radiation bending by the gradient of refractive index. As discussed below, except when analyzing clear sky measurements, the atmospheric emission term in (5) is often the only term required.

---

[1] T_{SKY} is derived phenomenologically by following the path of photons as they interact with particles. Similar to Boltzmann's transport equation, radiation transfer theory combines the scattering, absorption and emission of photons into a single first order integro-differential equation for radiation intensity.

Based on the radiation transfer equation, the mean temperature in (5) can be written as

$$T_M = \frac{\int_0^{\infty} T(z) \frac{d\tau(z)^{\text{Sec}\theta}}{dz} dz}{\int_0^{\infty} \frac{d\tau(z)^{\text{Sec}\theta}}{dz} dz} \qquad \text{(6a)}$$

$$\text{with} \qquad \tau(z) = e^{-\alpha(z)} = e^{-\int_0^z \gamma(z')dz'} \;. \qquad \text{(6b)}$$

The analysis resulting in (6a) is contained in Chapter 6 of the book "Atmospheric remote sensing by microwave radiometry" edited by Michael Janssen [7]. Physically, T_M is the atmospheric temperature $T(z)$ vertically averaged over the weighting function $-d\tau(z)^{\text{Sec}\theta}/dz$ where $\tau(z)$ is the transmittance function at height z above the ground. As mentioned above, $\tau = \tau(z = \infty)$ in (5) is the transmittance function from the ground to the top of the atmosphere, which is called total transmittance. The transmittance function in turn depends on the absorption coefficient per unit length, $\gamma(z)$, or opacity function $\alpha(z)$ in (6b) where $\alpha(z = \infty) = \alpha$ is the total opacity. Equation (6b) results from the Beer-Lambert law, which is strictly valid for low density diffuse media, as with the radiation transfer equation. Also, as a result of the Born-Oppenheimer approximation the absorption coefficient is the sum of that due to water vapor, oxygen, and even liquid water drops in the case of clouds and rain so $\alpha = \alpha_{H2O} + \alpha_{O2} + \alpha_{Liq}$ and therefore $\tau = \tau_{H2O}\,\tau_{O2}\,\tau_{Liq}$. In summary, although less fundamental than Maxwell's field equations, the results obtained based on the Beer-Lambert law, Born-Oppenheimer approximation and Radiation Transfer theory are accurate when analyzing radiometer measurements of gaseous atmospheres including clouds and rain. However, as discussed in Appendix A16, while these approximate models are applicable for diffuse media, they are insufficient when analyzing denser inhomogeneous media such as snow cover, aged sea ice and desert surfaces.

Figure 12 on the Right shows the calculated clear atmosphere total transmittance $\tau_{H2O}\,\tau_{O2}$ and individual components due to oxygen τ_{O2} and water vapor τ_{H2O} as a function of frequency for an atmosphere having 25 mm of water vapor. The calculations use the latest absorption models of oxygen and water vapor as developed by Dr. Philip Rosenkranz and described in Chapter 2 of the above referenced book [7]. From (6b), the weighting function $-d\tau(z)^{\text{Sec}\theta}/dz$ is $\gamma(z)\text{Sec}\theta\;\tau(z)^{\text{Sec}\theta}$ and found to be dominated by $\gamma(z)\text{Sec}\theta$ in Appendix A13. It is also shown to be largest at the surface, decreasing exponentially to zero as z approaches infinity. As such, T_M is mainly proportional to surface temperature and can be approximated as $\Gamma_M T_S$. The proportionality factor Γ_M is obtained by calculating T_M using an historical sample of temperature soundings from radiosonde observations (RAOB's) and correlating T_M with T_S. The resulting Γ_M values are 0.93, 0.94 and 0.96 at 4, 12 and 22 GHz respectively, with a standard error of about 3 K for T_M. As discussed in Appendix A13, the angular variation of Γ_M and T_M is largest for the small transmittance obtained at frequencies near the 22.23 GHz water vapor line. However, even near this frequency the cloud-free atmosphere transmittance is greater than 0.8 so Appendix A13 shows T_M increasing by only 1.2 K as θ increases from 0 to 70 degrees. In comparison, the sky brightness temperature (5) increases by 80 K due to the change in emissivity $[1 - \tau^{\text{Sec}\theta}]$. Therefore, as a good approximation, the angular variation of T_M is neglected compared to atmospheric emissivity when deriving the tipping curve calibration technique in Appendix A13.

At low frequencies, or at very high altitudes $\tau \approx 1$ so that (5) results in $T_{SKY} = T_{CB} = 2.7\ K$. In general, however $\tau < 1$ so $T_{SKY} > T_{CB}$ and one must determine T_{SKY} from measurements or model calculations. Under cloud free conditions, microwave radiation is absorbed by oxygen and water vapor. The measured transmittance due to water vapor and oxygen was first obtained by radiometer measurements using the tipping curve procedure[2] described by Dicke in his 1946 landmark paper (*Phys. Rev.* 70, 340–348). This **4th** and final calibration procedure discussed here uses clear sky radiometer measurements at two or more elevation angles to obtain the transmittance and calibrate radiometers. It has also been used to measure the cosmic background radiation. As noted in Table 1, this 4th calibration method is considered to be the most accurate, but is generally limited to frequencies having moderate atmospheric absorption. Details of the approach and application are given in Appendix A13 and used in Chapter 8 to calibrate the 20.5 GHz radiometer and compare it with the 1st calibration procedure in Table 1.

The atmospheric emission by water vapor, oxygen and the attenuation of cosmic radiation is shown in Figure 13. This Figure plots the simulated cloud free sky brightness temperature for a zenith viewing radiometer as a function of frequency, between 3 and 22 GHz, with Total Precipitable Water (*TPW*) or vertically integrated water vapor as a parameter. The *TPW* is the depth of water vapor in millimeters that would be accumulated if the entire vapor was condensed in a vertical column. The simulations use (5) with the transmittance and mean temperature calculated using the absorption models contained in the previously referenced book [7]. Three different vertical soundings of temperature and water vapor were used to calculate T_{SKY}.

The nearly flat frequency response shown in Figure 13 in the absence of water vapor ($TPW = 0$ mm) is due to the 2.7 *K* cosmic radiation in addition to non-resonant oxygen absorption below 50 GHz due to molecular collisions (*i.e.*, collisional broadening) whose strong resonant lines occur between 50 and 70 GHz. Atmospheric emission due to oxygen is comparable to the cosmic radiation of 2.7 *K* so that the brightness temperature increases to 5.0 *K* for the 4 GHz radiometer and 6.6 *K* for the 20 GHz radiometer whose center frequency is at 20.5 GHz. A much larger increase occurs due to water vapor emission which increases T_{SKY} as a function of *TPW* with the peak response occurring at the center of the water vapor line at 22.235 GHz. Water vapor emission is shown to be negligible at 4 GHz, but increases the brightness temperature an additional 5 *K* at 12 GHz, 55 *K* at 20.5 GHz and up to 90 *K* at 22 GHz for tropical atmospheres where the water vapor amount can reach 60 mm or more. It is therefore important to include this radiation when performing calibration using sky measurements.

---

[2] The procedure is based on (5) which can be written as $\ln\left[\dfrac{T_M - T_{SKY}(\theta)}{T_M - T_{CB}}\right] = \mathrm{Sec}\,\theta \ln \tau$. Differentiating with respect to *Sec θ* we obtain $\dfrac{d \ln[T_M - T_{SKY}(\theta)]}{d\, Sec\,\theta} = \ln \tau = -\alpha$ where α = opacity.

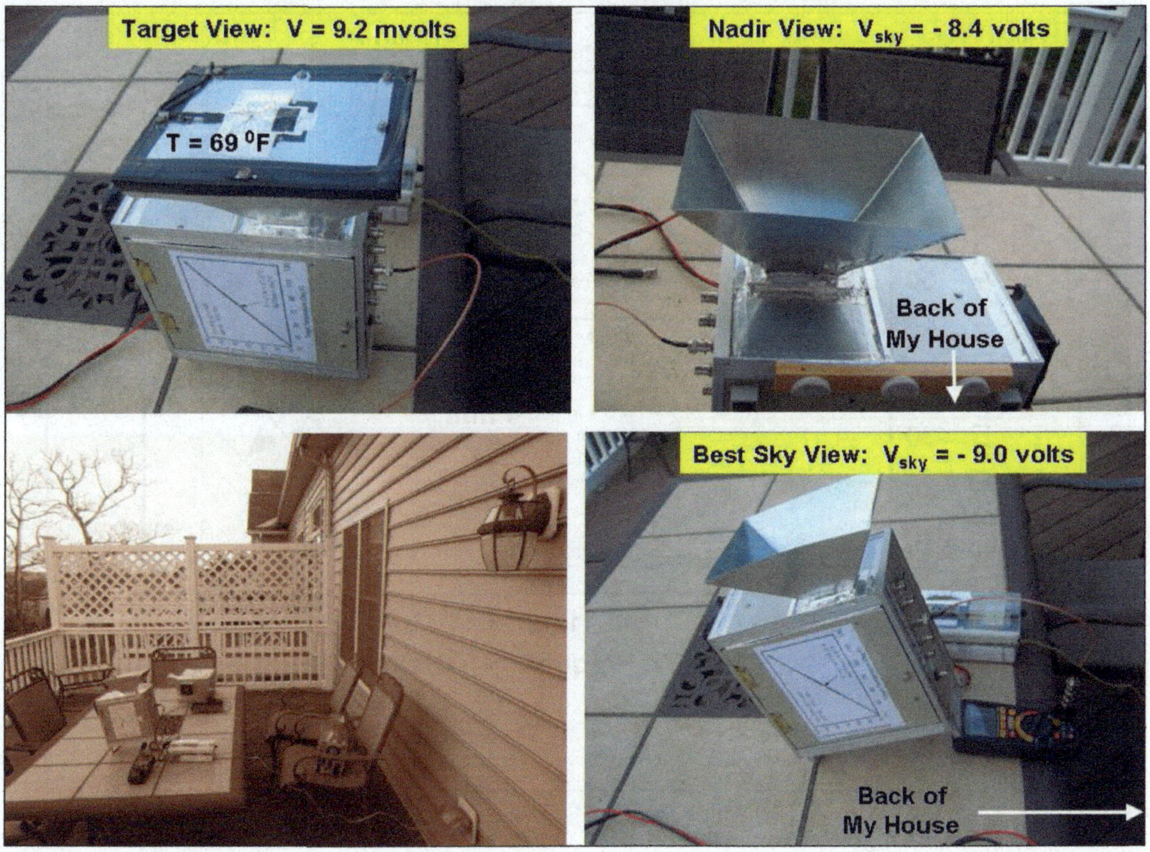

Figure 11 – The bottom-left shows the 4 GHz radiometer viewing clear skies. An enlarged picture is shown in the bottom-right. Calibration begins by placing the target over the antenna and setting the radiometer output to nearly 0 V by adjusting its fine offset (Top-Left). The top-right shows a sky measurement of -8.4 V at zenith viewing while the bottom-right shows it reduced to -9.0 V by rotating the antenna by 90^0 and directing the antenna away from the back of my house. The target temperature is 69^0 F so assuming the sky radiation is 2.7 *K* (-455 ^{0}F), the radiometer gain is (455+ 69)/9 = 58.22 ^{0}F/V or 32.34 K/volt. The radiometer parameters for these measurements are G_1 = 1930, G_2 = 1.7, G_d = 10.

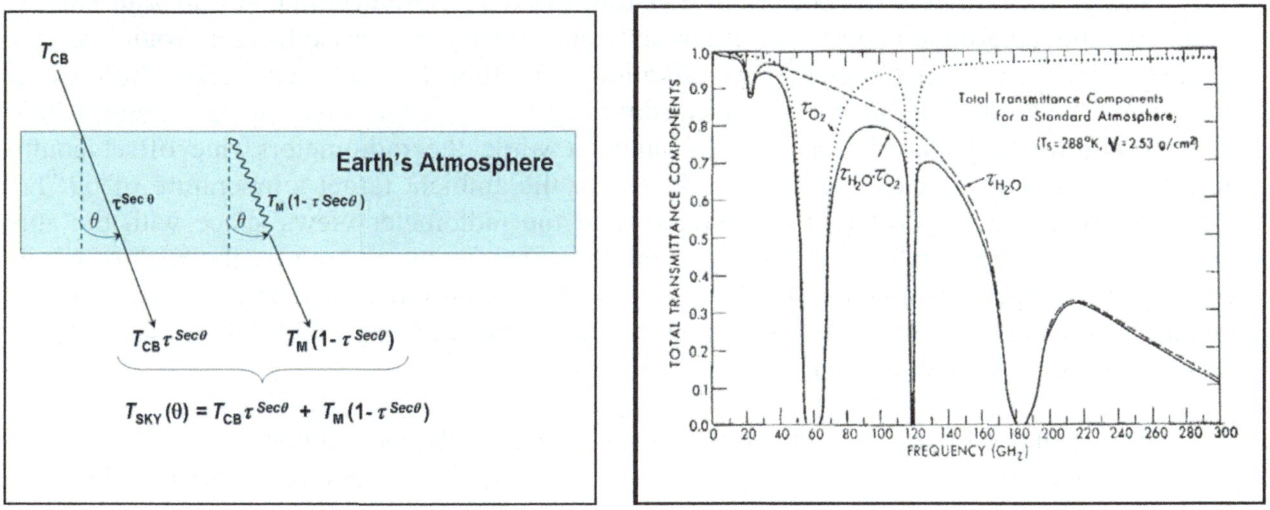

Figure 12 – (Left) Sky brightness temperature T_{SKY} received by a ground-based radiometer. The radiation consists of the cosmic background T_{CB} and atmosphere emitted radiation at mean temperature T_M with emissivity $1-\tau^{Sec\theta}$. The cosmic radiation is attenuated by the transmittance τ whose exponent Secθ accounts for the increased slant path at zenith angle, θ. The Right-most Figure shows the atmospheric transmittance due to oxygen (τ_{O2}) and water vapor (τ_{H2O}) as a function of frequency for a standard atmosphere having 25 mm of *TPW*. Also shown is the total transmittance $\tau = \tau_{O2}\,\tau_{H2O}$.

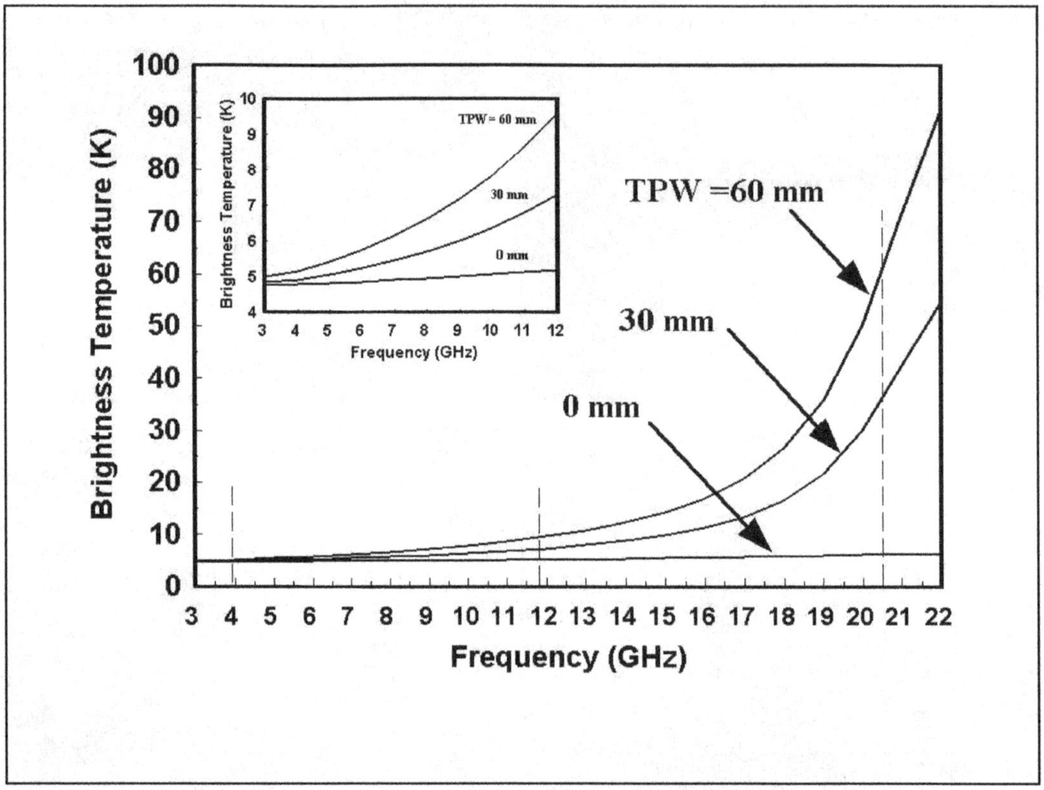

Figure 13 - Simulated brightness temperature for cloud free atmospheres as a function of frequency for different amounts of water vapor, *TPW*. The vertical lines identify the center frequencies of the 4, 12, 20 and 22 GHz radiometers which are at 3.9, 11.7, 20.5 and 22.2 GHz, respectively. The insert is an expanded plot for frequencies less than 12 GHz.

4.4 Clear Sky Calibration Measurements

Under cloud-free and dry atmospheric conditions it is relatively quick and easy to measure the offset and gain calibration parameters in equation (3). To assure accurate sky measurements, the antenna is moved around nadir to see if any antenna contributions arise from surrounding objects. Figure 11 (Bottom-Left) shows the sky radiation calibration measurements taken from my patio deck using the 4 GHz radiometer. The procedure begins as shown in the top-left figure, where the high emissivity target is placed over the antenna while the radiometers fine offset control is adjusted to produce a near zero output voltage for the ambient target temperature of 69 ^{0}F. The top-right figure shows the target removed so that the radiometer views space with the antenna directed at nadir. For this particular orientation, the antenna pattern seen by the back of my house contains the larger beamwidth of 28.7 degrees. This unfortunately results in maximum earth radiation reflected off the back of my house into the antenna Field Of View (FOV). The corresponding radiometer measurement is -8.4 volts.

The bottom right of Figure 11 shows the antenna rotated 90 degrees so that the smaller beamwidth of 25.5 degrees is directed toward my house. Furthermore, the antenna is slanted slightly away from the back of my house so the combined changes reduce the earth radiation scattered into the antenna FOV. As such, the radiometer measurement is decreased from -8.4 to -9.0 volts. Another means of reducing stray radiation is to surround the antenna by a large metal enclosure. Therefore, if the sky radiates at the cosmic background of 2.7 *K* (-455 ^{0}F), the gain is (455 + 69)/ 9 = 58.22 ^{0}F/volt or 32.34 K/volt. This gain decreases to 32.05 K/volt when including the 2 *K* (3.6 ^{0}F) increase in sky temperature due to oxygen emission. The radiometer parameters associated with these measurements is G_I= 1930, G_2= 1.7, G_d= 10 and τ_{Int} = 0.1 seconds. Nearly the same radiometric gain was obtained using the near-field calibration method in Table 1 and described before. Also, compared to sky measurements at 4 GHz, the narrow 16 degree beamwidth of the 12 GHz radiometer results in much smaller effects due to the surrounding earth radiation scattered into the antennas FOV. Calibration of the 12 GHz radiometer with nearly the same radiometric gain was obtained using the cold sky procedure and setting the instrument parameters as G_I = 286, G_2 =1.4 and G_d =10.

The next chapter describes the effect of gain variation for a Dicke and total power radiometer in addition to thermal noise effects. It also briefly discusses *RFI* detection and mitigation. This is followed by an experiment in Chapter 6 showing the sensitivity of a Schottky diode detector to temperature variation and its effect on the Dicke radiometer performance. The temperature effect is shown to be greatly pronounced when the radiometer views cold space. Therefore, a temperature compensated detector such as that shown in Figure A6 of Appendix A6 is used to reduce this effect. Unless specifically mentioned, all of the radiometers described here use the design shown in Figures 3 and 4, which use a temperature compensated detector.

5. Gain Variation, Noise and *RFI* Mitigation

Appendix A1 describes the total power radiometer I initially constructed. It is the simplest radiometer since it excludes the switch and synchronous demodulator shown in the Dicke radiometers of Figures 3 and 4. However, due to its high gain (120 dB), any slight temperature variation is shown to produce a large drift of the radiometer output. This drift can occur due to gain variations from self heating (see Section 4.2) as well as variations in the LNB noise Figure with temperature which varies in time and sets a practical limit on the maximum usable gain. As discussed in Chapter 3, the Dicke radiometers used here reduce the output drift due to such gain variations by switching the antenna input from the scene to a reference load and differencing the output signals using a synchronous demodulator. The requirement being that the gain is stable within the switching times, *i.e.*, $f_{switch}\,\tau_{Int} > 1$ where f_{switch} is the switching frequency and τ_{Int} is the integration time. A similar criteria is found in Appendix A15 using spectral analysis of the radiometer response, *i.e.*, $f_{switch} > f_c$ where f_c is the low-pass filter cutoff frequency. In this case, the filter suppresses the low frequency $1/f$ electronic flicker noise and wideband thermal noise. I found no problem using the lowest switching frequency and smallest integration time, *i.e.*, f_{switch} =173 Hz and τ_{Int} = 0.1 seconds. This is consistent with experiments that show the bulk of gain fluctuations is less than 1 Hz. In fact Dicke only used a 30 Hz switching frequency for his radiometers [1]. The following compares the noise characteristic of a total power and Dicke radiometer.

The random fluctuations in the output of a total power and Dicke radiometer is given by

$$\Delta T_b\big]_{Power} = \sqrt{\langle \Delta T_N^2 \rangle + \langle \Delta T_G^2 \rangle} = (T_A + T_N)\sqrt{\frac{1}{B\tau_{INT}} + \left(\frac{\Delta G(f)}{G}\right)^2} \tag{7a}$$

$$\Delta T_b\big]_{Dicke} = \sqrt{\langle \Delta T_N^2 \rangle + \langle \Delta T_R^2 \rangle + \langle \Delta T_G^2 \rangle} = \sqrt{\frac{(T_A + T_N)^2 + (T_R + T_N)^2}{B(\tau_{INT}/2)} + \left(\frac{\Delta G(f)}{G}\right)^2 (T_A - T_R)^2} \tag{7b}$$

with $$T_N = \frac{(1-\tau)T_0}{\tau} + \frac{T_{LNB}}{\tau} + \frac{1}{\tau G_{LNB}}\left[T_{Det} + \frac{1}{G_d}T_{Amp} + \frac{1}{G_d G_I}T_{Syn}\right]. \tag{7c}$$

Equations (7a) and (7b) define the minimum detectable change or NEΔT of each radiometer due to broadband thermal noise, as well as that due to amplifier gain variations ΔG(f), which has a low frequency 1/*f* electronic flicker noise power spectrum. The 1/*f* noise, also referred to as shot noise, depends on the total radiometer gain ($G = G_{LNB}\, G_d\, G_1 G_2$) shown in the block diagrams of Figures 3 and 4. On the other hand thermal noise is a function of the *IF* bandwidth B, the integration time τ_{int} of measurement and the total system noise temperature T_N which is referenced to the receiver's input after the antenna [4, pg 653]. Equation (7c) expresses T_N as a series of terms starting with the thermal radiation due to transmission loss τ by the connecting lines, switch and isolator, which when combined emit with emissivity (1 - τ) at average temperature T_0. This is followed by the LNB which radiates at its noise temperature T_{LNB} which is related to the receivers signal to noise ratio, SNR. The remaining terms in brackets is due to the detector, AC amplifier and synchronous demodulator. However, these contributions are generally small since (7c) shows that they are reduced by the high LNB gain.

As an example of radiometer noise measurements, Figure 14 shows the output voltage from a 12 GHz Dicke radiometer while viewing the high emissivity calibration target for 10 minutes at a temperature near the reference temperature T_R. The three top plots show the time series using an integration time τ_{int} of 5, 1 and 0.1 seconds. Each plot also shows the calculated standard deviation for each time series. To analyze the noise the bottom graph plots the standard deviation as a function of $\tau_{int}^{-1/2}$. Except for the intercept of 2 mv the straight line fit of the data has the same form as (7b) since $T_A \approx T_R$. As such, the random fluctuation of measurements corresponds to that of thermal noise. However, in addition to the standard deviation, which is the second moment of the probability distribution function (PDF), higher order moments of the distribution are generally needed to more precisely identify the Gaussian distribution corresponding to instrumental noise. As discussed next, statistical measurements of the PDF has also been used for *RFI* detection and mitigation.

Since Gaussian noise also results from the natural thermal emitted radiation by the earth's atmosphere and surface, any non-Gaussian noise such as *RFI* can be identified using higher order statistical moments of the measurements. A radiometer using this approach to mitigate interference is described by Dr. Chris Ruf [8]. The paper demonstrates the technique using an L-Band radiometer centered at 1412 MHz with a 24 MHz bandwidth. To identify and remove interference the bandwidth is subdivided into 8 high isolation filters of 3 MHz bandwidth. Histograms are then obtained using 128 bins for each subband, where only the subband measurements having Gaussian distributions are considered interference-free, with the others discarded. Also, the second moment of the Gaussian distributions or standard deviation of the interference-free measurements is proportional to the thermally emitted radiation by the radiometer and the thermal radiation seen by the antenna. Therefore, the standard deviation of the time series within each *RFI*-free subband provides an alternative means of measuring the brightness temperature without the use of square law detectors. This point will be expanded upon in Section 6.1.

While this digital radiometer is most agile, it is much more difficult to construct than the analog technique developed by Dicke which minimizes the radiometers low frequency drift, 1/*f* noise and gain variations. Also, as in the design of the 4 GHz Dicke radiometer, *IF* filters are often adequate to attenuate *RFI* signals at specific frequencies. Furthermore, filters based on temporal information as well as polarization can be used to remove *RFI* for a given channel. Lastly, relationships among different channels can also detect and mitigate *RFI*. As an example, Figure 51 in Section 8.7 shows the 20.5 and 22.2 GHz brightness temperature measurements highly correlated and related by the equation $T_b(20) = 2.33 + 0.623\ Tb(22)$ with a standard error of 0.82 K. While a smaller error would be better, it can still be used as a discriminate function to filter much of the *RFI* observed at

20.5 GHz (see Figure 33) since the 22.2 GHz channel resides in a frequency protected region allocated for radio astronomy by the FCC (Federal Communication Committee).

Returning back to (7a) and (7b), it is noted that the NEΔT from a total power radiometer is half that of a Dicke radiometer when assuming no gain variations and $T_A = T_R$, *i.e.*, $\Delta T_b]_{Dicke} = 2 \Delta T_b]_{Power}$. This factor of two larger noise for a Dicke radiometer results from the half integration time used when measuring T_A. In addition to the larger noise its sensitivity is half that of a total power radiometer since $V]_{Dicke} = kGB\,(T_A - T_R)/2$ while $V]_{Power} = kGB\,T_A$ where k is Boltzman's constant. Therefore, $(\Delta V/\Delta T_A)_{Dicke} = 1/2\,(\Delta V/\Delta T_A)_{Power}$. Unfortunately, the lower thermal noise and higher sensitivity of a total power radiometer is offset by its larger drift and noise due to gain variations compared to a Dicke radiometer. In fact, the gain variation of a Dicke radiometer can be virtually eliminated using a null balancing or noise injection radiometer which uses feedback to minimize the factor $T_A - T_R$ in (7b) by adding noise of sufficient amplitude to T_A. However, this requires additional electronics and was not used since the unbalanced Dicke radiometer was found to provide sufficiently high stability and noise reduction. This stems from the fact that the modulation provided by the Dicke switch reduces radiometer drift. Also, synchronous demodulation suppresses low frequency flicker noise by shifting the noise spectrum to a higher frequency as explained in Appendix A15. Incidentally, a general device that uses modulation followed by amplification and synchronous demodulation is called a phase lock amplifier or simply a "lock-in amplifier". It is not only used in radiometers, but also in other electronic instruments to recover a signal originally below the noise floor.

In summary, the reduced drift and gain variations of a Dicke radiometer greatly offsets the smaller thermal noise of total power radiometers. In fact, to operate properly a total power radiometer must be continuously calibrated less than a minute to minimize any drift and gain variations (see Figure A1-4 in Appendix A1). As described in Chapter 10, and shown in Figure 59, this is best obtained using satellite radiometers whose antenna is rapidly scanned to view earth, followed by unrestricted views of cold space and its on-board warm calibration target. In this case the drift and gain variations can be reduced sufficiently so ΔT_b in (7a) is dominated by thermal noise.

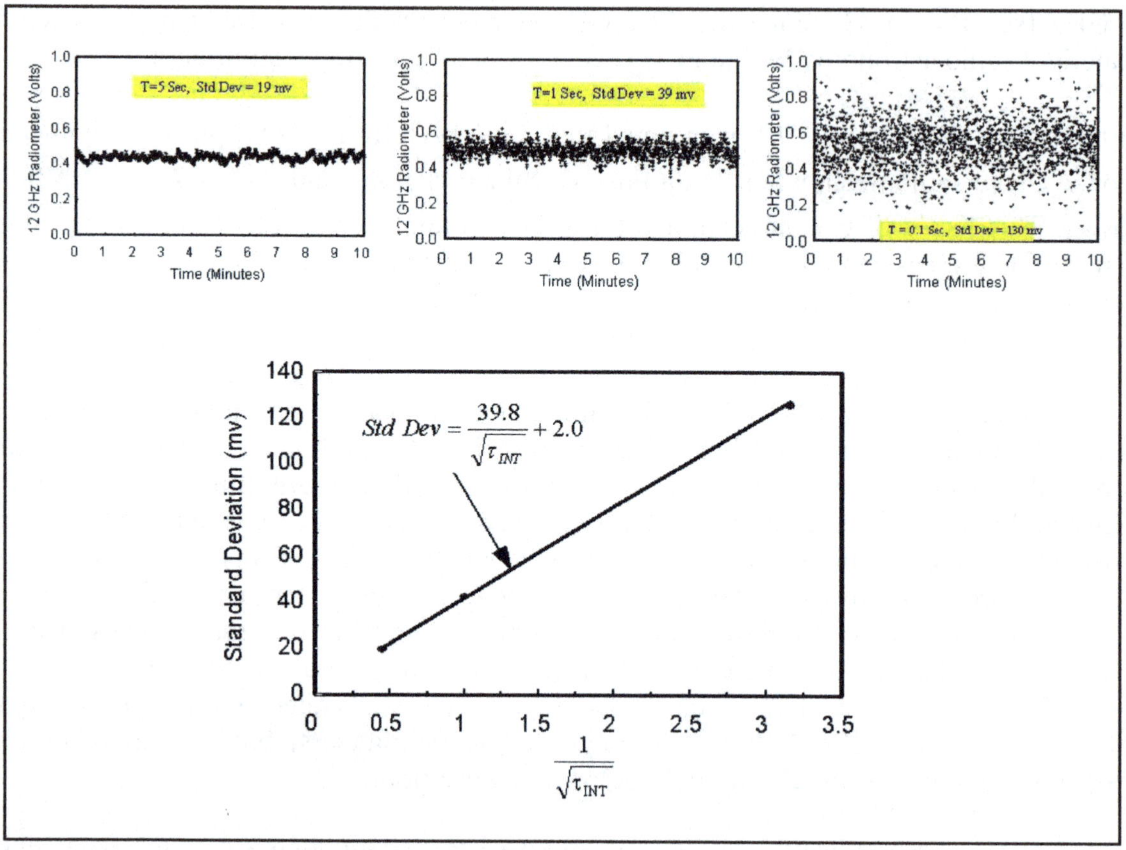

Figure 14 – The time series in the top plots show the 12 GHz Dicke radiometer measurements when viewing a calibration target at integration times τ_{INT} of 5.0, 1.0 and 0.1 second. Each plot also shows the calculated standard deviation from the time series. The bottom graph plots the standard deviation as a function of $1/\sqrt{\tau_{INT}}$ where the straight line fit corresponds to thermal noise fluctuations.

6. Detector Response

In addition to the LNB, the detector is critical in determining the radiometer performance. Not only does the detector define the linearity of the radiometer, it can also affect the overall gain, bandwidth and temperature dependence of the radiometer. These issues are discussed in this chapter as well as in Appendix A6 and A18.

A dual packaged Schottky diode series detector element, model HSMS 2862 was chosen because of its high sensitivity at microwave frequencies. When used with an optimally designed input impedance matching network the diode sensitivity, γ_{det}, is specified as 50 mv/μw at 1 GHz, decreasing to 25 mv/μw at its highest frequency of 6 GHz. This decreased sensitivity with frequency results from the diode's intrinsic capacitance which is specified to be about 1 pf. Furthermore, as described in Chapter 8 (see Figure 29), stray capacitance and inductance reduces the sensitivity and frequency response when the Schottky diode is connected in a more complete circuit that includes a difference amplifier for temperature compensation and a multiplexer to supply DC power to the LNB and extract its *IF* signal. As such, I found the detectors sensitivity and frequency response to be different for each unit constructed. Instead of obtaining a broadband detector with high sensitivity its response is found to peak anywhere between 1 and 2 GHz with sensitivity between 5 to 50 mv/μw due to these stray parameters. Such parasitic element effects are only minimized in commercially produced detectors, where the diode element and its associated components are placed on the well designed printed circuit board shown in Figure 16 to provide better impedance matching with a broadband response. This detector is mentioned in Appendix A6 and used in Appendix A14 to construct dual as well as a single frequency radiometer.

The following describes another important detector characteristic, namely its temperature dependence. However, before discussing the temperature effect, a brief overview is given of the detector's square law response. The bottom-left of Figure 15 shows the detector connected in a voltage doubler circuit along with pictures of the opened and enclosed unit used in an early 4 GHz radiometer. For illustration, the bottom-right of the Figure displays a representative detector response, showing a Log-Log plot of the input power against the output voltage. Note that to ensure a square law response, so the radiometer output is linearly proportional to power or brightness temperature, the diode input power should be kept below -15 dBm with its output voltage below 100 mv. This is generally not a problem since the detector signal is below 100 mv as shown in Figure 5. Other detector issues are discussed next as well as in Section 6.1.

While the detector's frequency response and input power level affect the overall gain and linearity the radiometers sensitivity is also found to be temperature dependent. The temperature dependence is due to the characteristic of Schottky diodes which although consisting of a metal semiconductor junction, its current-voltage characteristic can be described by the p-n junction diode equation [4, pg 510]. If the diodes series resistance is neglected, $I = I_S\left[\exp(V/\eta V_{th}) - 1\right]$ where I_S is the diodes saturation current and $V_{th} = kT/q$ is the thermal voltage with k being Boltzman's constant and q being the electron charge. Also, the quality factor η typically varies from 1 to 1.6 depending on the material and fabrication process. The temperature dependence, T, appears in the thermal voltage as well as the saturation current which is proportional to $T^2\exp(-V_b/V_{th})$ where V_b is the surface barrier potential and can result in poor performance if not accounted for. Also, as derived in Appendix A18, when the exponential quantity is expanded in a series, the diodes response extends beyond 1st order in power so equation (3) becomes $T_b = I + SV + \mu S^2(V - V_c)(V_W - V)$ where the μ parameter results from the 2nd order contribution.

In contrast to Schottky diodes, the highly doped tunnel diodes consist of a p-n junction, and are much less temperature dependent since the detectors operate at zero-bias where the tunnel current dominates the diode operation. This current is characterized by the peak voltage V_p and peak current I_p which is insensitive to temperature and given by $I_t = I_p(V/V_p)\exp(1 - V/V_p)$. The detector operates at voltages up to the peak voltage so its performance hardly depends on temperature excursions. Furthermore, some tunnel diode detectors operate as backward diodes to increase their frequency response and sensitivity. More will be said in the next section about the use of tunnel diode detectors.

Temperature effect on the radiometer output is obtained using the equations in the block diagrams of Figures 3 and 4, *i.e.*,

$$V)_{Dicke} = \tfrac{1}{2}kG'B\,\gamma_{det}\,[T_A - T_R] \qquad (8a)$$

$$\frac{\Delta V)_{Dicke}}{\Delta T_{det}} = \tfrac{1}{2}kG'B\,[T_A - T_R]\frac{\Delta\gamma_{det}}{\Delta T_{det}} \qquad (8b)$$

$$\text{where} \quad G' = G_{LNB}\,G_{Amp}\,G_{Syn}\,. \qquad (8c)$$

From (8b), the temperature variation of a Dicke radiometer is shown to occur predominately from changes in the detector sensitivity γ_{det} . Furthermore, the temperature variation occurs primarily when viewing cold space since then $T_A << T_R$. Conversely, the temperature variation is much less when viewing warmer ground temperatures since then T_A is comparable to T_R. To demonstrate these features, experiments were performed using an earlier 4 GHz radiometer which uses the uncompensated detector shown in Figure 15. As shown on the books cover page, the radiometer is mounted on top of the 12 GHz radiometer and placed on a movable cart that is elevated so that the antenna views space through my basement glass patio door. During the experiment the detector temperature is heated by about 25 ^{0}F for a short time using a small incandescent light. For the first 90 minutes the high emissivity target is placed over the antenna as in Figure 9, and then removed for about 60 minutes so that the antenna views space.

The two top plots in Figure 17 show the radiometer voltage (Left) and detector temperature (Right) as a function of time. Note the abrupt decrease in radiometer voltage when the detector is heated while the antenna views space at the time of 125 minutes. No change is seen at the earlier time of 75 minutes when the antenna views the high emissivity target. The bottom left Figure shows the measurements viewing space on expanded scales, while the bottom right Figure shows the corresponding voltage plotted against detector temperature. Using least squares regression analysis, a linear equation is obtained between the voltage and detector temperature that has a slope of -0.1 Volts / ^{0}F. Also, using the detector temperature as input, the voltage derived from this equation is shown to accurately reproduce the true measurements in the bottom left figure. What is also noteworthy is how well the radiometer voltage follows the detector temperature even though the temperature sensor is mounted outside of the detector case as shown in Figure 15 (Top-Right). In fact, much of the difference between the two measurements seen in the bottom-right Figure is due to the time lag between the outside temperature change and detector response. A similar example of this time lag between the measured temperature and detector temperature response will be displayed in the next section when testing a temperature compensated detector.

The results shown in Figure 17 are in qualitative agreement with equation (8b), which gives the slope of the radiometer output voltage versus detector temperature, *i.e.*,

$$\frac{\Delta V)_{Dicke}}{\Delta T_{det}} = \tfrac{1}{2} k G' B [T_A - T_R] \frac{\Delta \gamma_{det}}{\Delta T_{det}} \quad .$$

Note that as in the measurements, the slope is negative when viewing space since then $T_A < T_R$, and becomes negligibly small when viewing the high emissivity target since then $T_A \cong T_R$. It should be mentioned that I initially observed the detector problem by viewing space while warming the detector with my fingers as shown in Figure 15. I originally thought that the temperature stability issue was due to the LNB amplifier whose gain can be temperature dependent. However, after heating the LNB in the same manner as the detector and finding no effect, I concluded that the issue was solely due to the detector. This was different than that found using the 20 GHz radiometer. As mentioned previously in Section 4.2, the LNB used for this higher frequency radiometer results in a noticeable gain variation that decreases with temperature due to self heating, where this effect can be reduced using passive cooling as indicated in the section.

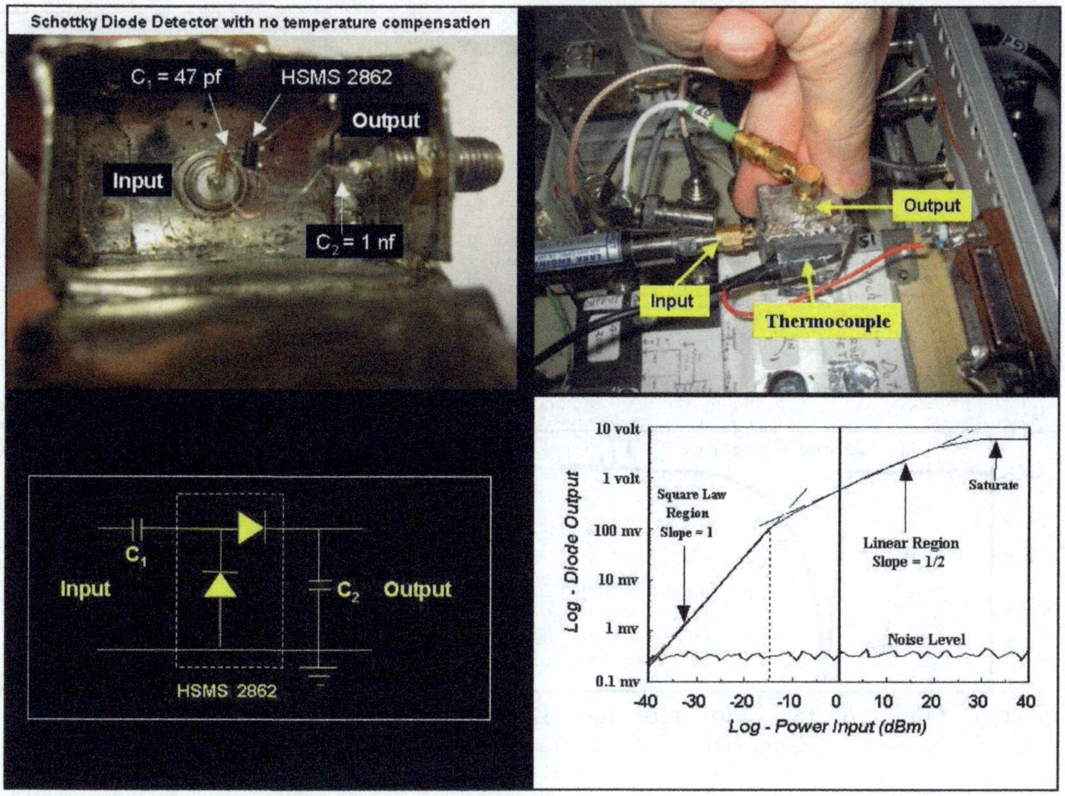

Figure 15 - The HSMS 2862 Schottky diode chip contains two elements connected together to double the sensitivity. The detector circuit is shown in the bottom-left, while the open unit (top left) shows the labeled components of the detector that is wire connected. This uncompensated detector was used in an early 4 GHz radiometer. The detector's temperature is monitored using a thermocouple which is attached to its outer case (top right). The bottom-right shows the generic detector response on a Log-Log plot. It shows the regions for which a diode displays a square law and linear response. Note that to assure a square law response the input power must be below -15 dBm so that the diode's output voltage is less than 100 mv.

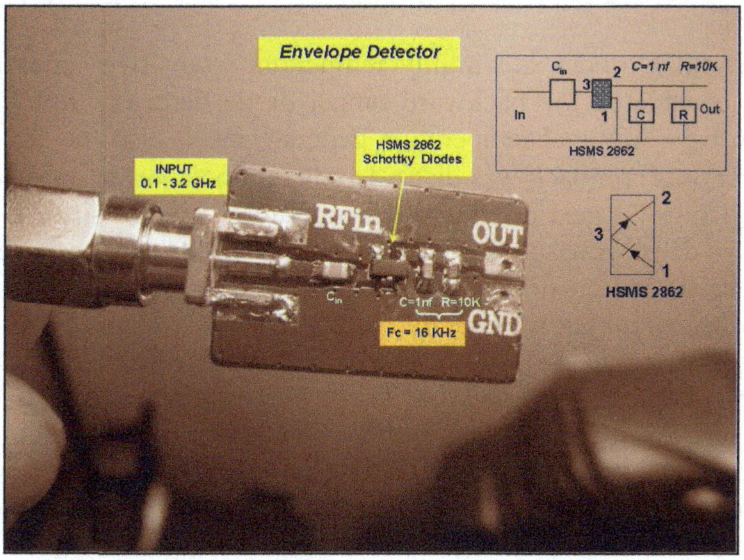

Figure 16 – Commercially available envelope detector using a single Schottky diode element. It has a square law response with a sensitivity of 3 mv/mw from 0.1 to 3.2 GHz with a temperature variation of about 0.1 mv/^{0}F. This wideband response is obtained using the optimally designed component layout on a printed circuit board rather than the homebuilt unit shown in Figure 15. This detector was used in the 22 GHz radiometer discussed in Section 8.6 and dual frequency radiometer described in Appendix A14 and shown in Figure A14 -3.

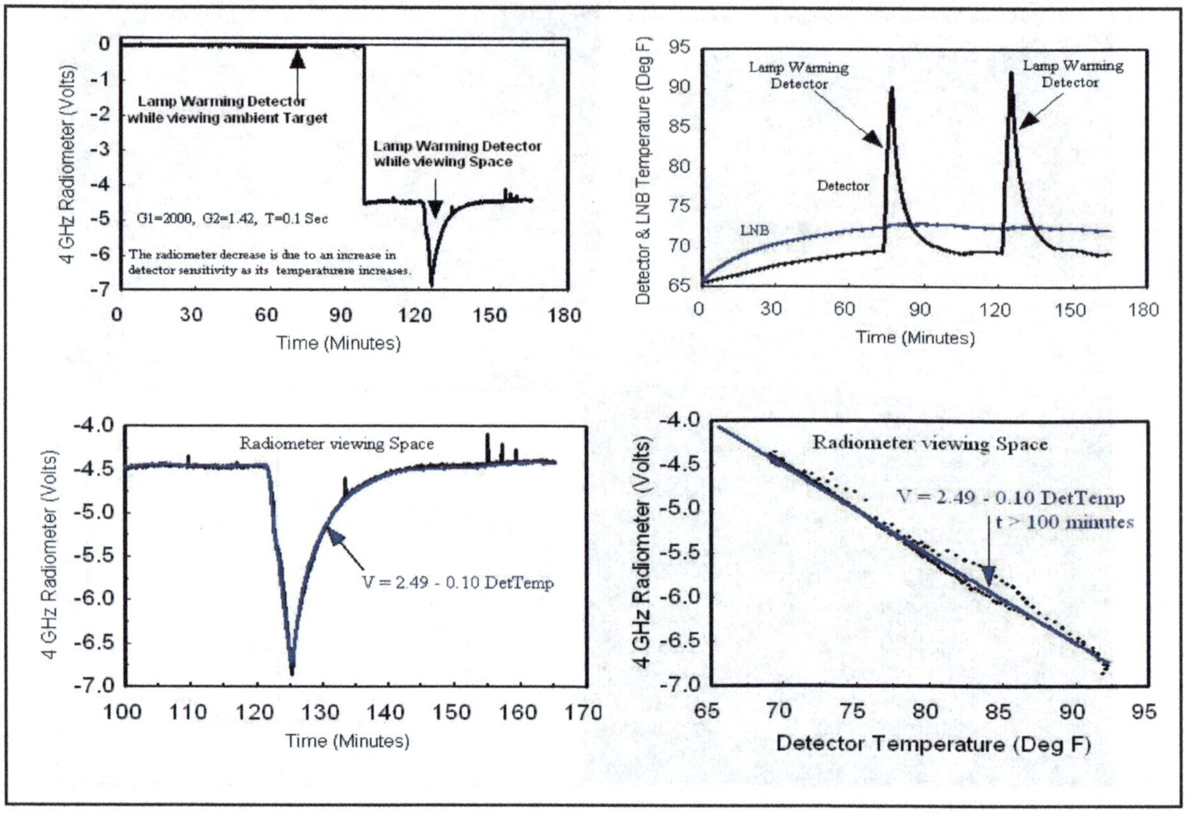

Figure 17 - Earlier 4 GHz radiometer with an uncompensated Schottky diode detector displays the detector temperature effect. The top-left Figure shows the radiometer voltage when the antenna views the high emissivity target followed by the space view. The top-right Figure shows the detector and LNB temperatures. The bottom-left Figure shows the space viewing measurements while the bottom right Figure plots these voltages against detector temperature.

6.1 Detector Temperature Compensation

Upon realizing Schottky diodes are highly temperature sensitive, alternative approaches were considered. I first considered the backward tunnel diode detector, which was mentioned in the previous section to have minimal temperature dependence. It also has lower noise than Schottky diodes [9]. However, this approach was discarded due to their lower sensitivity of around 1 $mv/\mu w$ and its lack of availability. I next considered a quadrature detector since it eliminates the need for diodes. A quadrature detector requires a power divider and 90^0 phase shifter to convert the LNB's *IF* signal to in-phase (I) and quadrature-phase (Q) components, after which the I and Q quantities are squared to produce an output representing total power. As mentioned at the end of this section, this I/Q technique is used in Software Defined Radio (SDR) based radiometers, but was not considered due to its complexity. Likewise, the elimination of square law detectors using digital techniques to measure thermal noise as described in Chapter 5 and at the end of this section was also not considered. As such, I ended up using the conventional and simpler temperature compensated circuit in Figure A6 of the Appendix. The circuit also uses the readily available dual packaged matched pair of Schottky diodes, which are connected to a difference amplifier. One diode element serves as a reference while the other is connected to the LNB's *IF* output. The difference output then approximately cancels temperature effects since both diodes operate at about the same temperature.

Figure 18 shows the finalized 4 GHz radiometer with the temperature compensated detector while Figure 19 (Bottom left) shows the detectors overall response. The detector response was determined by connecting a signal generator to the input and measuring the output in millivolts (mv) as the input power was increased from 0.1 microwatts (– 40 dBm) to 1 milliwatt (0 dBm). The frequency was set at 1.4 GHz, which is near the center of LNB *IF* bandwidth. Figure 19 (Bottom right) shows that for input power levels between -30 dBm (1 μw) to -12 dBm (63 μw), a linear relationship exists between input power and output voltage with a maximum error of +/- 0.2 dBm, *i.e.*,

$$P(\mu w) = -0.6 + 0.2\, V(mv) \qquad (9a)$$

$$V(mv) = 3.0 + 5.0\, P(\mu w)\,. \qquad (9b)$$

Equation (9b) was obtained by inverting (9a), and shows a detector sensitivity of 5 mv/μw for input power within the above mentioned 18 dBm range. This detector sensitivity includes the difference amplifier gain of 10. Also, for completeness, the two plots on the top of Figure 19 compare the above equations with the measurements over a 250 mv output range. From these sets of measurements it is evident that to insure a linear detector response, the input signal strength must be such that the output detector voltage is less than about 200 mv or the input power is less than 40 μw. The output voltage must also be greater than about 8 mv, corresponding to input power greater than 1 μw. Also, not to be overlooked is the detectors frequency response, which can define the overall response of the radiometer. This point was mentioned previously and shown in Figure 29 of Chapter 8 for the 20 GHz radiometer detector.

The 4 GHz radiometer response is again measured after substituting the temperature compensating detector as shown in Figure 18. For this experiment, the 4 GHz radiometer is subjected to the same type of measurements as before. Since the temperature compensated detector has a different gain than the uncompensated detector, the improved radiometer was calibrated to have the same radiometric gain as the earlier unit. As shown in Figure 20 (Top Right), the detector temperature is now increased by 55 ^{0}F using the small incandescent lamp. Also, unlike before, the radiometer voltage now increases (Figure 20 Top Left) rather than decreases as its temperature increases. Of greater importance, Figure 20 (Bottom) shows the radiometer temperature sensitivity is +0.01 V/^{0}F whereas before it was –0.10 V/^{0}F using the uncompensated detector. This factor of 10 improvement in stability using a temperature compensated detector was implemented for the 12 GHz radiometer (see Figure 3) as well as the 4 GHz Dicke radiometer of Figure 4. Also note in Figure 20 a small wave-like displacement between the radiometer response and the detector temperature change. This delayed radiometer response in the top-left and bottom plots occurs because the temperature sensor is mounted to the detector case, which is at a different location, and therefore at a slightly different temperature than the detector element. A similar effect was mentioned when discussing Figure 17, where again a time lag was evident between the two measurements

As mentioned above, the SDR approach can also be used to construct radiometers. Rather than use hardware, this approach uses software to perform the operations of mixers, detectors, filters, amplifiers and demodulators. To achieve this, the LNB down-converted *IF* output between 1 and 2 GHz is first separated into its in-phase (I) and quadrature-phase (Q) components using for example a power splitter with a 90^0 phase shift at one port. The I/Q outputs are then sampled using high speed analog to digital converters (ADC's), and the detected power is obtained by squaring the I and Q quantities using computer software. All of the other waveforms and operations shown in Figures 3 and 4 can also be obtained using software, which includes that of the synchronous demodulator, integrator and offset. Also, a digital *IF* filter can be used to precisely define the radiometer bandpass. This is particularly useful for *RFI* mitigation and when constructing the high spectral resolution temperature sounders discussed in Chapters 10 and 11. Lastly, Chapter 5 describes a different digital approach that uses statistical analysis of the *IF* signal for thermal noise detection as well as *RFI* mitigation. However, in general I felt the analogue Dicke radiometers developed here had sufficient precision and stability, so that although more flexible, the hardware needed to construct digital radiometers was unnecessary for ground-based radiometers.

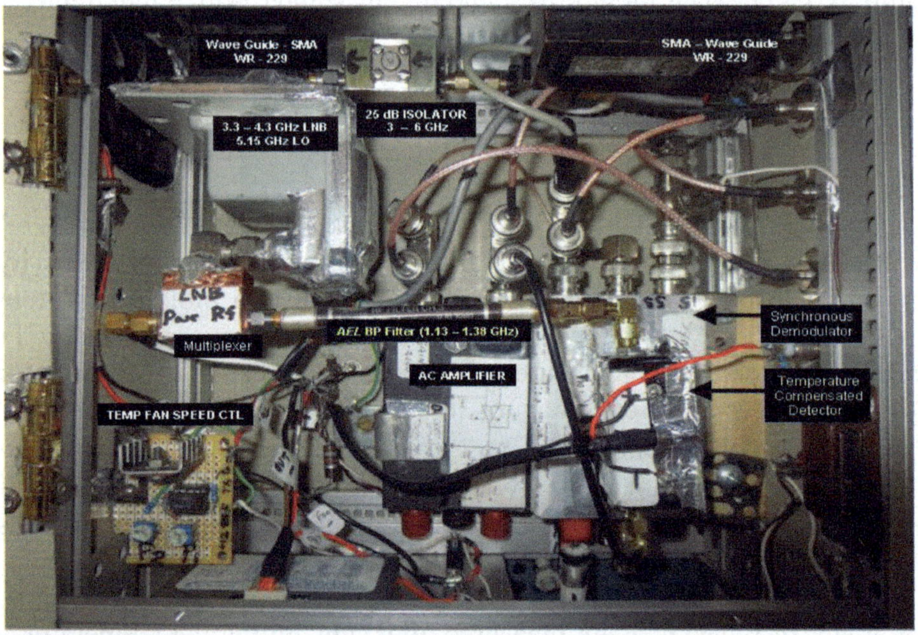

Figure 18 - Finalized 4 GHz radiometer has a temperature compensated detector. The labeled components are shown here as well as in the block diagram of Figure 4.

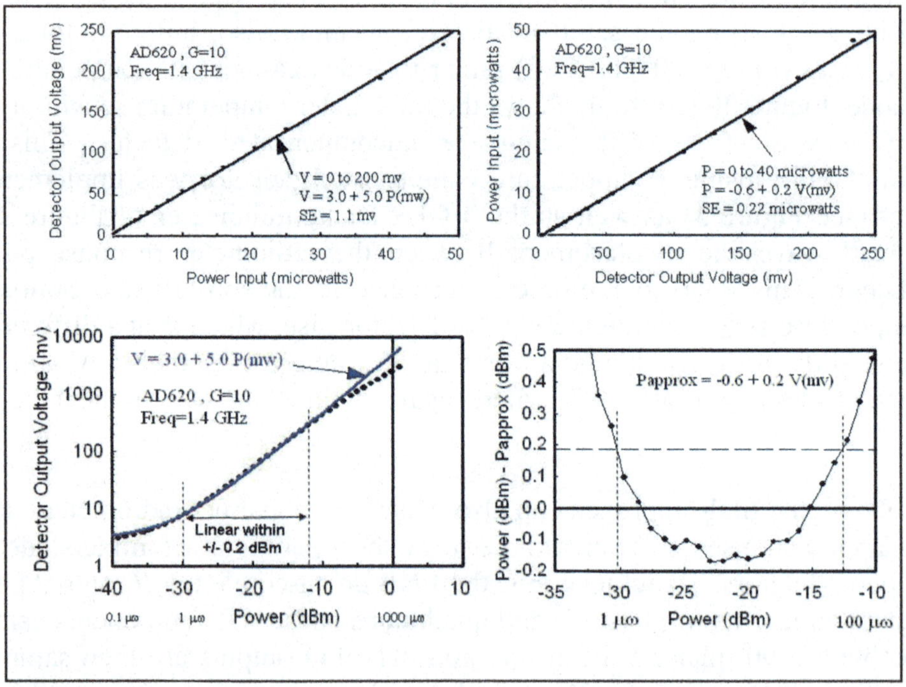

Figure 19- Measurements of 4 GHz radiometer detector at 1.4 GHz with the AD620 amplifier gain set to 10. The top-left shows the detector output voltage as a function of input power, while the adjacent figure shows the inverse relationship. The bottom-left shows the detector output voltage varies linearly with input power over a -30 to -12 dBm range, with a +/- 0.2 dBm accuracy (bottom-right). However, the bottom figures show the detector is greater than the power law at very low levels and closer to linear at high levels.

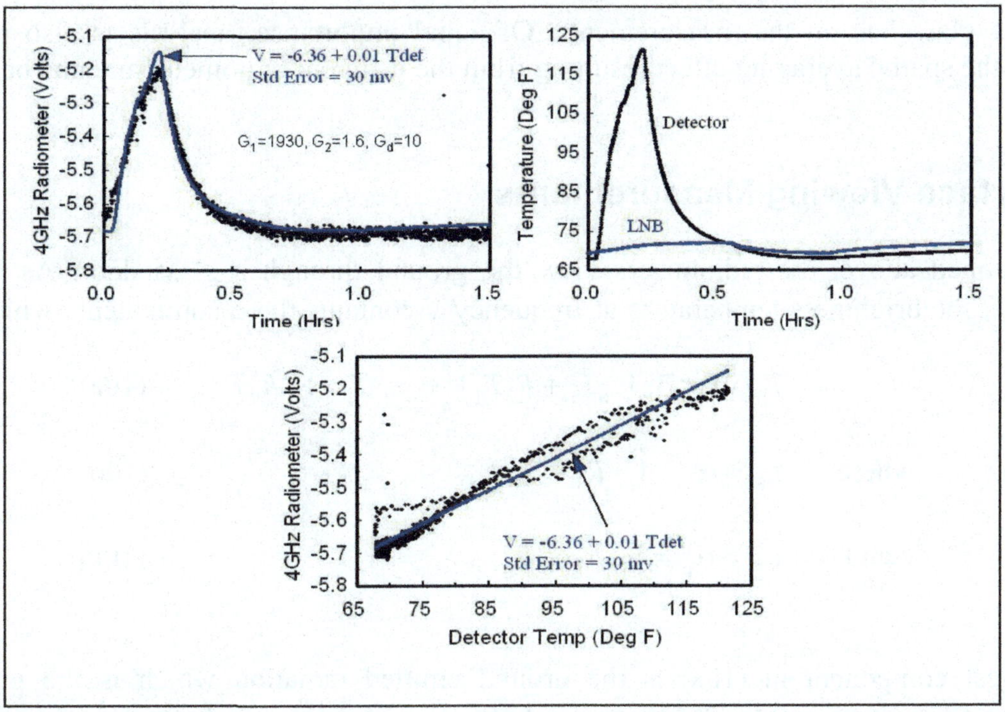

Figure 20 – The 4 GHz radiometer with a temperature compensated detector displays a reduced temperature effect compared to the uncompensated detector in Figure 17. As in Figure 17, the radiometer views space while the detector temperature is increased by 55 ^{0}F using an incandescent lamp. The top-left shows the radiometer voltage increasing with detector temperature (top-right) with a sensitivity of + 0.01 V/^{0}F (bottom). This is 10 times less than the uncompensated detector whose temperature sensitivity is -0.10 V/^{0}F.

7. Radiometer Applications

Much time was spent in constructing radiometers with different components and testing their performance. This was followed by experiments to measure their response and demonstrate their use in earth remote sensing. In this chapter, examples of sky and ground viewing measurements are described using the 4 and 12 GHz radiometers. This is extended in Chapter 8 to include measurements at higher frequencies using the 20.5 and 22.2 GHz radiometers. The measurements are also compared with analytical models to help explain the observations.

As seen from the books cover, the radiometers are mounted above each another and placed on a cart that can be elevated to view the sky or ground at elevation angles up to 25 degrees. Note also that the measurements were done in my basement with the radiometer viewing outdoors through a glass patio door. Incidentally, this measurement could not be done using infrared radiometers since glass (also Styrofoam) is opaque[3]. Fortunately, glass has very little absorption at microwave frequencies although it partially reflects radiation. Therefore, while it is a poor substitute for a radome it shields the radiometers from the environment so it is particularly useful when measuring rain. A better alternative, although not as convenient, would be to put the radiometers in a transparent less reflecting Styrofoam enclosure. Since this was not done, analysis is used to show

[3] While Styrofoam is essentially transparent in the microwave region it is measured to be opaque at infrared wavelengths around 0.65 microns and 10.5 microns.

the effect glass has on the measurements. Of equal importance, analysis is also performed to examine the spatial averaging effect resulting from the different radiometer antenna beamwidths.

7.1 Surface Viewing Measurements

As mentioned above, the radiometers view the ground through a glass door. As illustrated in Figure 21, the brightness temperature at frequency ν contains three components which combined become,

$$T_b(\nu) = \Im_g\,[\varepsilon_S T_S + R_S T_d] \;+\; \varepsilon_g\, T_g \;+\; R_g T \qquad (10a)$$

where $\varepsilon_S = \alpha_S = 1 - R_S - \Im_S$ (10b)

and $\varepsilon_g = \alpha_g = 1 - R_g - \Im_g$. (10c)

The largest component in (10a) is the ground emitted radiation which is the product of its emissivity ε_S and temperature T_S. From Kirchoff's law of thermal radiation, the emissivity equals the absorption coefficient α_S , which from energy conservation is given by (10b) where R_S is the reflection and $\Im_S$ is the transmission coefficient. As seen by the radiometer this radiation $\varepsilon_S T_S$ is attenuated by the glass door where $\Im_g$ is its transmission coefficient, which is mainly due to reflection. Equation (10a) also contains the downwelling atmospheric radiation T_d which is reflected by the ground and attenuated by the glass door before reaching the radiometer. Not included is any sky radiation seen indirectly when viewing the ground, such as by antenna side lobes. As discussed later in Section 8.3, T_d is largest for rain emission, but even then its reflected contribution is small over low reflectivity land surfaces. Although glass absorption is very small, for completeness equation (10a) also includes the glass emitted radiation which is the product of its emissivity ε_g and temperature T_g. As with the ground emissivity, the glass emissivity equals its absorption coefficient α_g in (10c) where R_g is the reflection and $\Im_g$ is the transmission coefficient. The last term in (10a) is the thermal radiation in my basement at temperature T which is reflected by the glass door and viewed by the radiometer.

Many surfaces are highly absorbing so that $\Im_S = 0$ and from (10b) $\varepsilon_S = 1 - R_S$. As such, the emissivity for flat homogeneous surfaces such as water and smooth soils can be obtained using the reflection coefficient derived from the Fresnel equations[4] with the dielectric constant, viewing angle and polarization as input parameters. In contrast to these surfaces, the glass door has negligible absorption so $\varepsilon_g \cong 0$ and $\Im_g \cong 1 - R_g$. However, as discussed in Section 7.2 and Appendix A10 the glass door reflection coefficient requires extensive analysis since it contains two glass panes separated by an air gap. Even more complex models are required for rough surfaces such as oceans, and inhomogeneous media such as snow and aged sea ice. The reflection coefficient and emissivity for these surfaces require solutions of Maxwell's equations that include surface scattering in the case of oceans and volume scattering by the ice grains in snow and air

---

[4] Fresnel reflection coefficients; $(R_S)_{VH} = |r_{VH}|^2$ where $r_V = -\mathrm{Tan}\,(\theta - \theta')/\,\mathrm{Tan}\,(\theta + \theta')$, $r_H = \mathrm{Sin}\,(\theta - \theta')/\,\mathrm{Sin}\,(\theta + \theta')$ and $\mathrm{Sin}\theta' = \mathrm{Sin}\theta / \sqrt{\varepsilon}$. However, most surfaces do not appear perfectly smooth at high frequencies so the reflectivity is altered by surface roughness. Therefore, $(R_S)_V$ is less and $(R_S)_H$ is greater than from the Fresnel equations.

bubbles formed in aged sea ice due to brine depletion. The emissivity of sea ice is discussed more fully in Section 12.3.

Although analytical models can provide physical insight, actual emissivity measurements are required using ground based, aircraft flown and satellite-launched radiometers. Such far-field observations are particularly needed at high frequencies where models are most deficient as the wavelength approaches the scale of inhomogeneities and surface roughness. Other aspects of emissivity modeling are discussed in Appendix A16. Lastly, Appendix A17 describes the difficulty in using near-field laboratory techniques to measure emissivity. For reference, Figure 22 shows examples of emissivity spectra. As stated in the Figure legend, these plots use multifrequency observations fitted to an empirical function. Note that the highest emissivity occurs for vegetated land, new sea ice and melting snow which is about 0.95. The lowest emissivity occurs for open water which is about 0.4. Also note that the emissivity spectrum has a positive slope for water and wet surfaces which absorb microwave radiation, while having a negative slope for surfaces containing ice grains and air bubbles that scatter radiation. Both the magnitude and slope of emissivity are used when developing algorithms to identify surfaces and atmospheric features from satellite observations [10]. These emissivity characteristics are also discussed in Section 12.3 when developing a dual frequency algorithm to identify ice type and measure sea ice concentration.

As mentioned above, at microwave frequencies the absorption and emissivity of glass is negligibly small so that $\varepsilon_g = \alpha_g \cong 0$ and $\Im_g \cong 1 - R_g$. Furthermore, the downwelling reflected radiation $R_S T_d$ is generally negligible over vegetated land so equation (10a) reduces to

$$T_b(\nu) = \left[1 - R_g(\nu)\right] \varepsilon_S(\nu) T_S \; + \; R_g(\nu) T \; . \qquad (11)$$

Therefore, the brightness temperature change ΔT_b due to variations in emissivity $\Delta\varepsilon_S$ and surface temperature ΔT_S becomes

$$\Delta T_b(\nu) = \left[1 - R_g(\nu)\right] \left[\, \varepsilon_S(\nu)\, \Delta T_S + T_S \,\Delta\varepsilon_S(\nu) \,\right] \qquad (12)$$

where this equation is used next to analyze the following two experiments.

1 - <u>Surface Temperature Variation</u> :

In the first experiment shown in Figure 23 (Top-Left), the 4 and 12 GHz radiometers view the patio whose temperature decreases over 5 hours beginning at 7 pm on August 8, 2013. Both radiometers are assumed to view the same temperature change so the measurement ratio based on (12) is

$$\left.\frac{\Delta T_b(\nu_1)}{\Delta T_b(\nu_2)}\right]_{\varepsilon_S = \text{Constant}} = \left[\frac{1 - R_g(\nu_1)}{1 - R_g(\nu_2)}\right] \frac{\varepsilon_S(\nu_1)}{\varepsilon_S(\nu_2)} \; . \qquad (13a)$$

The bottom-left of Figure 23 shows both radiometer voltages decreasing steadily as the surface temperature decreases. Upon plotting the 4 against 12 GHz measurements on the bottom-right, a slope of 3.1 is obtained using least squares regression analysis. Since the emissivity over vegetated land is about 0.95 at both frequencies, the slope of 3.1 is the glass transmission coefficient ratio at the two frequencies.

2 - **Surface Emissivity Variation** :

In the second much later experiment shown in Figure 24, the 4 and 12 GHz radiometers view the patio on January 15, 2017 at 4 pm. The patio was sprayed with water for 3 minutes to lower its emissivity from about $0.95(\varepsilon_0)$ to $0.37(\varepsilon_W)$ as indicated in Figure 22. The spraying was then stopped and the wetness area slowly decreased due to runoff. From (12), the measurement ratio is

$$\left.\frac{\Delta T_b(\nu_1)}{\Delta T_b(\nu_2)}\right]_{T_S=Constant} = \left[\frac{1-R_g(\nu_1)}{1-R_g(\nu_2)}\right]\frac{\Delta\varepsilon_S(\nu_1)}{\Delta\varepsilon_S(\nu_2)} \quad . \qquad (13b)$$

The emissivity is expressed as $\varepsilon_S(\nu) = \varepsilon_0 - [\varepsilon_0 - \varepsilon_W] f_W(\nu)$ where $f_W(\nu)$ is the fractional wet area seen by the radiometers. This simple expression of emissivity using fractional areas is applicable to far-field measurements and fully discussed in Appendix A17. It is particularly important to use this type of expression for satellite measurements to include spatial averaging by their large field of view. This is also described in Section 12.3 with regard to the measurement of sea ice concentration. Therefore, the emissivity ratio in (13b) is $\Delta f_W(\nu_1)/\Delta f_W(\nu_2)$, which is unity if both radiometers view the same area. In that case the measurement ratio (13b) is the glass transmission coefficient ratio. However, the bottom right plot in Figure 24 shows the slope of T_b to be 1.7 while it was 3.1 in Figure 23. This smaller slope of 1.7 results from the emissivity ratio, which is less then unity since $\Delta f_W(4) < \Delta f_W(12)$ due to the antenna beamwidth of 27^0 at 4 GHz compared to 16^0 at 12 GHz.

3 - **Insertion Loss Measurement** :

For a more definitive study done prior to the 1st experiment, the glass transmission ratio was obtained using the insertion loss measurements described in Appendix A11. In that experiment the radiometer output voltage is measured after opening and closing the glass patio door. The reflection coefficient at 4 and 12 GHz is then determined to be 0.20 and 0.64, respectively. As such, the transmission coefficient ratio between 4 and 12 GHz is $(1-0.20)/(1-0.64) = 2.2$. This more direct measurement is less than the 3.1 slope in Figure 23 and greater than the 1.7 slope in Figure 24. In fact the insertion loss measurement of 2.2 is nearly the average of the other two measurements. To best understand how the glass door effects the radiometer measurements, model analysis is performed. As shown in the next Section 7.2, the transmission coefficient of the glass door varies as a function of frequency as well as the glass door parameters. In fact, as discussed in Appendix 10, the door acts as a filter with characteristics similar to a Fabry-Perot Etalon or interferometer whose resonant frequencies and spectral shape depend on the glass dielectric constant, thickness, separation and incident angle.

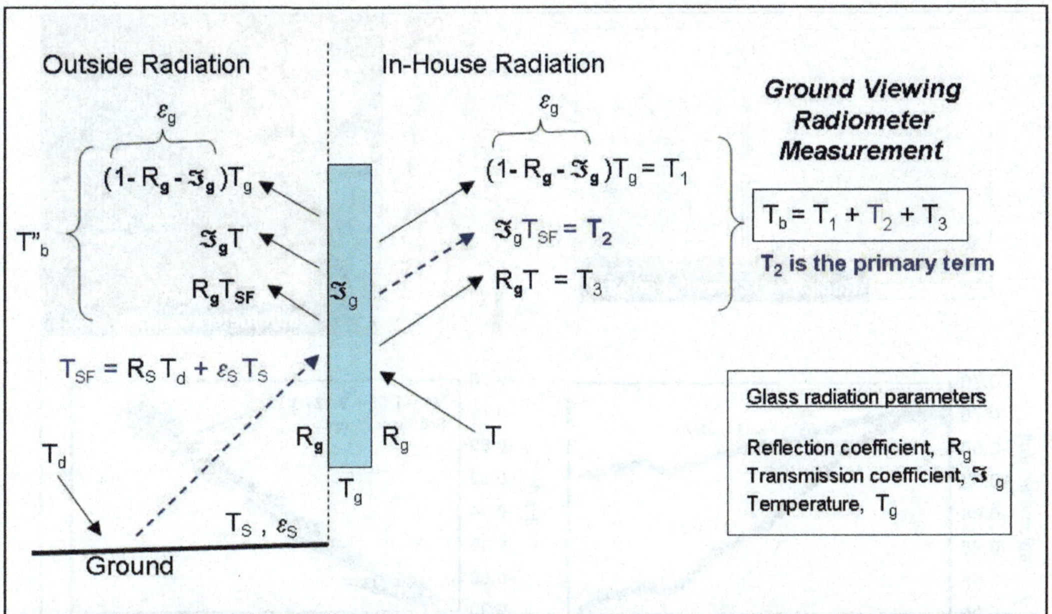

Figure 21 - Schematic of different brightness temperature, T_b, contributions seen inside my house by a radiometer viewing the ground through a glass window. The primary radiation seen outside my house is the ground emitted radiation, which is the product of emissivity, ε_S, and surface temperature, T_S. It is highlighted in blue. Also shown is the downwelling atmospheric radiation, T_d, reflected by the ground R_ST_d which is added to ε_ST_S and shown by the dashed line. Inside my house these two terms are attenuated by the glass door. It is denoted as T_2, and highlighted in blue, and shown by the dashed line. Two other quantities seen inside my house is the emitted, T_1, and reflected radiation T_3 by the glass.

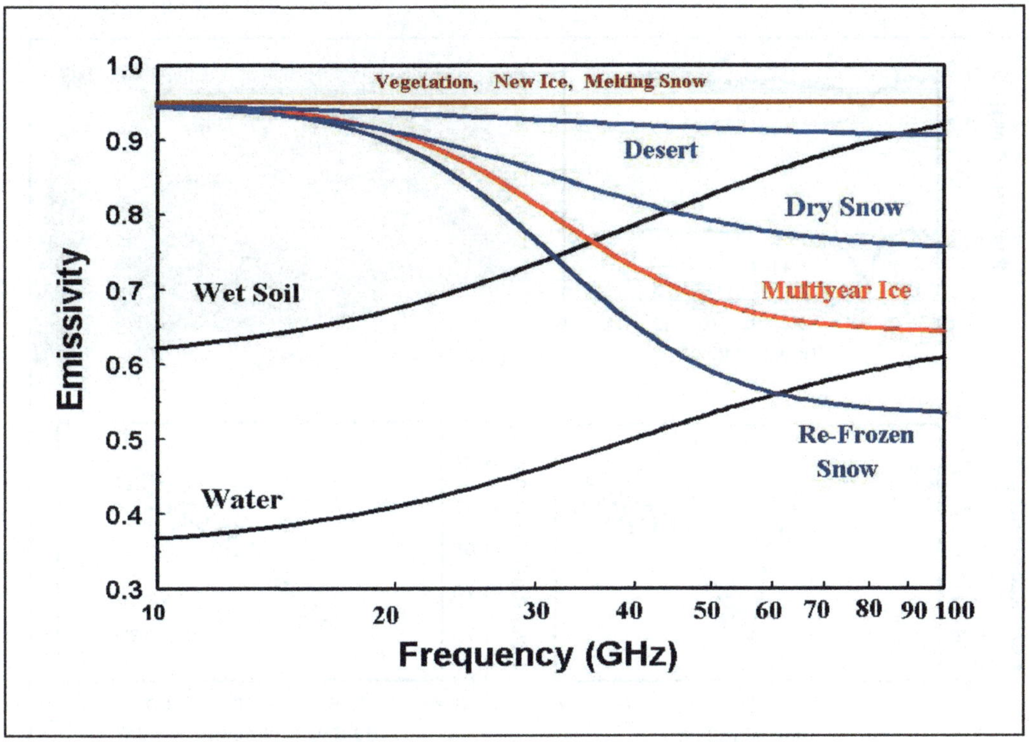

Figure 22 – Emissivity obtained using the empirical function $\varepsilon_S = \left[\varepsilon_0 + \varepsilon_\infty(\nu/\nu_0)^n\right] / \left[1 + (\nu/\nu_0)^n\right]$ whose parameters $\varepsilon_0, \varepsilon_\infty, n, \nu_0$ are obtained from representative measurements at nadir viewing. Note that the highest emissivity is for vegetated land, new sea ice and melting snow while the lowest is for water and wet soil. Multiyear sea ice and re-frozen snow also has very low emissivity at high frequency ν with a negative slope. Surface identification is obtained from satellite radiometer measurements using the magnitude and slope of emissivity with frequency.

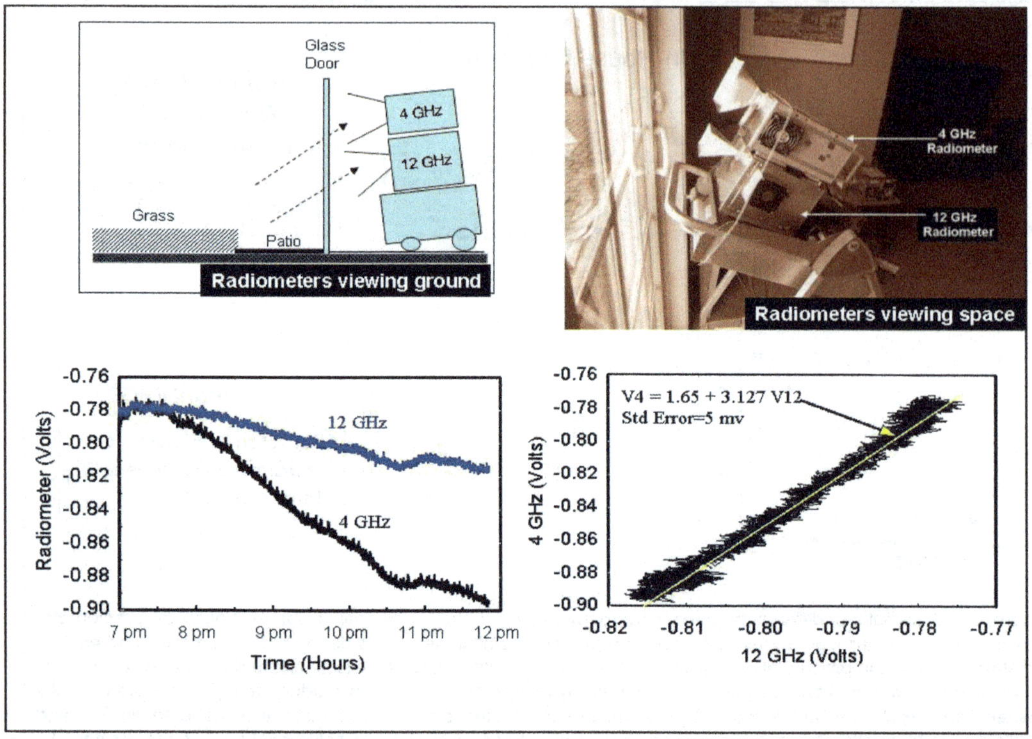

Figure 23 - The 4 and 12 GHz radiometers are placed on a movable cart to view the ground (Top Left) and sky (Top-right) from my basement. Radiometer voltages are plotted as function of time (bottom-left) when viewing the ground and a best fit linear equation is obtained (bottom-right) and shown in yellow.

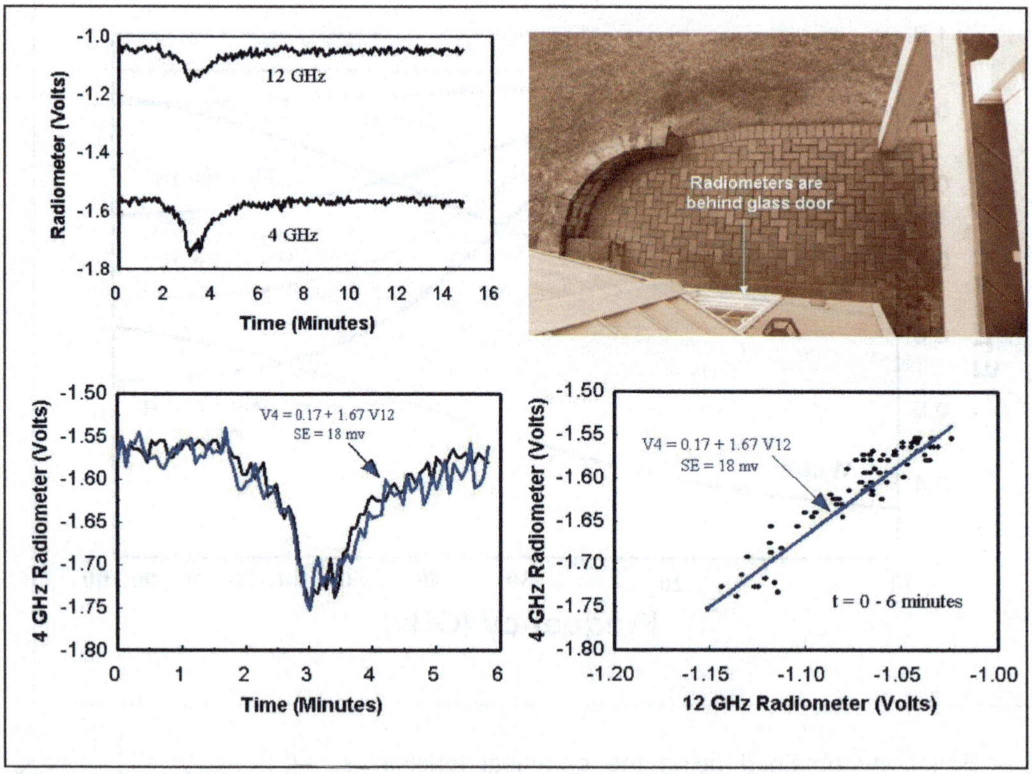

Figure 24 – The top-right picture shows the patio and grass area after being watered for 2 minutes. The top left shows the 4 and 12 GHz radiometer measurements before and after watering. An expanded plot of the 4 GHz data is shown in the bottom-left. Also shown is a plot of the 4 GHz estimate using a best fit linear equation relating the 4 to 12 GHz measurements. The radiometer measurements together with the best fit equation is also plotted in the bottom right.

7.2 Glass Door Reflectance

As noted in Appendix A2, the antenna receives vertical polarization to reduce surface reflections when viewed at large elevation angles. The dielectric constant of glass is reported to be about 6 so its reflection coefficient is quite large. In fact, R_g is computed to be 0.16 at vertical polarization for a 20 degree viewing angle using the above mentioned Fresnel equations for a smooth dielectric interface. However, a more realistic model must include wave interference due to multiple reflections within a glass sheet. Using this model, Figure A10-1 of Appendix A10 shows large variations in reflectance even for very small changes in the frequency and glass parameters. Similar calculations were also performed by Dr. Philip Rosenkranz who I have been collaborating with and whose results are summarized in the following Tables. The Tables below also compare the model calculations with the above mentioned insertion loss measurements of reflection coefficient.

Table 2 lists the calculated reflection coefficient at normal incidence for a 2.5 mm glass slab. Note that the reflection coefficient is 0.20 at 4 GHz but now increases to 0.51 at 12 GHz at normal incidence. Also, for a 20^0 viewing angle the calculated reflection coefficient is reduced to 0.17 at 4 GHz and 0.47 at 12 GHz. These calculations compare much better with the insertion loss measurements than the fixed reflectance of 0.16 for a single glass interface using the Fresnel coefficients. Furthermore, the glass door actually contains two glass panes separated by an air gap or spacer. The reflectance then varies depending on the separation distance as well as the glass thickness. In fact, the model calculations as well as Figure A10-1 show it is possible to obtain a wide range of reflectance by fixing the dielectric constant to 6 and just varying these two parameters. Since the results are sensitive to frequency, the power reflection coefficients were averaged over the radiometer bandwidths because thermal noise is uncorrelated at different frequencies. As an example, Table 3 shows the calculated reflection coefficient for a glass separation of 5 mm with a glass thickness of 4 mm. Note that for these particular parameters the 12 GHz calculations are nearly the same as the measurements while the 4 GHz results are now different. There is also little difference at the two viewing angles. Unfortunately, the glass door dimensions were not available to obtain more exact comparisons.

Table 2: Reflectivity for a 2 mm glass sheet.

	Measurements*	**Calculation****	**Calculation*****
4 GHz	$R_g\ (0^0) = 0.20$	$R_g\ (0^0) = 0.20$	$R_g\ (20^0) = 0.17$
12 GHz	$R_g\ (0^0) = 0.64$	$R_g\ (0^0) = 0.51$	$R_g\ (20^0) = 0.47$

* Insertion loss measurements at 0^0 incident angle (See Figure A11 of Appendix).
** Calculated at 0^0 incident angle used in the insertion loss measurements.
*** Calculated at 20^0 incident angle used in ground and sky viewing measurements

Table 3: Reflectivity for two 4mm glass sheets separated by a 5 mm air gap.

Frequency	Measurements*	Calculation**	Calculation***
4 GHz	$R_g\ (0^0) = 0.20$	$R_g\ (0^0) = 0.11$	$R_g\ (20^0) = 0.13$
12 GHz	$R_g\ (0^0) = 0.64$	$R_g\ (0^0) = 0.67$	$R_g\ (20^0) = 0.65$

* Insertion loss measurements at 0^0 incident angle (See Figure A11 of Appendix).
** Calculated at 0^0 incident angle used in the insertion loss measurements.
*** Calculated at 20^0 incident angle used in ground and sky viewing measurements.

7.3 Sky Viewing Measurements

While the previous section studied changes due to temperature and emissivity variation, this section examines changes due to atmospheric variations. Figure 23 (top-right) shows the 4 and 12 GHz radiometers viewing space through the glass patio door. Analogous to the ground viewing experiment, whose brightness temperature components are shown in Figure 21, the sky viewing components are depicted in Figure 25. Combining the components, the brightness temperature is

$$T_b(\nu) = \Im_g\, T_d + \varepsilon_g\, T_g + R_g T \qquad (14a)$$

$$\text{where}\quad T_d = (1 - \tau^{\mathrm{Sec}\theta})T_M \qquad (14b)$$

$$\text{and}\quad \varepsilon_g = 1 - R_g - \Im_g\ . \qquad (14c)$$

The first term on the right side of (14a) is the downwelling atmospheric radiation T_d attenuated by the glass door transmission coefficient $\Im_g$. This component is denoted as T_2 in Figure 25 and is the largest contribution unlike that of surface observations. Omitting the cosmic radiation term in (5), T_d is given by (14b) where τ is the atmospheric transmittance and T_M is the mean temperature in (6a). The second term in (14a) is the thermal emission by the glass door which is the product of its glass emissivity ε_g and temperature T_g. It is designated as T_1 in Figure 25 with the emissivity given by (14c). The right-most term in (14a) is the in-house reflected radiation at temperature T, where R_g is the glass reflection coefficient. This term is depicted as T_3 in Figure 25. Lastly, (14a) assumes no ground radiation is observed indirectly by way of the antenna side lobes.

Substituting (14b) into (14a) and neglecting the very small glass emissivity,

$$T_b = (1 - R_g)\,(1 - \tau^{\mathrm{Sec}\theta})T_M + R_g T \qquad (15a)$$

$$\text{where}\quad \tau = \tau_{H20}\, \tau_{02}\, \tau_{Liq}\ . \qquad (15b)$$

As mentioned in Section 4.3, the atmospheric transmittance (15b) is the product of the water vapor, oxygen and liquid water transmittances. The liquid water component τ_{Liq} is due to the water drops in clouds and rain. Its transmittance is obtained by summing the energy absorbed by each drop. At frequencies where the drops are smaller than the wavelength each drop radiates as dipoles. Its absorption then varies as the 2nd power of drop size to wavelength based on the Rayleigh model. Also, due to their small fractional volume the radiation from each drop is considered unperturbed from its neighbors. Collective effects such as wave interference is therefore neglected. Using this Born approximation, which is applicable for a sparse group of particles (see Appendix A16) the liquid water transmittance can be written as

$$\tau_{Liq}(\nu) = e^{-Q/Q(\nu)} \qquad \text{(16a)}$$

$$\textit{with} \qquad Q(\nu) = 41.5\,\frac{\nu^2 + \nu_0^2}{\nu^2\,\nu_0} \qquad \text{(16}b\text{)}$$

$$\text{and} \qquad \nu_0 = 160\,e^{7.2[1-287/T_{\text{CLD}}]}\,. \qquad \text{(16}c\text{)}$$

The derivation leading to (16a) is based on analysis by Herbert Goldstein in Section 8.6 of the Radiation Lab. book [11]. Due to Rayleigh's model the resulting transmittance τ_{Liq} in (16a) is independent of drop size but only on the total liquid water content Q in millimeters. It is analogous to the *TPW* described in Section 4.3, where Q is the depth of liquid that would be accumulated if all drops were compressed in a vertical column. Equation (16a) also contains a frequency dependent parameter $Q(\nu)$ given by (16b) and a temperature dependent relaxation frequency ν_0 resulting from Debye's dielectric constant of water. The parameter ν_0 is given by (16c) where T_{CLD} is the cloud temperature. Table 4 lists the values of $Q(\nu)$ and ν_0 at three cloud temperatures for the 4, 12, 20 and 22 GHz radiometers whose measured center frequencies are also listed.

Table 4: Cloud transmittance parameters $Q(\nu)$ and ν_0 at different temperatures.

Radiometer	Center Frequency	T_{CLD} = 270 K	T_{CLD} = 275 K	T_{CLD} = 280 K
4 GHz	ν = 3.9 GHz	ν_0 = 102 GHz $Q(\nu)$ = 278 mm	ν_0 = 117 GHz $Q(\nu)$ = 319 mm	ν_0 = 134 GHz $Q(\nu)$ = 365 mm
12 GHz	ν = 11.7 GHz	ν_0 = 102 GHz $Q(\nu)$ = 31.3 mm	ν_0 = 117 GHz $Q(\nu)$ = 35.8 mm	ν_0 = 134 GHz $Q(\nu)$ = 40.8 mm
20 GHz	ν = 20.5 GHz	ν_0 = 102 GHz $Q(\nu)$ = 10.5 mm	ν_0 = 117 GHz $Q(\nu)$ = 11.9 mm	ν_0 = 134 GHz $Q(\nu)$ = 13.5 mm
22 GHz	ν = 22.2 GHz	ν_0 = 102 GHz $Q(\nu)$ = 9.03 mm	ν_0 = 117 GHz $Q(\nu)$ = 10.2 mm	ν_0 = 134 GHz $Q(\nu)$ = 11.7 mm

Since $\nu < \nu_0$, $Q(\nu) \cong 41.5\ \nu_0/\nu^2$. Also, for clouds and light rain $Q < 1$ mm so $Q\,/\,Q\,(\nu) < 0.1$ and (16a) approximately becomes,

$$\tau_{Liq}(\nu)^{Sec\theta} \approx 1 - [Q/Q(\nu)]\ Sec\theta \quad \text{where} \quad Q\,(\nu) \cong 41.5\ \nu_0/\nu^2\ . \qquad (17)$$

In the experiment described below the skyward viewing radiometers observe changes in cloud and rain water ΔQ_ν where the subscript ν denotes the frequency of the particular radiometer. From (15a), (15b) and (17) the observed brightness temperature change ΔT_b is

$$\Delta T_b(\nu) = [1 - R_g(\nu)]\ [\tau_{H20}(\nu)\tau_{02}(\nu)]^{Sec\theta}\ [\Delta Q_\nu\,/\,Q(\nu)]\ T_M\ Sec\theta \qquad (18)$$

so
$$\frac{\Delta T_b(\nu_1)}{\Delta T_b(\nu_2)} \approx \left[\frac{1 - R_g(\nu_1)}{1 - R_g(\nu_2)}\right] \left[\frac{\tau_{H20}(\nu_1)\tau_{02}(\nu_1)}{\tau_{H20}(\nu_2)\tau_{02}(\nu_2)}\right]^{Sec\theta} \left(\frac{\nu_1}{\nu_2}\right)^2 \frac{\Delta Q_{\nu_1}}{\Delta Q_{\nu_2}}, \qquad (19)$$

where $\Delta Q_{\nu_1} = \Delta Q_{\nu_2}$ if both radiometers have the same antenna beamwidth so they view the same scene. For $\nu_1 = 11.7$ GHz, $\nu_2 = 3.9$ GHz and $\theta = 20^0$ the transmittance ratio term in (19) is calculated to decrease from 0.99 to 0.97 as the *TPW* increases from 0 to 40 mm. More important is the liquid water absorption ratio $(\nu_1/\nu_2)^2$ which is 9.0 at the two frequencies. This factor would be the dominant term in (19) if not for the glass transmission coefficients $1 - R_g(\nu)$.

Using the insertion loss measurements described in Section 7.1 on page 39, the ratio of transmission coefficients is $1/2.2 = 0.45$ so combining all these factors, equation (19) becomes

$$\frac{\Delta T_b(12)}{\Delta T_b(4)} = 4\ \frac{\Delta Q_{12}}{\Delta Q_4}\ . \qquad (20)$$

Equation (20) is used to analyze the 12 and 4 GHz radiometer measurements of clouds and rain. As an example, Figure 26 shows the measurements obtained on June 12, 2014 when two heavy rain events occurred between 5 and 7 pm. The top-left plot shows two large voltage spikes at 12 GHz. Even the 4 GHz radiometer responds to the rain although with smaller increases. The figure also shows a period of moderate rain between the two heavy rain events. Since the non-precipitating clouds observed prior to 5 pm have smaller drops with less water content they produce a much smaller increase at 12 GHz with no change at 4 GHz. The top-right shows an expanded plot of the first heavy rain event which includes moderate rain. For comparison, the bottom left shows the measurements plotted against each other for the moderate rain period, while the bottom right includes the first heavy rain event. The moderate rain plot has a slope of 8.6 between measurements while the plot containing heavier rain has a slope of 12.4. Interestingly, as shown in Figure A12 of Appendix A12, nearly the same large slope was obtained for another severe storm event on February 24, 2016. However, other rain events have slopes between 10 and 15. The reason for the different slopes is discussed next.

The slopes in Figure 26 combined with (20) result in $\Delta Q_{12}/\Delta Q_4 = 12.4/4 = 3.1$ for heavy rain and $\Delta Q_{12}/\Delta Q_4 = 8.6/4 = 2.1$ for moderate rain. These different values of $\Delta Q_{12}/\Delta Q_4$ is attributed to the larger spatial averaging by the 4 GHz antenna beamwidth of 27^0 compared to the 12 GHz beamwidth of 16^0. To estimate the maximum spatial averaging effect we consider an opaque rain feature within the antenna beamwidth so the 12 to 4 GHz brightness temperature ratio is proportional to the ratio of viewing areas, *i.e.,* $(27^0/16^0)^2 = 2.8$. This value is comparable with the 3.1 factor obtained for heavy rain. Also, for widespread rain the spatial averaging effect is smaller, which is consistent with the smaller ratio of 2.1 for moderate rain.

In summary, the spatial averaging effect defined by $\Delta Q_{12}/\Delta Q_4$ when combined with the glass transmission ratio in (19) results in T_b ratios comparable to the measurements. It is also noteworthy that the 12 GHz radiometer detects rain similar to radar. However, unlike the Rayleigh absorption measured by radiometers which depends on the 2nd power of drop size to wavelength, the Rayleigh scattering cross section measured by radar depends on the 4th power of drop size to wavelength for small spherical drops. As such, the radiometer measurement is proportional to the liquid water content Q while radar measurements are not as directly related to Q. However, radar has the added advantage of being able to measure the range and fall velocity of rain drops, which can not be achieved using radiometers. It must also be mentioned that in addition to frequency, the polarization signals resulting from non-spherical raindrops can provide additional information on rainfall detection when using radar as well as radiometry.

Lastly, a relationship can be obtained between the liquid water measured by radiometers and rain rate. As mentioned previously, Q is the depth of liquid that would be accumulated if all drops were compressed in a vertical column. If we think of rainfall like water from a sprinkler (see Appendix A1) then the vertically integrated, area-averaged liquid water measured by a radiometer is the amount of rain drops accumulated on the ground over the time it takes them to reach the ground, *i.e.,*

$$\int_0^{\tau_d} \hat{R}\, dt = Q \qquad (21)$$

where $\hat{R}$ is the rain rate in mm/hr and τ_d is the descent time in hours.

Since rain is highly random, varying both spatially and temporally, rain rate has been related statistically to parameters such as the drop-size distribution, fall velocity and their vertical extent. While many of these variables can be inferred using different types of radar, even when utilized together they generally provide a less direct measure of rainfall than the liquid water obtained from radiometers. In fact, the area average rainfall obtained from radiometer measurements is probably more representative than that measured locally by rain gauges. However, due to the complex nature of rain, the best way to characterize precipitation remotely is to use ground-based radar and radiometer measurements.

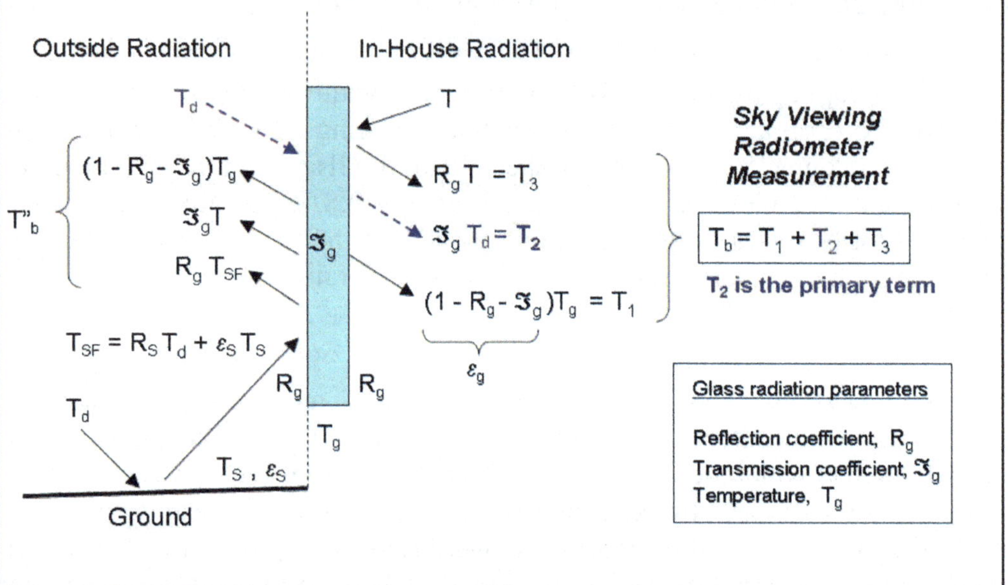

Figure 25 - Schematic of different brightness temperature, T_b, contributions seen inside my house by a radiometer viewing the sky through a glass window. The primary radiation seen outside my house is the downwelling atmospheric radiation T_d. Upon entering my house this radiation is attenuated by the glass door and denoted as T_2. It is highlighted in blue and shown by the dashed line. This downwelling radiation is also reflected by the ground $R_S T_d$ and added to the surface emitted radiation $\varepsilon_S T_S$. The other two quantities seen inside my house is the emitted, T_1, and reflected radiation T_3 by the glass.

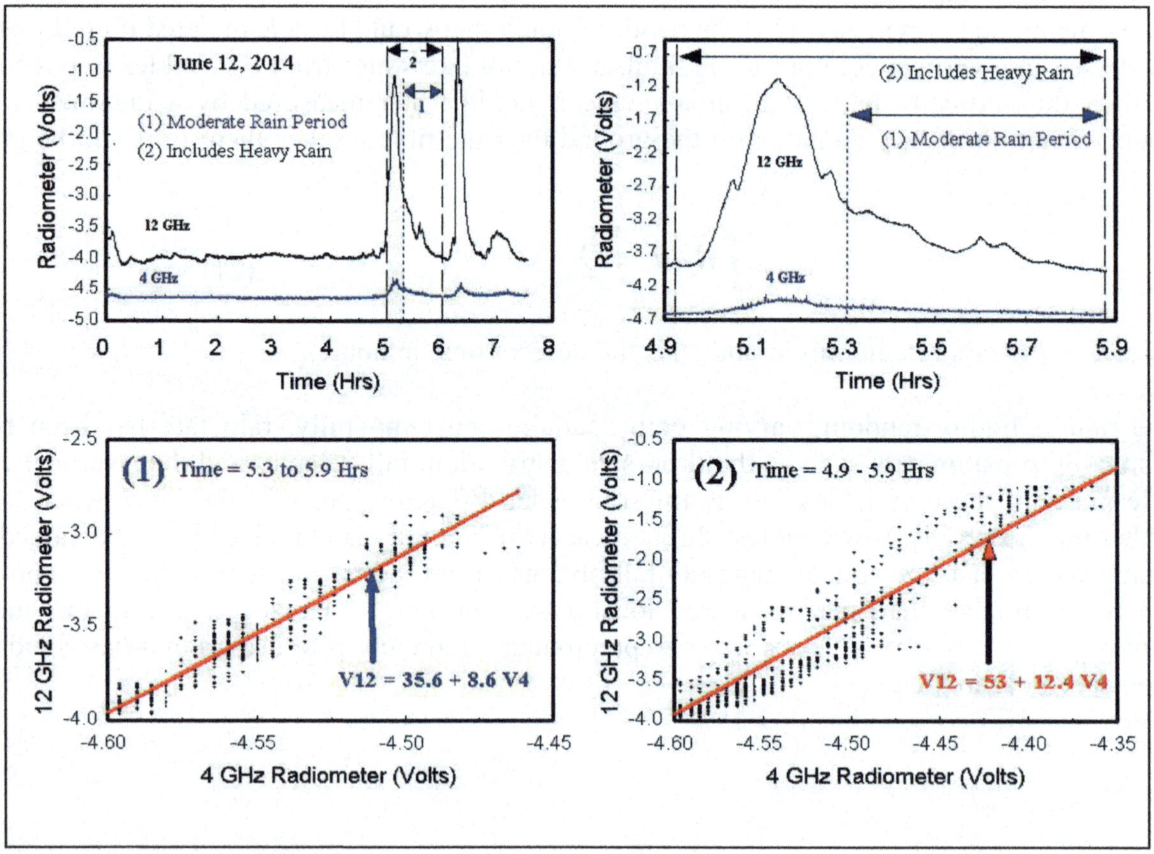

Figure 26 - The top left shows the 4 and 12 GHz measurements over 8 hours beginning at 12 pm on June 12. Two heavy rain events proceeded by clouds occur between 5 and 7 pm, with the 12 GHz output increasing from -4 to less than -1 volt while the 4 GHz output increases by only 0.2 volts. The top right shows an expand view of the 1st rain event which consists of heavy and moderate rain. During this period the bottom right shows the 12 GHz increasing by a factor of 12.4 compared to 4 GHz. However, the bottom left shows the 12 GHz only increased by a factor of 8.6 for the widespread moderate rain.

8. Water Vapor Radiometers

This chapter describes my highest frequency Dicke radiometers together with some measurements, analysis and simulations. The first such radiometer was built using the Norsat 9000C Ka band LNB shown in Figure 6 which amplifies frequencies between 20.2 and 21.2 GHz at linear polarization. Its higher frequency makes the radiometer more sensitive to atmospheric absorption by clouds and rain than the two lower frequency radiometers. Figure 13 also shows it to be more sensitive to water vapor than the lower frequency radiometers at 4 and 12 GHz. As explained next, the center frequency is at 20.5 GHz although it will sometimes be referred to as a 20 GHz radiometer. For comparison, I also discuss radiometers constructed at 21.1 and 22.2 GHz, and compare their measurements with the 20.5 GHz radiometer.

8.1 Radiometer at 20 GHz

The Norsat LNB has an *LO* of 19.25 GHz with a down converted *IF* between 0.95 to 1.95 GHz. Therefore, the corresponding *RF* is between 20.2 and 21.2 GHz. As with the lower frequency LNB's in Figure 6 it has waveguide input and coax output at the *IF* port. Also, the LNB gain is 55 dB with a noise figure of 1.3 dB or 100 *K* noise temperature. This is 7 times greater than the 4 and 12 GHz LNB noise temperatures of 14 *K*. The radiometer noise is then proportionally larger so a 1 second integration time is required for its noise to be comparable to the lower frequency radiometers which generally use a 0.1 second integration time. Its higher frequency also makes the 20 GHz radiometer more expensive to construct due to its higher LNB and waveguide adapter cost. The LNB also requires a current of 200 ma versus 100 ma for the lower frequency units. As mentioned in Section 4.2, this LNB also has a larger gain decrease with temperature. Lastly, to obtain low front end loss the radiometer requires a well matched antenna cable in addition to low VSWR waveguide adapters, etc., at this frequency.

While the Norsat Ka-Band LNB is more expensive than the C- and Ku-band units, I was able to find a model 9000C on eBay for only $50. The block diagram of the 20 GHz radiometer is similar to the 12 GHz radiometer shown in Figure 3. Also, Figure 27 shows the radiometer lid opened to display the components. Its smaller size antenna has 20 dB gain, while the larger 12 and 4 GHz antennas have gains of 19 and 15 dB, respectively. Calibration was performed as shown in Section 8.4 using the tipping curve and near-field methods listed in Table 1. The resulting calibration equation is nearly the same as the lower frequency radiometers by setting the AC amplifier, detector and DC amplifier gains to G_1=1000, G_d=10, G_2=3.2, respectively. As such, the total radiometer gain which includes the LNB is 145 dB. This is larger than the 12 GHz radiometer gain of 130 dB and less than the 4 GHz radiometer gain of 158 dB.

While waveguide adapters and isolators are available for the 18 to 22 GHz Ka frequency band, I was unable to find a pin diode switch on eBay at frequencies beyond 18 GHz. I therefore used the same General Microwave (M862B) pin diode switch used for the lower frequency radiometers. This switch is specified to operate between 0.1 and 18 GHz with 2 dB insertion loss and 45 dB isolation. However I measured different frequency characteristics beyond 18 GHz depending on the unit. For example, Figure 30 shows little degradation between 18 and 20.8 GHz for this switch. Also, at these higher frequencies the Figure shows a 3 dB insertion loss when the pin diodes are switched on and more than 45 dB isolation when the pin diodes are powered off. However, beyond 20.8 GHz the insertion loss of the switch increases to 12 dB at 21.5 GHz. Such high loss not only requires larger gain but also increases the radiometer noise due to increased front end loss as indicated by equation (7c). In comparison, Figure A14-2 only shows a 3 dB insertion loss beyond 21 GHz for a different M862B switch. In general, to obtain the lowest insertion loss at high frequencies, mechanical switches are used. However, they require larger activation power than pin diode switches so they were not used. Also the switching time must be small compared to integration time. This is no problem for pin diode switches with can operate in nanoseconds compared to microseconds for mechanical switches.

Unlike the lower frequency units, which are far removed from the 22.235 GHz water vapor line, the 20 GHz radiometer requires a more precise frequency determination. The radiometers frequency response is shown in Figure 29 next to the detector. These measurements were obtained using an *RF* sweep generator whose attenuated output of -87 dBm is connected to the pin diode switch input. The sweep generator frequency was varied between 20.15 to 21.15 GHz and the normalized radiometer output voltage is plotted in dBV or 20 $Log_{10}V$ as a function of equivalent *IF* frequency (0.9 to 1.9 GHz). However, unlike the LNB which covers a 1 GHz bandwidth from 20.2 to 21.2 GHz, the radiometer output peaks at 20.5 GHz with a 250 MHz bandwidth. As explained next, this narrower bandwidth centered at 20.5 GHz occurs primarily due to the detector.

In a similar experiment, the detector was excited using a sweep generator that operates at lower frequencies with its power level set to -25 dBm. As with the radiometer, the normalized detector output voltage in dBV is plotted in Figure 29 as a function of *IF* frequency. The detector circuit shown in the Figure has its amplifier gain set to 10 by setting R_G to 5.5 K. Its response peaks at 1.26 GHz with a sensitivity of 18.6 mv/μw for input power less than 20 μw, or output voltages less than 370 mv. For some unknown reason this sensitivity is much larger than the 5.0 mv/μw measured in Figure 19 for the 4 GHz radiometer detector. Note also that the detectors response has a 230 MHz bandwidth with a shape similar to the radiometer spectrum. As such, the radiometer response is mainly due to the detector which inadvertently narrows the bandwidth beyond that of the LNB. Also, as mentioned above, the increased insertion loss of the pin diode switch beyond 20.8 GHz also contributes to the response.

Unfortunately, the Ka band radiometer detects *RFI* at certain times when viewing space. An example of the interference seen by the 20 GHz radiometer is shown in Figure 31, where the *RFI* begins at 9 pm on May 3, 2017. Unlike the occasional spikes observed at 4 GHz in Figure A12 of Appendix A12, the 20 GHz interference exhibits long-period stepwise jumps displaying similar noise variations before and during *RFI*. Compared to the 12 GHz measurements, the 20 GHz radiometer displays abrupt offsets of 0.1 to 0.3 volts that persist throughout the evening until the next day at 12 pm. Other observations show similar interference beginning at later times. As explained next, different methods were tried to identify *RFI*.

As discussed in Section 8.3, the *RFI* only occurs when viewing space so I originally felt it was due to TV broadcast from geostationary satellites. However, the antenna was not directed toward Direct-TV geostationary satellites and the *RFI* only occurred at certain times. Also, the interference was not observed for the radiometers operating within the C and Ku broadcast bands. As such, I concluded that the *RFI* is probably due to non-stationary low orbiting satellites or to localized sources in my area. I next tried to identify the *RFI* frequencies using spectrum analyzer measurements at the *IF* output. However I could not isolate any signals coincident with the time of interference. Also, unlike the 4 GHz radiometer, I was unable to reduce the *RFI* using different *IF* filters. Lastly, I considered performing statistical analysis of the time series as discussed in Chapter 5 to identify and mitigate *RFI*. This technique however requires a number of narrow band filters, which I did not have, to partition the *IF* signal into small subbands and digitally obtain the histogram distribution. The distribution would then be used to determine if the *RFI* differs significantly from Gaussian thermal noise. However, as stated above, the noise fluctuations in Figure 31 appear similar before and during *RFI* so this approach may not work.

After constructing the 20.5 GHz radiometer, I found that the FCC allocates a narrow protective band between 21.2 to 21.4 GHz for radio astronomy and space research although it excludes mobile services. I therefore constructed a 2nd radiometer using another LNB but shifted it's *LO* from 19.25 GHz to a maximum of 19.93 GHz by reducing its DRO size and tuning it. Different views of the LNB are shown in Figure 28 as well as its DRO and tuning screw. This 680 MHz increase of the *LO* increased the radiometer *RF* frequency to the edge of the protective band at 21.2 GHz as defined by the detector's peak response at 1.26 GHz. Unfortunately, such DRO modification reduces its LNB performance by altering its gain and temperature stability. Also, as shown in Figure 30 this pin diode switch has an 8 dB insertion loss at 21.2 GHz. This required a doubling of the detector amplifier gain from 10 to 20. The switch's larger insertion loss also increased the NEΔT so a 5 second integration time was needed compared to 1 second for the 20.5 GHz radiometer. However, after all of this modification I only observed a small reduction in interference so it was not worth the effort. By the way, a much improved 21.2 GHz radiometer is described in Appendix A14. It uses a higher frequency Norsat LNB (9000D) and a General Microwave pin diode switch with lower insertion loss beyond 20 GHz. It's notable however that Figure A14-4 of Appendix A14 also shows that this 21.2 GHz radiometer displays the same small *RFI* reduction when compared with the 20.5 GHz measurements. *Furthermore, no interference is observed for the 22.2 GHz radiometer measurements since its frequency is allocated mainly for radio astronomy use by the FCC.*

Figure 27 - The top lid of the 20.5 GHz radiometer is opened to show the components. Except for the Norsat LNB, the components are similar to the 12 GHz radiometer shown in Figure A4.

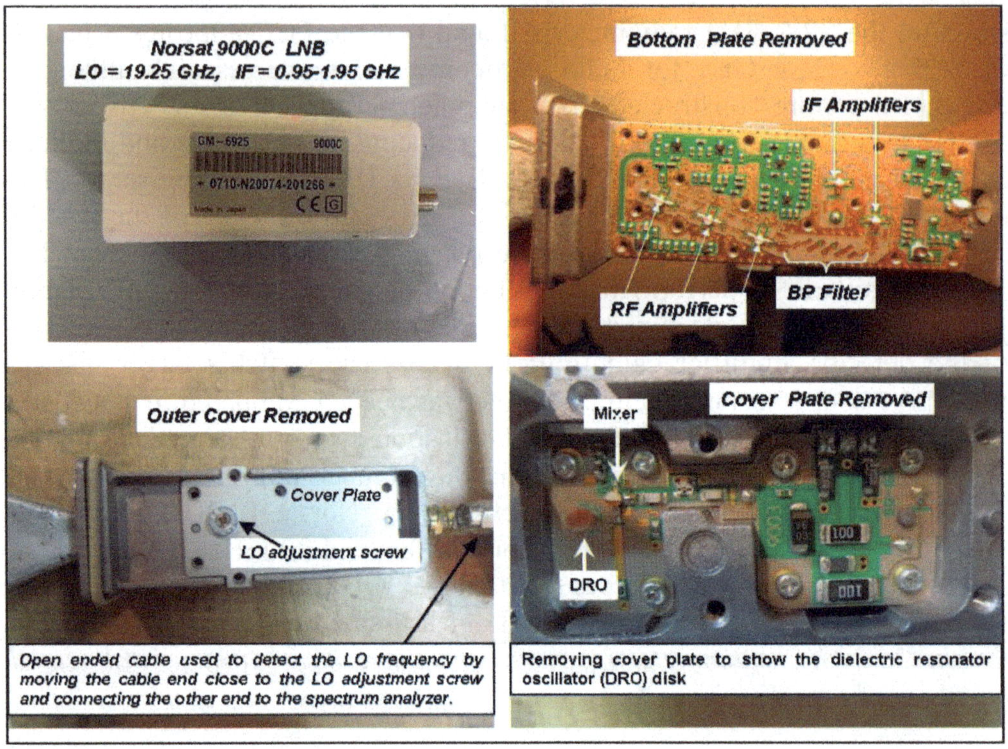

Figure 28 – The top and bottom-right show the inside view of the Norsat 9000C LNB used in the 20.5 and 21.2 GHz radiometers. Microstrips on a printed circuit board connect the components. The top-right shows the cover plate removed to view the *RF* amplifiers, filter and *IF* amplifiers illustrated in Figure 6. The bottom-left shows the *LO* tuning screw to set the dielectric resonator oscillator (DRO) frequency. Its cover plate is removed on the bottom-right to view the DRO.

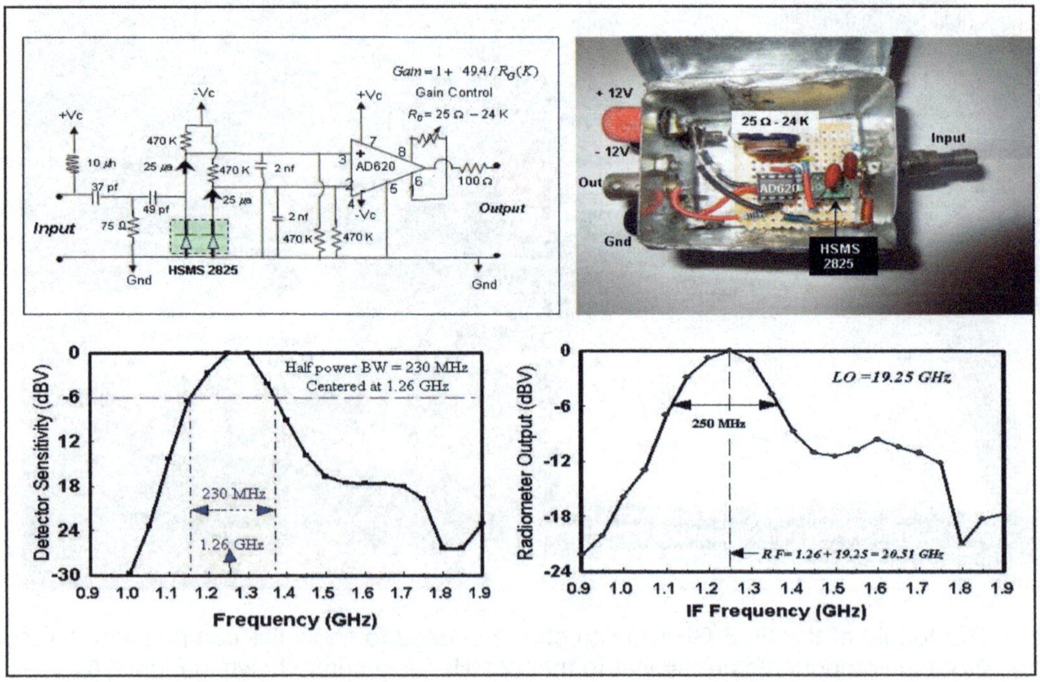

Figure 29 - The 20 GHz radiometer uses a temperature compensated detector (top-right) similar to the other radiometers. Its circuit (top-left) uses an AD620 difference amplifier with its gain set to 10 by setting R_G to 5.5 *K*. Note that its input contains a multiplexer circuit to power the LNB while passing the *IF* signal to the detector. Also observe that the normalized detector sensitivity (bot-left) peaks at 1.26 GHz with a 230 MHz bandwidth. Due to the detector, the normalized radiometer response (bot-right) peaks at 1.26 GHz (*IF*) + 19.25 GHz (*LO*) = 20.51 GHz with a 250 MHz bandwidth.

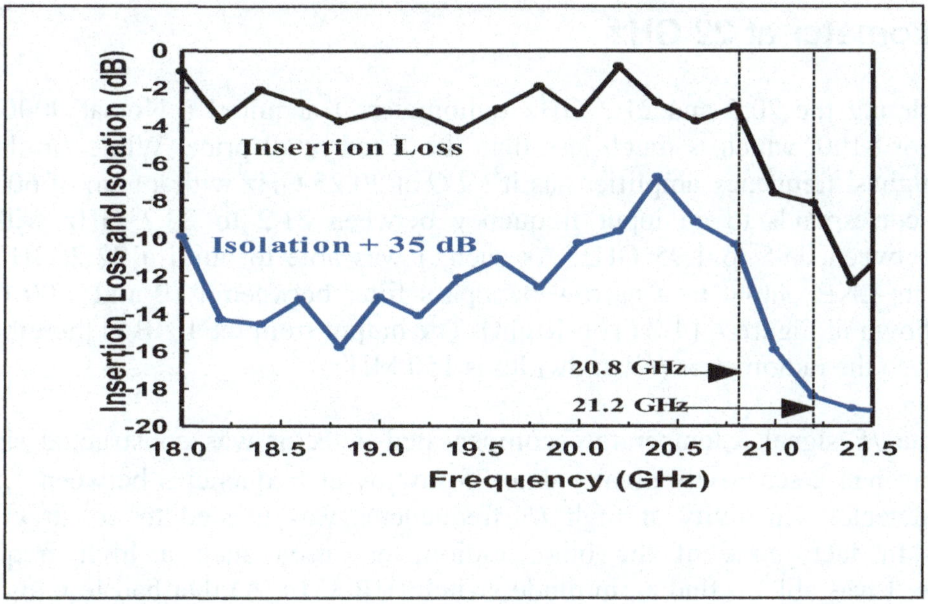

Figure 30 - General Microwave pin diode switch (M862B) measurements of insertion loss and isolation at frequencies beyond 18 GHz. It is specified to operate between 0.1 and 18 GHz with 2 dB insertion loss and 45 dB isolation. The measurements show little degradation between 18 and 20.8 GHz. However, beyond 20.8 GHz its insertion loss decreases to 8 dB at 21.2 GHz and continues decreasing.

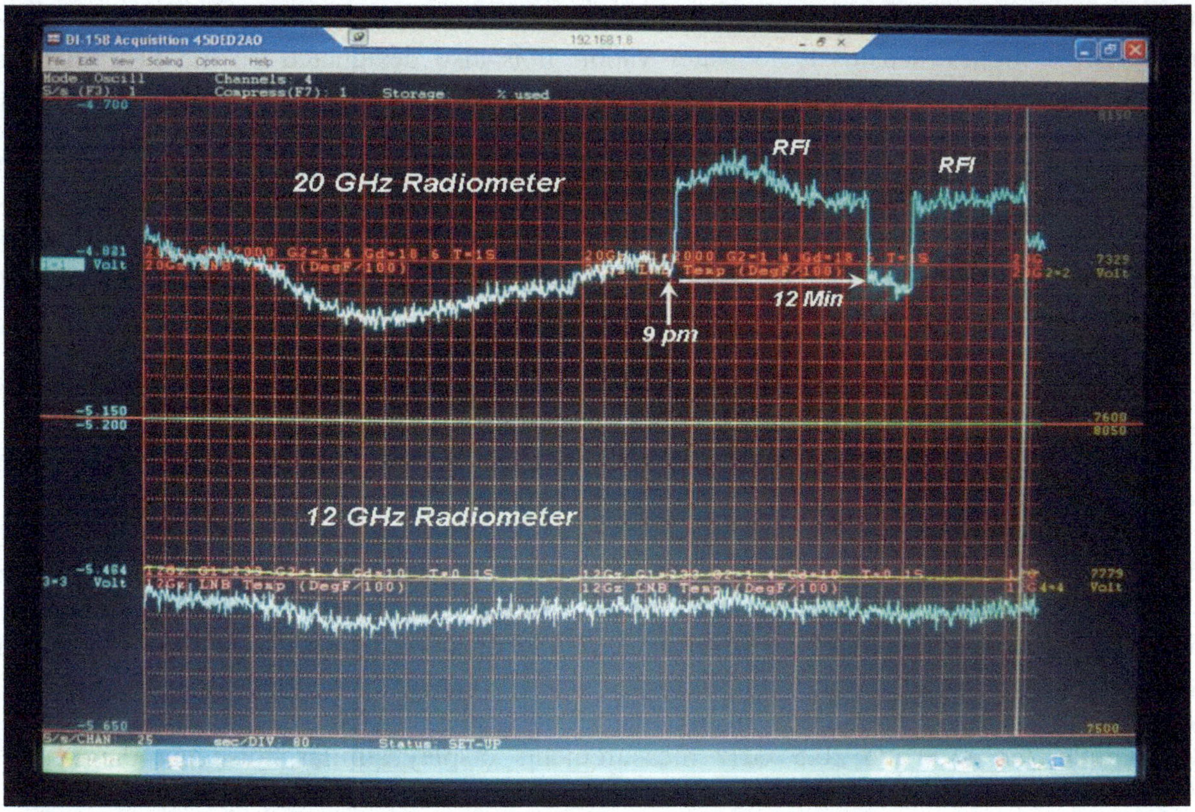

Figure 31 – Cloudy sky measurements using the 20 GHz and 12 GHz radiometers on May 3, 2017. This oscilloscope picture was taken a few minutes prior to and after 9 pm when abrupt jumps lasting 12 minutes occur due to *RFI*. Integration times for the 20 and 12 GHz radiometer are 1.0 and 0.1 second, respectively. The radiometer output voltages are shown using the same vertical range of 0.45 volts, with both horizontal time scales covering the same 53 minute period.

8.2 Radiometer at 22 GHz

Upon completing the 20.5 and 21.2 GHz radiometers I acquired a Norsat 9000D LNB from a distributor for $100, which is much less than the listed retail price. While similar to the 9000C LNB, this highest frequency amplifier has it's *LO* at 20.25 GHz with a gain of 60 dB. The 1 GHz higher *LO* corresponds to an input frequency between 21.2 to 22.2 GHz with its *IF* output frequency between 0.95 to 1.95 GHz. As such, I was able to build a 22.2 GHz radiometer by connecting its LNB output to a narrow bandpass filter between 1.80 and 2.00 GHz with sharp cutoffs as shown in Figure A14-2 (Top-Right). The output from the LNB is therefore between 1.80 to 1.95 GHz so the radiometers *IF* bandwidth is 150 MHz.

To detect the *IF* signal a temperature compensated detector was constructed similar to that in Figure 29 but had a sensitivity greater than 25 mv/μw at frequencies between 1.5 and 2.2 GHz. This high detector sensitivity at high *IF* frequencies was needed to construct the 22.2 GHz radiometer. In fact, none of the other radiometers used such a high frequency detector. Furthermore, I was able to find a pin diode switch (HP 33142A) that had low insertion loss at 22 GHz. The radiometer is shown in Figure 32 with its components mounted on a metal baseplate which was later enclosed in a metal cabinet. To set the different amplifier gains and approximately calibrate the radiometer, clear sky measurements were taken during the winter when the water vapor was low. The corresponding gains were found to be G_1=2000, G_d=10 and G_2=2.8 for a radiometer output of about -9 volts, which is increased to near zero volts when viewing an ambient temperature calibration target.

Since the radiometer frequency is centered near the 22.235 GHz water vapor line, Figure 13 shows it to have the highest sensitivity to water vapor. It is also designated by the FCC to be in a protective region used mainly for radio astronomy. As an example, Figure 33 shows the effect of *RFI* on the 22.2 GHz radiometer measurements. This digitally recorded data also shows measurements obtained using the 20.5 and 11.7 GHz radiometers while viewing overcast skies through the glass patio door on December 28 between 7:30 and 11:00 PM. The three radiometers were placed above each other similar to that shown on the book's cover page. During this time period the 20.5 GHz radiometer measurements display sporadic jumps due to strong *RFI* while none is seen for the 22.2 and 11.7 GHz measurements.

To obtain similar instrumental noise the 22 GHz radiometer uses a 1 second integration time while the 20 and 12 GHz radiometers have a 0.1 second integration time. Such noise is similar to that seen in Figure 33 around 10:30 PM when no *RFI* is evident at 20.5 GHz. The 10 times larger integration time required for the 22.2 GHz radiometer is attributed to its LNB noise figure of 1.5 dB. In comparison the LNB noise figure is 1.3 dB at 20.5 GHz and only 0.3 dB for the 11.7 GHz radiometer. As shown by equation (7b), the 22 GHz radiometer noise is also larger due to its smaller bandwidth of 150 MHz compared to 250 MHz for the other radiometers. Lastly, the pin diode switch insertion loss is likely larger at 22.2 GHz than at 20.5 GHz. In contrast to the larger noise at high frequencies, the 11.7 GHz measurements display very little noise due to its small LNB noise figure. Another example of the smaller noise at 11.7 GHz was shown in Figure 31. In this example, the 11.7 and 20.5 GHz measurements display similar noise but the 11.7 GHz radiometer only uses a 0.1 second integration compared to 1 second at 20.5 GHz.

Another comparison between radiometer measurements was recorded the following day on December 29 when light rain occurred between 9:30 to 11:00 AM. The measurements are plotted in Figure 34 and show no *RFI* during this time period for the 20.5 GHz measurements. Beginning at about 10:30 AM, all of the measurements respond to the rain. However, while all radiometers show an increase due to rain and clouds, the 20.5 GHz measurements display the largest variation, while the 22.2 and 11.7 GHz are very similar with a slope of 1.1 between the measurements. These different variations are also seen in the bottom of Figure 34 when plotting the measurements against each other. The following explains the much larger signal observed due to rain and clouds at 20.5 GHz compared to 22.2 GHz.

The ratio of the 20.5 and 22.2 GHz measurements is analyzed using equation (19). Calculations of the transmittance ratio in (19) for $\nu_1 = 22.2$ GHz, $\nu_2 = 20.5$ GHz and $\theta = 20^0$ is found to decrease from about 0.99 to 0.90 as the water vapor increases from 0 to 40 mm. Also, the liquid water absorption ratio $(\nu_1/\nu_2)^2$ is 1.1 so the combined terms is nearly unity. Also, the two radiometers have nearly the same antenna beamwidth so $\Delta Q_{\nu_1} / \Delta Q_{\nu_2}$ is also near unity. Therefore, the rain measurements should be about the same without the glass door. The slope of 0.53 between measurements must therefore result from the glass transmission coefficient ratio in (19). This slope is similar to the 1/2.2 = 0.45 slope mentioned on page 39 from the insertion loss measurements for $\nu_1 = 11.7$ GHz and $\nu_2 = 3.9$ GHz. However, as shown in Appendix A10 the transmission coefficients vary greatly with frequency and glass dimensions. As such the slopes can not be compared using models without having the exact glass dimensions, which I did not have.

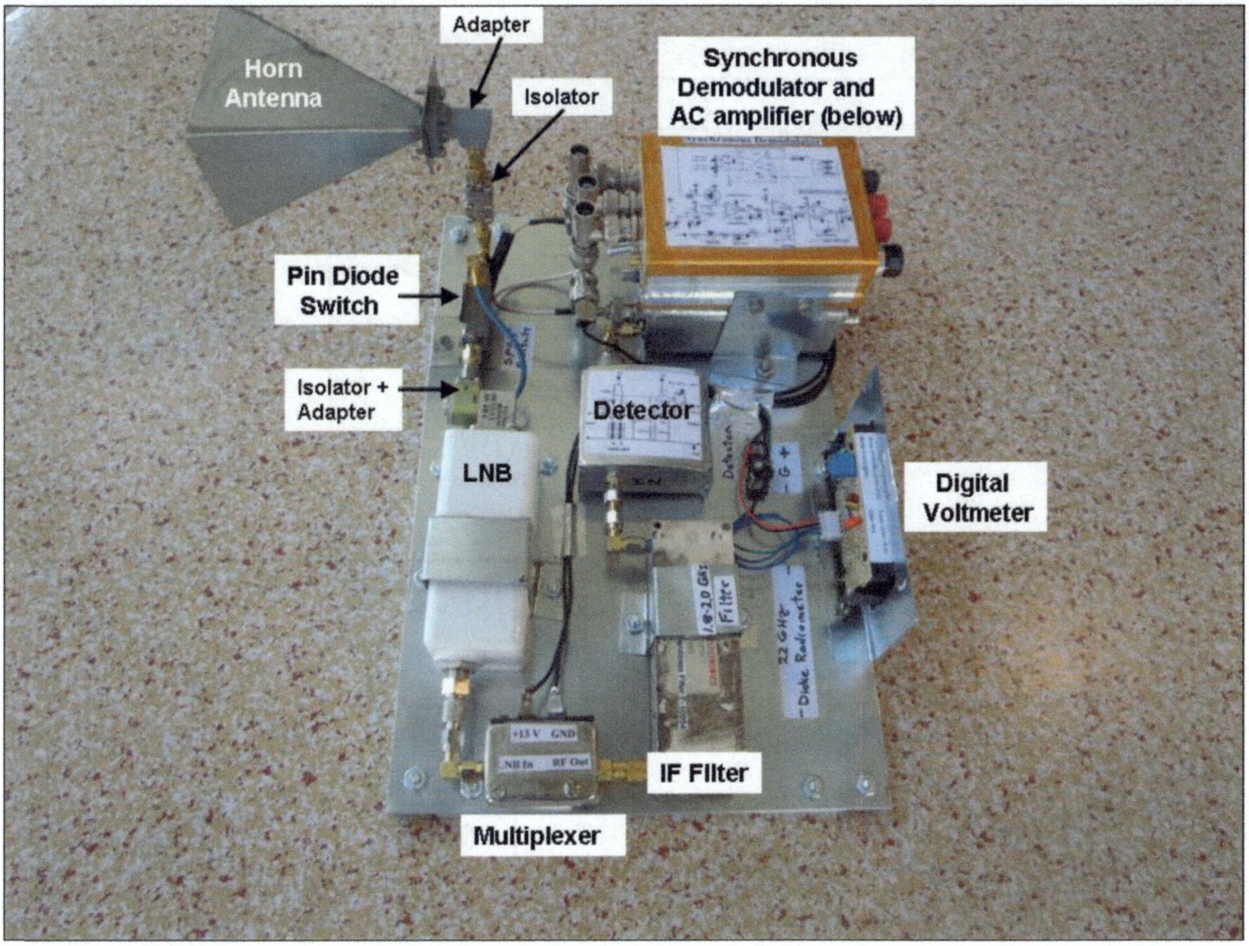

Figure 32 - Shown is my highest frequency, 22.2 GHz, radiometer prior to being enclosed in a cabinet. All components are mounted on a metal baseplate. Starting at the front end is the horn antenna, which is connected in-turn to the isolator, pin diode switch, Norsat 9000D LNB, multiplexer, *IF* filter, detector, AC amplifier and lastly the synchronous demodulator.

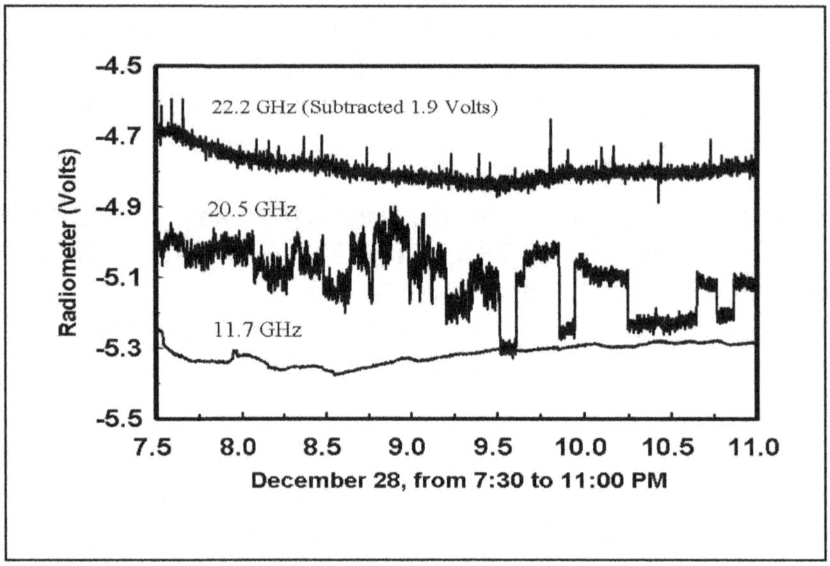

Figure 33 - The 22.2, 20.5 and 11.7 GHz radiometers view overcast skies on December 28, 2019 from 7:30 to 11 PM. Only the 20.5 GHz radiometer displays sporadic *RFI* at this time. Integration times are 0.1 second for the 11.7 and 20.5 GHz radiometers and 1 second for the 22.2 GHz radiometer. For display purposes the 22.2 GHz measurements are offset by 1.9 volts (see Text).

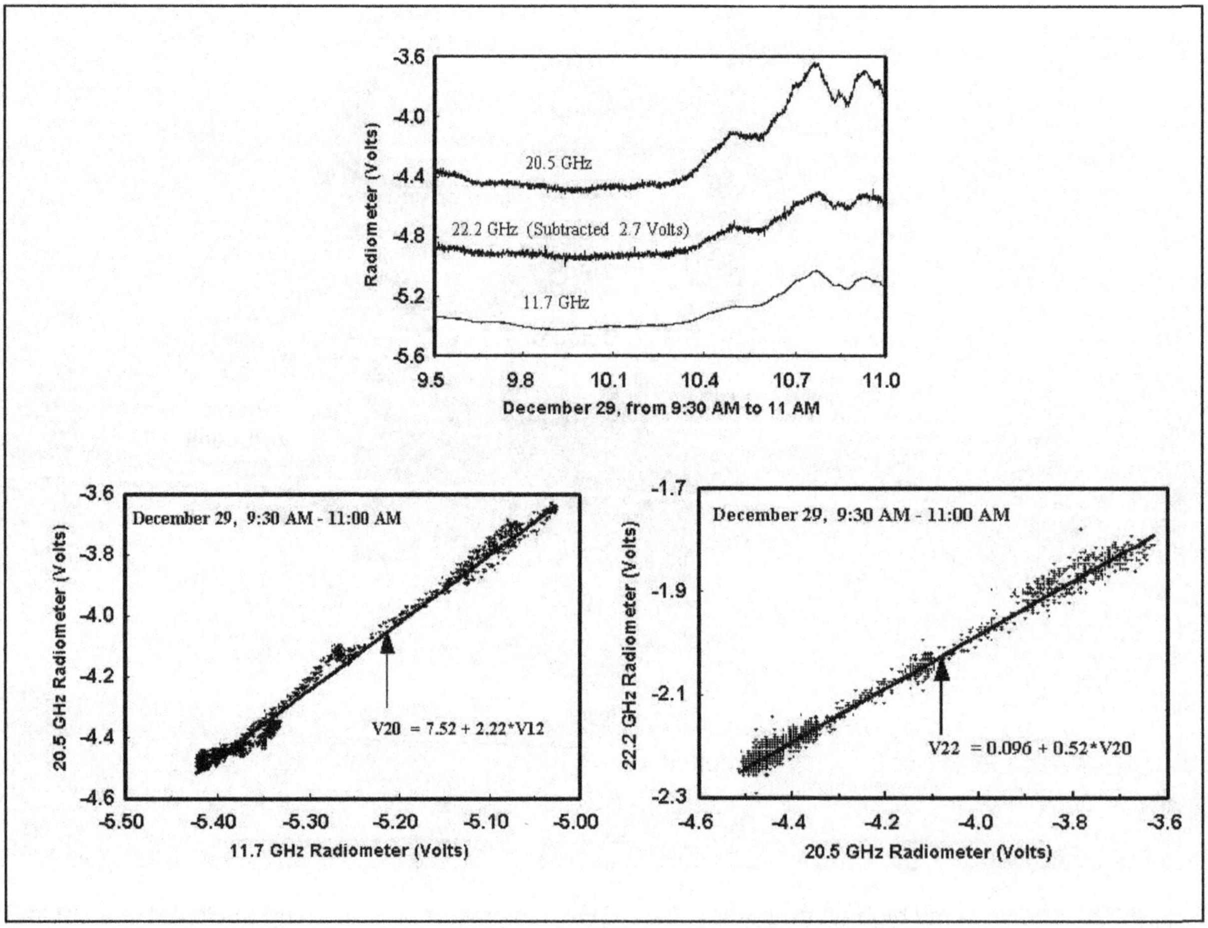

Figure 34 - The 22.2, 20.5 and 11.7 GHz radiometers view clouds and light rain on December 29, 2019 between 9:30 to 11 AM. Unlike Figure 33, no *RFI* occurs at this time at 20.5 GHz (top plot). The bottom right shows the 22.2 GHz against 20.5 GHz measurements. Its slope of 0.52 is due to the glass reflectivity. Similarly, the bottom left shows the 20.5 GHz plotted against 11.7 GHz measurements. This slope of 2.22 is due to the glass reflectivity as well as the cloud absorption ratios in equation (19).

8.3 Cloud and Rain Measurements

1 - Sky viewing cloud measurement

In the absence of interference the 20.5 GHz radiometer provides high quality atmospheric measurements when viewing space. As an example, Figure 35 shows the 20 GHz and 12 GHz radiometer measurements of clouds seen through the glass patio door on May 3, 2017 between 12 pm and 9 pm (Top-left). Throughout this 9 hour period no *RFI* is seen except for its small onset at 9 pm. The top-right picture taken at about 4 pm shows a large visually opaque (*i.e.*, deep) cumulous cloud while the bottom-left is a plot of the 20 GHz against 12 GHz cloud measurements over the 9 hour period. During this time the radiometers view varying amounts of cloud liquid water within the FOV which appears as abrupt variations. Also, the slope between the 20 and 12 GHz radiometer measurements is shown to be 2.0 in Figure 35. This slope will be compared next with ground viewing measurements as well.

2 - Ground viewing (indirect) rain measurement

As shown in Figure 35, *RFI* is observed to start at 9:30 PM on May 3 when viewing space. To help identify the direction of *RFI* the radiometer cart is tilted downward so the antenna views the ground rather than space. However, while no interference is seen when viewing the ground, a small increase of 0.168 volts is seen for the 20 GHz measurements at 10:20 PM along with a smaller increase of 0.084 volts at 12 GHz. Both voltage increases were found to occur during a brief light rain event. These measurements are shown in Figure 36 and analyzed below.

Neglecting any sky radiation contribution from antenna sidelobes, the increased measurements shown in Figure 36 must be due to the downwelling radiation T_d by rain which is reflected by the ground. As such, the brightness temperature obtained using (10a) and (14b) is

$$T_b(\nu) = [(1 - R_g)\varepsilon_S T_S + R_g T] + T_b' \qquad (22a)$$

$$\text{where } T_b' = (1 - R_g) R_S [1 - (\tau_{H20}\tau_{02}\tau_{Liq})^{\mathrm{Sec}\,\theta}]\, T_M \,. \qquad (22b)$$

The dominant term in (22a) is the glass-attenuated surface emitted radiation $(1\text{-}R_g)\varepsilon_S T_S$. Equation (22a) also contains T_b' which is the downwelling atmospheric radiation reflected by the ground and attenuated by the glass door. This term, given by (22b) is responsible for the small perturbations seen in Figure 36. It increases the brightness temperature due to the liquid water transmittance. However, as seen in Figure 22 the emissivity over vegetated land is about 0.95 so the ground reflectivity $R_S = 1\text{-}\varepsilon_S$ is only 0.05. Therefore, to obtain a significant brightness temperature increase τ_{Liq} must be small or the liquid water content must be large.

To further analyze the measurements in Figure 36, equations (22b) and (17) are used to obtain the ratio of 20 to 12 GHz brightness temperature change ΔT_b due to liquid water variations ΔQ_v , viz.,

$$\frac{\Delta T_b(\nu_1)}{\Delta T_b(\nu_2)} \cong \frac{1 - R_g(\nu_1)}{1 - R_g(\nu_2)} \left[\frac{R_S(\nu_1)}{R_S(\nu_2)}\right] \left[\frac{\tau_{H2O}(\nu_1)\tau_{O2}(\nu_1)}{\tau_{H2O}(\nu_2)\tau_{O2}(\nu_2)}\right]^{Sec\,\theta} \left(\frac{\nu_1}{\nu_2}\right)^2 \frac{\Delta Q_{\nu_1}}{\Delta Q_{\nu_2}} \qquad (23)$$

where $\nu_1 = 20.5$ GHz and $\nu_2 = 11.7$ GHz .

Since the land reflectivity is nearly the same at the two frequencies, the brightness temperature ratio (23) is the same as the sky viewing equation (19). This similarity is also supported by the fact that the slope of 2.0 when viewing clouds in Figure 35 is identical to that when viewing the reflected radiation due to rain in Figure 36, *i.e.*, $\Delta T_b(20)/\Delta T_b(12) = 0.168/0.084 = 2.0$.

To obtain larger reflected sky radiation, a 2nd ground viewing experiment was performed on May 5, 2017 when heavier rain occurred. Figure 37 (Top) shows the time series over a 9 hour period starting at 12 AM. For more detail, the bottom Figure shows an expanded plot of the measurements around 8 AM during the heavy rain period. As in Figure 36, the radiometers measure abrupt increases due to rain, whose ground reflected radiation is larger than before. Furthermore, the 20 GHz measurement saturates at 0.65 volts, while the 12 GHz measurement increases to 0.55 volts. The slope $\Delta T_b(20)/\Delta T_b(12)$ now becomes $0.65/0.55 = 1.2$ which is smaller than the 2.0 slope found in the previous light rain case due to saturation at 20 GHz.

This indirect observation of rain by way of ground reflection is a demonstration of the high radiometric sensitivity to detect very small signals. A much larger signal is obtained using direct (skyward) observations. Instead of (22b) the brightness temperature increase due to rain is based on (15a) so T_b' becomes $(1-R_g)[1-\tau^{\mathrm{Sec}\,\theta}]\,T_M$. The absence of the ground reflectivity term results in a 20 fold increase in T_b' compared to (22b) assuming a land reflectivity of 0.05. Compared to the indirect result, the much larger direct measurement of rain occurs due to the large contrast between cold space and the rain emitted radiation.

In an analogous situation, satellite radiometers operating below 20 GHz rarely detects rain over land due to the very small difference between the radiation emitted by land and that of rain. However, these same lower frequency radiometers readily detect both rain and clouds over oceans since the sea surface emissivity is only about 0.45 so a large contrast exists between the ocean and atmospheric emission. Furthermore, as discussed in Chapter 12 and displayed in Figure 69, satellite radiometers operating at high frequencies (*e.g.*, 85 GHz) detect rain over land not by emission, but due to the scattering of upwelling radiation by the ice particles formed aloft in rain clouds as part of the precipitation process. Ice has very low absorption at microwave frequencies so it mainly scatters radiation. The back-scatter due to ice particles therefore reduces the brightness temperature measurements below that emitted by land as well as ocean surfaces. This contrast due to precipitating clouds is further enhanced using vertical polarization since then the surface emissivity is highest, resulting in the largest brightness temperature difference due to ice-scattering over land as well as over lower emissivity oceans.

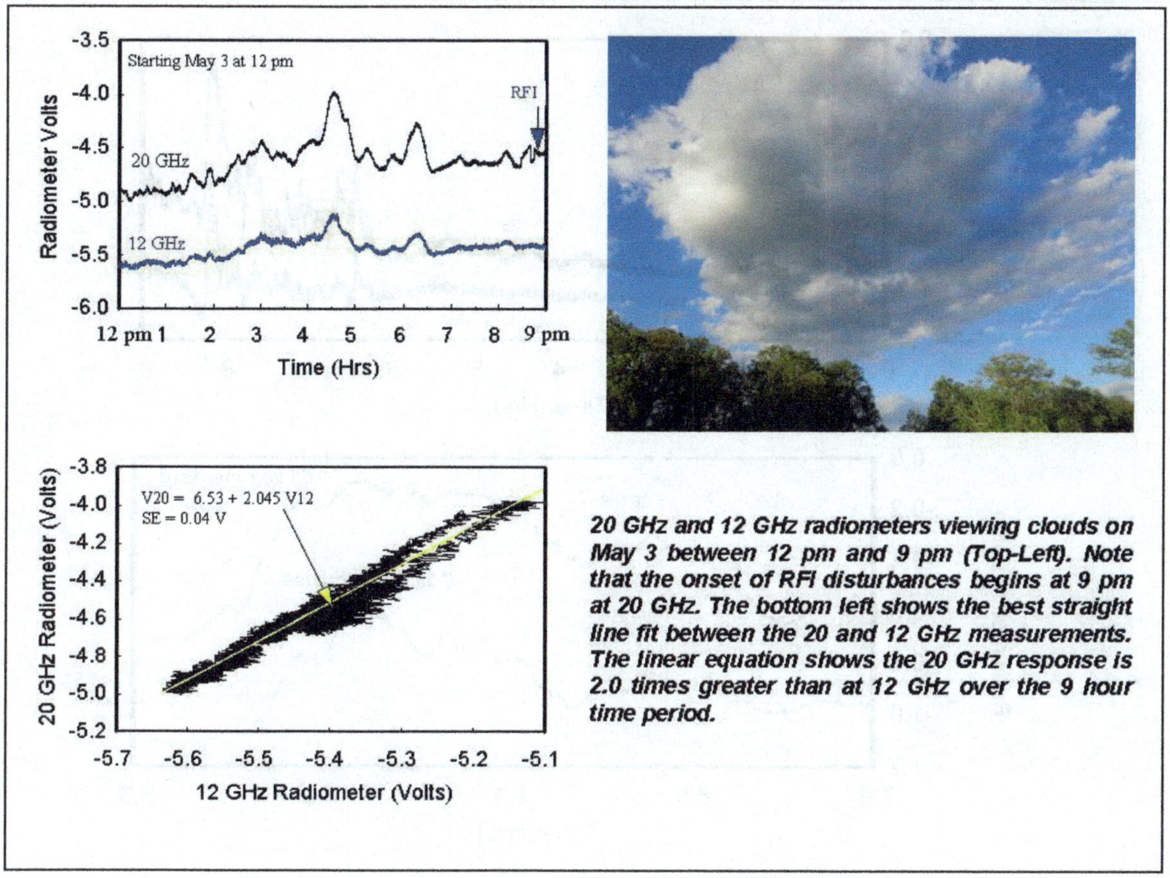

Figure 35 - Clouds measured using the 20 and 12 GHz radiometer on May 3, 2017 between 12 and 9 pm (Top-left). The cloud picture on the top-right is at 4 pm, while the bottom-left plots the 20 GHz against 12 GHz measurements over the full 9 hour time period. The measurement ratio of 2.0 is the product of the cloud transmittance and glass transmission ratios in equation (19).

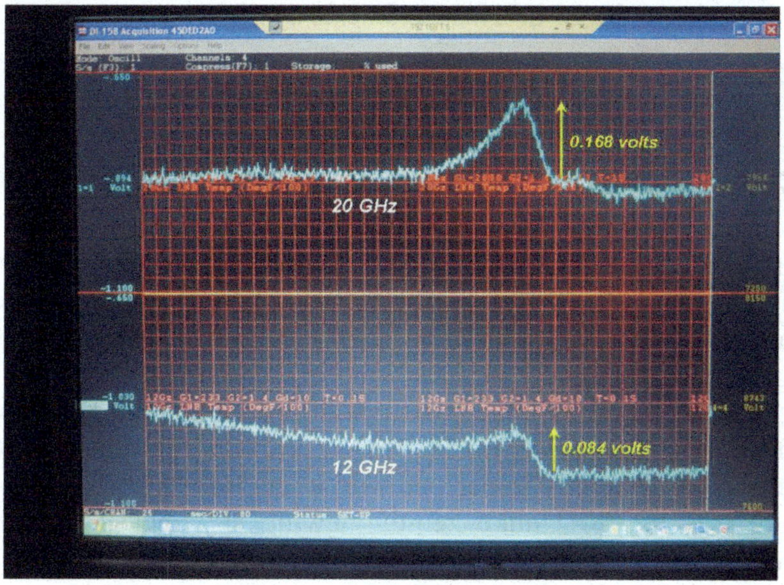

Figure 36 – Surface viewing rain response on May 3, 2017 at 10:20 PM using the 20 GHz and 12 GHz radiometers. The integration time is 1 second for the 20 GHz radiometer and 0.1 second for the 12 GHz radiometer. Both measurements are viewed on my oscilloscope using a 0.45 volt vertical scale and 53 minute horizontal time scale. The 0.168 volt increase at 20 GHz and 0.084 volt at 12 GHz results from rain emitted radiation reflected by the ground to the antennas. This ratio of 0.168/ 0.084 = 2.0 is the product of cloud transmittance and glass transmission ratios in equation (23) at the two frequencies.

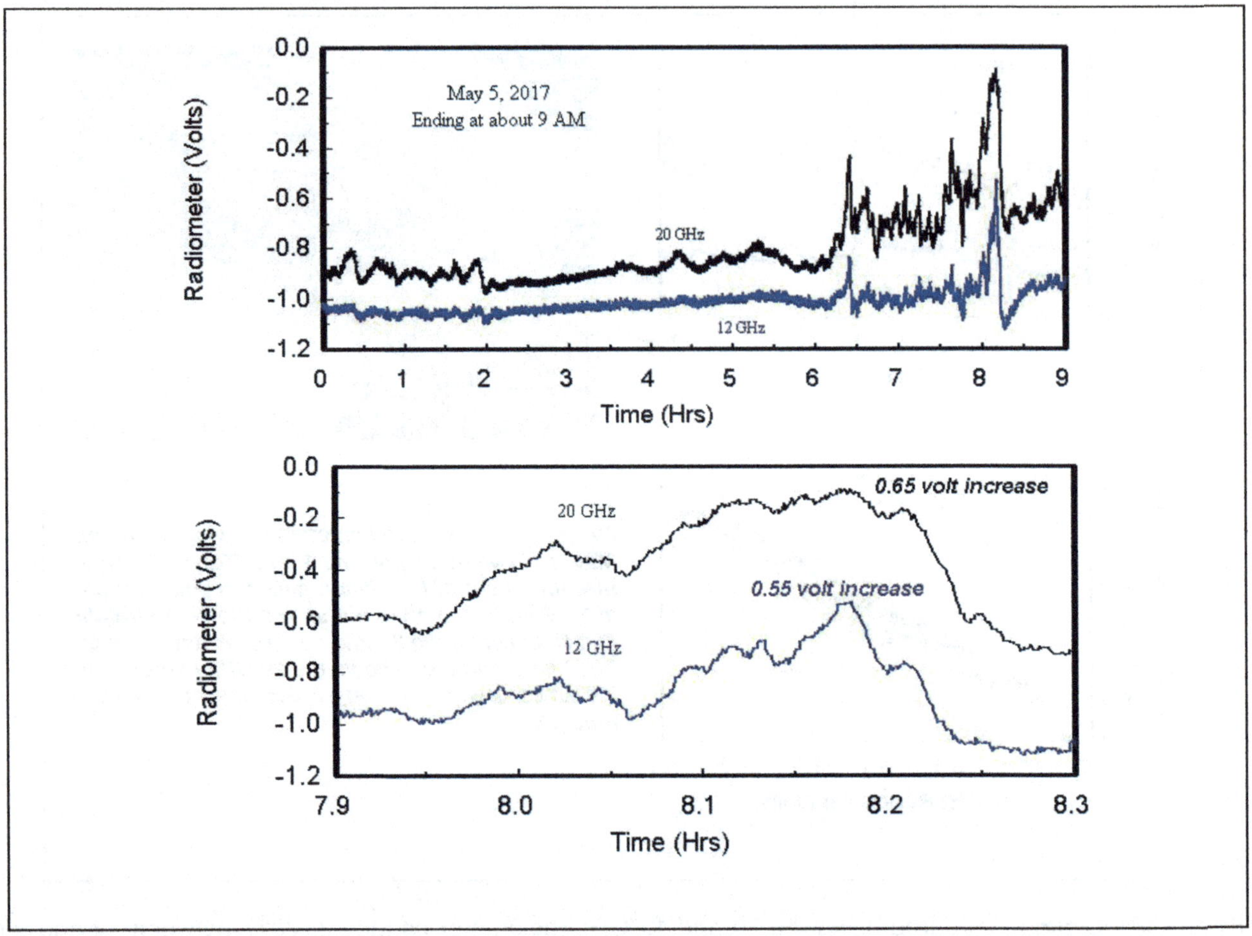

Figure 37 - Ground viewing measurements by the 20 and 12 GHz radiometers on May 5, 2017 over a 9 hour period starting at 12 AM. The top plot displays the total time period while the bottom is an expanded plot during the most intense rain period around 8 AM. As in Figure 36, the integration time is 1 second for the 20 GHz radiometer and 0.1 second for the 12 GHz radiometer. The ratio of the 20 GHz to 12 GHz measurements during this intense rain period is 1.3. This ratio is less than the 2.0 value found for light rain in Figure 36 due to saturation of the 20 GHz measurements by heavy rain.

8.4 Tipping Curve Calibration Measurements

Clear sky calibration of the 20 GHz radiometer requires atmospheric corrections due to water vapor and oxygen absorption. As mentioned in Section 4.4, such corrections can be obtained using the modeled calculations in Figure 13 based on the RAOB observations. Alternatively, this section applies the tipping curve calibration procedure listed in Table 1 and described in Appendix A13. In summary, it uses angular scan measurements to determine the atmospheric absorption as well as the slope and intercept parameters in the linear calibration equation (3). The first such measurements were made from my upper patio deck on July 9, 2018 at about 2 PM when the surface temperature was 92 ^{0}F. Unlike zenith viewing sky measurements at 4 and 12 GHz described in Sections 4.3 and 4.4, the 20 GHz radiometer views the sky at different zenith angles θ using the flat copper reflector in Figure 38. However, as shown later a nonlinear increase in radiometer voltage occurs at low elevation angle due to blockage by my house and trees. Therefore, to obtain the best view of space the reflector was scanned only at zenith angles between 0^0 and 70^0 in 10^0 steps so that Secθ varies between 1.0 and 2.9. Figure 39 plots the radiometer voltage as a function of $\mathrm{Sec}\theta$ where the procedure was repeated four times for consistency. The following outlines the steps used for calibration.

As discussed in Section 4.3, the linear brightness temperature equation is

$$T_b = \mathrm{I} + \mathrm{S}\,\mathrm{V} \tag{24}$$

where I is the offset, S is the radiometric gain and V is the radiometer output voltage. Figure 39 shows the best straight line fit of the measurements is

$$\mathrm{V} = -8.38 + 1.26\ \mathrm{Sec}\theta \tag{25}$$

so that upon substituting (25) into (24) the brightness temperature is

$$T_b = (\mathrm{I} - 8.38\mathrm{S}) + 1.26\ \mathrm{S}\cdot\mathrm{Sec}\theta. \tag{26}$$

As explained in Section 4.3, the sky brightness temperature is

$$T_b = \tau^{Sec\theta}\, T_{CB} + (1 - \tau^{Sec\theta})\,T_M \tag{27}$$

so that for $Sec\theta = 0$, $T_b = T_{CB} = 2.7\,K$. Therefore, setting $Sec\theta = 0$ in (26) we can write[5],

$$T_{CB} = \mathrm{I} - 8.38\mathrm{S}. \tag{28}$$

Also, when viewing the high emissivity warm target at temperature T_W = 307 K (92 ^{0}F), the radiometer voltage is 94 mv so from (24),

$$T_W = \mathrm{I} + 0.094\mathrm{S}. \tag{29}$$

Subtracting (29) from (28) the radiometric gain is,

$$\mathrm{S} = \frac{T_{\mathrm{W}} - T_{\mathrm{CB}}}{8.38 + 0.094} = 35.91\ \ K/\mathrm{Volt} \tag{30}$$

and from equation (29) the offset is

$$\mathrm{I} = T_{\mathrm{W}} - 0.094\mathrm{S} = 303.62\ \ K. \tag{31}$$

Substituting (30) and (31) into (24), the calibration equation becomes

$$\boxed{\mathrm{T_b} = 303.62 + 35.91\ \mathrm{V}} \tag{32}$$

---

[5] If the calibration parameters I and S are known *a-priori* then (27) can be used to derive T_{CB}. This approach was in fact used by Dicke, and later by Penzias and Wilson, to accurately measure T_{CB}. Conversely, as done here, (27) can be used to calibrate the radiometer since T_{CB} is now known with very high accuracy.

and the calibrated tipping curve (26) becomes

$$T_b = 2.7 + 45.175 \cdot \mathrm{Sec}\theta \ . \tag{33}$$

It should be noted that the radiometric gain of 35.91 *K*/Volt in (32) depends on the extrapolated cosmic background measurement of -8.38 volts as obtained by setting $Sec\theta = 0$ in (25). Also, as discussed next the atmospheric opacity depends on the tipping curve slope of 1.26 in (25).

As explained in Appendix A13, the atmospheric opacity is obtained using equation A13-19,

$$\alpha = -\frac{d\ln\left[T_M - T_b(\theta)\right]}{d\,\mathrm{Sec}\theta} \quad \text{with} \quad T_M \approx 285\,K\,, \tag{34}$$

with $T_b(\theta)$ obtained by applying the calibration equation (32) to the voltage measurements of Figure 39. The natural logarithm term is then plotted against $Sec\theta$ in Figure 40 to obtain the best fit line,

$$\ln\left[T_M - T_b(\theta)\right] = 5.71 - 0.229\,Sec\theta \ . \tag{35}$$

Therefore, upon applying (35) to (34) the opacity becomes 0.229 nepers so the transmittance is

$$\tau = e^{-0.229} = 0.80\,. \tag{36}$$

The cloud-free atmospheric transmittance is the product of the oxygen and water vapor components, *i.e.*,

$$\tau = \tau_{O2}\tau_{H2O} = e^{-\alpha_{O2}(\nu)}e^{-TPW/W(\nu)} \tag{37}$$

where the water vapor opacity is proportional to *TPW* so its transmittance can be parameterized using the frequency dependent quantity $W(\nu)$. Neglected is the small effect of the water vapor vertical distribution on transmittance and its frequency dependence. This dependence is discussed in Section 8.7 when describing the *TPW* retrieval accuracy. At $\nu = 20.5$ GHz the calculated values of α_{O2} and W are listed in Table 5 on page 68 as 0.014 and 270 mm, respectively. Therefore, from equation (37), $TPW = -W\,[\ln\tau + \alpha_{O2}] = 58$ mm for $\tau = 0.80$. However this *TPW* value approaches the upper limit found in tropical atmospheres and is therefore too large.

The large opacity of 0.229 contained in equation (36) results from the slope of 1.26 seen in Figure 39. This large slope is likely due to blockage of the sky radiation at low elevation by objects such as trees. Furthermore, the thermally emitted radiation by trees can be leaked into the feed horn as the reflector rotates to large zenith angles. Such leakage or spillover at large viewing angles is a result of sidelobes in the antenna pattern. As discussed next, one way to reduce the angular variation of the antenna pattern is to use a conical scanning reflector. Better yet, any antenna pattern variation caused by the reflector rotation is eliminated by mounting the radiometer on a tripod as described below.

To minimize the antenna pattern variation with scan angle, radiometers use a circularly polarized corrugated feed horn with the reflector rotated in azimuth about the feed horn axis [12]. As such, the antenna pattern is nearly unchanged as the reflector rotates in azimuth. My feed horn is linearly polarized so the brightness temperature varies with azimuth angle, *i.e.*, $T_b = T_V\,Cos^2\phi + T_H\,Sin^2\phi$, where T_V and T_H are the vertical and horizontal polarized components, respectively. Fortunately, the sky radiation (27) is un-polarized so $T_b = T_V = T_H = \tau^{\mathrm{Sec}\phi} T_{CB} + (1 - \tau^{\mathrm{Sec}\phi}) T_M$ where the zenith angle now becomes the azimuth angle. Alternatively, rather than use a reflector to view the sky, the radiometer can be scanned in elevation and azimuth by mounting it on a tripod. As shown in Figure 41, this approach was used by Dicke for sky measurements and will be used here. By the way, as an interesting side-note, when Dicke developed his radiometers in the 1940's there were no isolators and no Eccosorb target available. In fact, he constructed a target out of cloth imbedded with conductive graphite material, which was wedge-shaped and mounted on a board. Furthermore, as shown in Figure 41, this "shaggy dog" target was shaken randomly when obtaining near-field measurements. This random motion reduces the time-averaged coherent *LO* radiation, which is leaked out of the antenna and reflected by the target back into the radiometer (see Chapter 4, page 15).

Figure 42 shows the 20 GHz radiometer mounted on a tripod whose elevation can be varied from 0^0 to 60^0 so that the zenith angle varies from 90^0 to 30^0, respectively. However, while no problem occurs when viewing the sky at high elevation, various objects can obscure the sky at low elevation. The area outside my patio is surrounded by trees. Also, for privacy, a plastic lattice panel exists between my house and my neighbors. Since this panel produces blockage, the tripod is raised on a patio table to elevate it as shown in Figure 42. The best orientation to view space is determined by moving the tripod and scanning it in elevation and azimuth to find the location having the lowest radiometer voltage measurement. Using this procedure, Figure 42 shows the best sky view at low elevation while Figure 43 shows the cloud free measurements obtained at this location on August 26 when the surface temperature was 89^0 F.

The left plot in Figure 43 displays the full range of radiometer measurements as a function of Secθ for zenith angles between 35^0 and 80^0. Note that the voltage increases linearly for $\theta < 70^0$, after which the response becomes nonlinear for larger angles. This nonlinear behavior is due to blockage of the sky radiation by trees in addition to its thermal emission. Also, as explained in Appendix A13, the tipping curve procedure also becomes less accurate for zenith angles exceeding 70^0. Such effects for $\theta > 70^0$ is also displayed by comparing the best fit quadratic equation with the linear equation obtained for $\theta < 70^0$ or Sec $\theta < 3$, *i.e.*,

$$V = -8.10 + 1.105\ \mathrm{Sec}\theta\ . \qquad (38)$$

For further analysis, Figure 43 on the right shows an expanded plot for zenith angles less than 70^0. It also plots the tipping curve (38), and for comparison shows the previous tipping curve equation (25) with its larger slope of 1.26. This earlier tipping curve was obtained using the setup in Figure 38 which results in greater blockage due to the privacy panel and radiation leakage from surrounding objects. Unlike the nonlinearity seen at large zenith angles, the slope difference for smaller zenith angles appears as a smaller effect in Figure 43. Furthermore, upon repeating the previous analysis using the new tipping curve (38), the calibration equation is

$$\boxed{T_b = 310.6 + 38.0\ V}\ . \qquad (39)$$

Also, after substituting (38) into (39) the calibrated tipping curve equation now becomes

$$T_b = 2.7 + 42.0\ \mathrm{Sec}\theta . \qquad (40)$$

Finally, upon applying (40) to (34) the opacity is 0.192. Also, the transmittance becomes 0.825 with $TPW = -W\,[\ln\tau + \alpha_{O2}] = 48$ mm. This *TPW* is 10 mm less than that obtained using the setup in Figure 38, which contains more error sources. This smaller amount of water vapor is more reasonable so that the calibration equation (39) should be more accurate than that of (32).

For comparison with the above results, an additional set of tipping curve measurements was taken on April 3, 2019. During this early spring period the trees were bare and the water vapor amount is much less than during the summer period. Figure 44 (Left) shows a plot of the radiometer measurements for this day in addition to the natural logarithm plot (Right). Both quantities are plotted as a function of air mass (Secθ). Due to less obstruction by trees a smaller elevation angle is obtained so the maximum zenith angle is 80 degrees (Sec $\theta = 5.76$) with none of the nonlinear effects seen in Figure 43. Also, the resulting calibration equation becomes

$$\boxed{T_b = 293 + 38.89\ \mathrm{V}} \qquad (41)$$

which has a slope or radiometric gain slightly larger than that of equation (39). This calibration should be more accurate than that obtained from Figure 43 (Right) since it is obtained using a wider range of data. Furthermore, the opacity given by the slope of the logarithmic plot is 0.0955 so that the resulting *TPW* is now only 22 mm, which is less than half that obtained on August 26, 2018.

Lastly, for comparison the 20 GHz radiometer was calibrated on August 7, 2021 using near-field calibration procedure described in Section 4.1. Using this method, the target was cooled below 10^0 F (260 K) until it warmed up to room temperature (299 K) for over an hour. It was then heated up to 125^0 F (324 K) and slowly cooled down to room temperature. Figure 45 shows the range of temperature (Top-Left) and radiometer measurements (Top-Right) used for calibration. It also shows the derived calibration equation (Bottom-Left). Note that although the temperature range is much smaller than the tipping curve experiment, the resulting calibration equation obtained from these measurements is

$$\boxed{T_b = 299 + 39.57\ \mathrm{V}} \qquad (42)$$

whose radiometric gain is nearly the same as in (41). This similarity between the two calibration methods attests to the linearity and long term stability of the radiometer. Comparing equations (39), (41), and (42) the gain increased from 38.0 to 38.9 to 39.57. This 4 % increase in gain can be attributed due to slight changes in the instrumental parameters and differences in measuring techniques.

The above comparison demonstrated very good agreement between the tipping curve and laboratory calibration measurements. Furthermore, one also expects similar agreement using the cold sky measurements indicated in Table 1 on page 14. In a somewhat different vain, the question arises as to whether it is also possible to use the near-field laboratory procedure to measure the emissivity of any arbitrary surface rather than the more traditional far-field measurement approach? Unfortunately, as explained in Appendix A17, laboratory measurements of emissivity are difficult to obtain accurately. In essence, the *emission from a non-unity emissivity surface is seen differently by an antenna when viewed in the near-field and far-field.* Stated differently, only the high emissivity calibration target radiates virtually unperturbed by antenna-coupling when seen in the near-field. As such, it is necessary to know the "coupling factor" when determining the emissivity using near-field radiometer measurements. This parameter is related to the fractional area *f* referred to in Appendix A17.

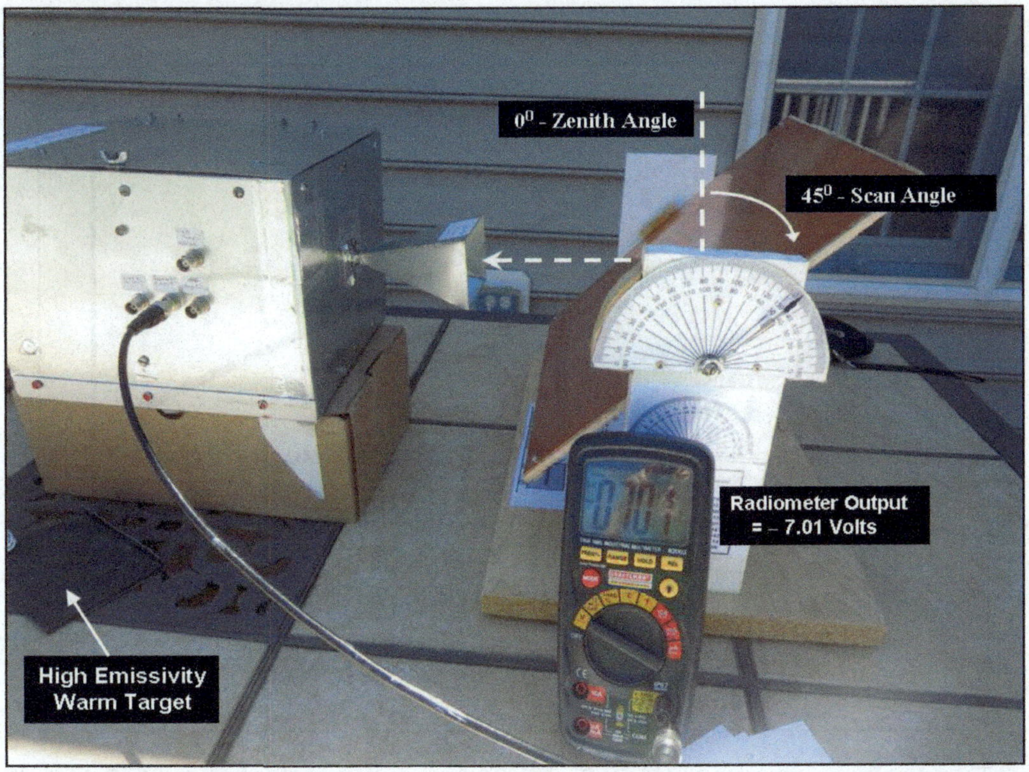

Figure 38 -Tipping curve measurements of the 20 GHz radiometer on July 9, 2018. The radiometer antenna views space using a flat copper reflector. The zenith angle is varied between 0^0 to the largest unobstructed slant angle of 70^0. The picture shows the upward viewing scan angle of 45^0, corresponding to a zenith angle of 0^0 and radiometer voltage of -7.01 volts.

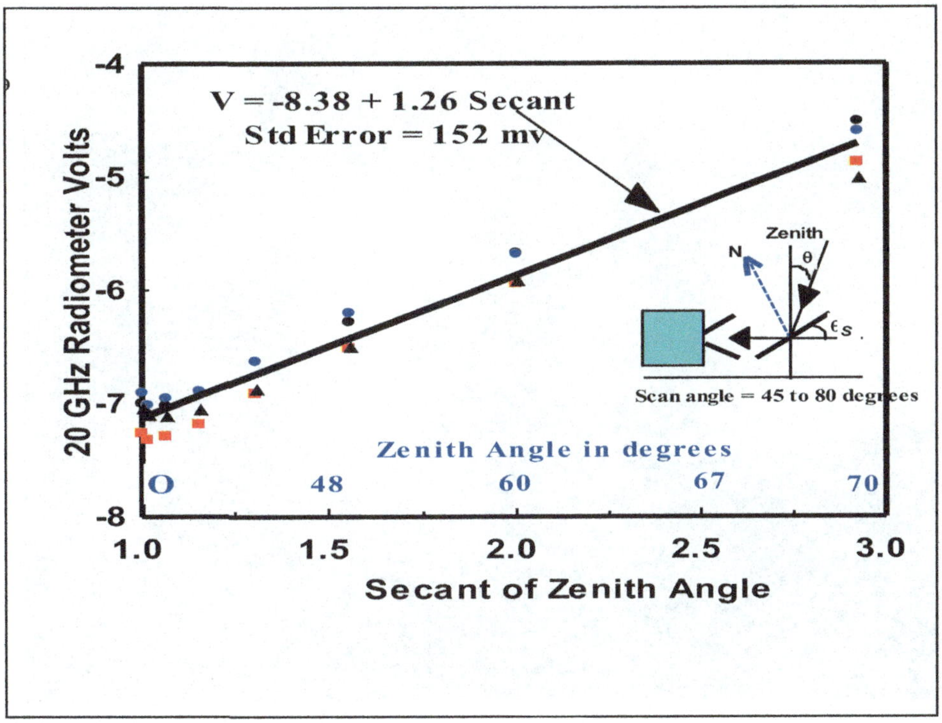

Figure 39 - Tipping curve measurements on July 9, 2018 for the 20 GHz radiometer using the setup shown in Figure 38. For a consistency check, the series of eight angle measurements was repeated four times. The plot shows the data and straight line fit between the radiometer voltage and secant of zenith angle, θ. The resulting calibration equation based on the plotted data is given by equation (32).

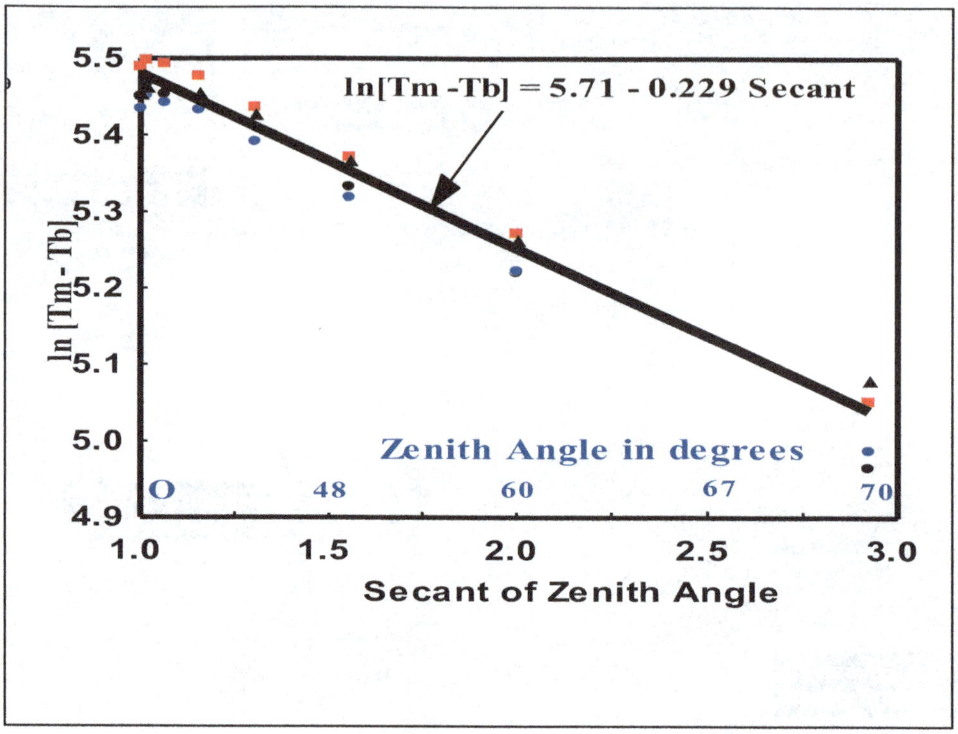

Figure 40 - Tipping curve measurements in Figure 39 is used to calibrate the 20 GHz radiometer. The resulting calibrated brightness temperature given by equation (32) is used to obtain the atmospheric opacity by plotting ln (T_M -T_b) as a function of *Sec* θ. The straight line fit shown above is then applied to equation (34) to obtain the opacity, which is the negative slope of 0.229.

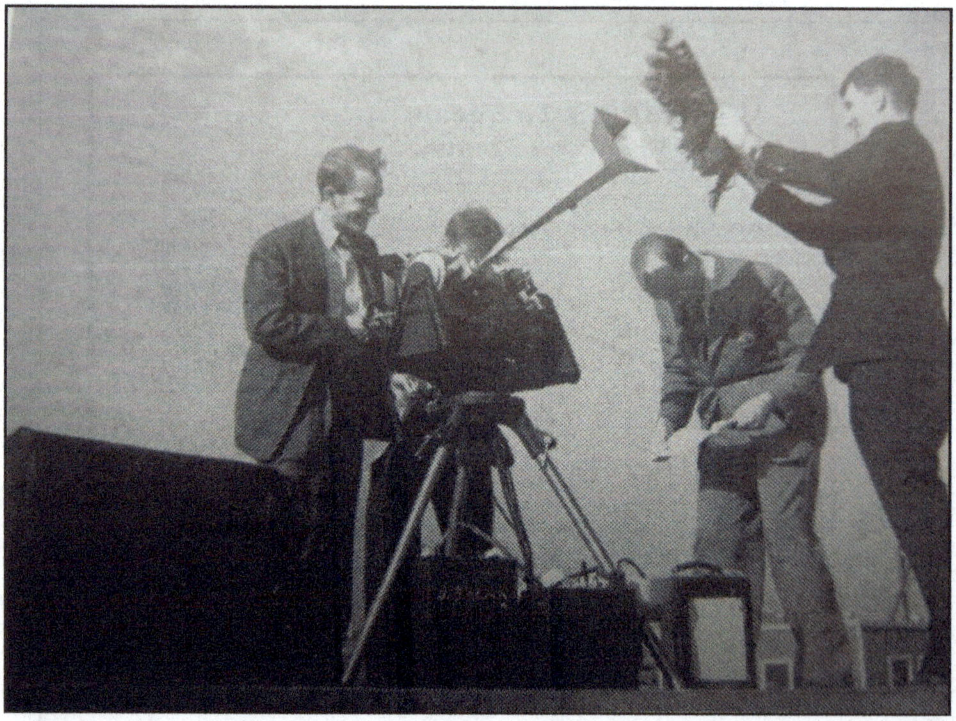

Figure 41 - Tipping curve absorption measurements by Dr. Robert Dicke along with his associates. Starting on the Left is E. Beringer, R. Kyhl, A. Vane and R. Dicke. It shows Dicke holding up a "shaggy dog" absorber in front of one of his radiometers while a chart recorder on the ground plots the measurements. This picture is in the book "Five Years at the Radiation Laboratory, MIT, by C. Newton, T.E. Peterson and N. J. Perkins (Eds.)," *The Andover Press, Ltd*., 205 pgs., 1946.

Figure 42 - Clear sky tipping curve measurements for the 20 GHz radiometer made August 26, 2018. Unfortunately, blockage of the sky radiation at low elevation (large zenith angle) occurs by surrounding trees, the house and privacy panel. To minimize blockage the radiometer is mounted on a tripod that is elevated on a table. The right picture shows the view seen by the radiometer when the antenna views at low elevation. Observations for zenith angles greater than 70^0 are obstructed by the trees and house.

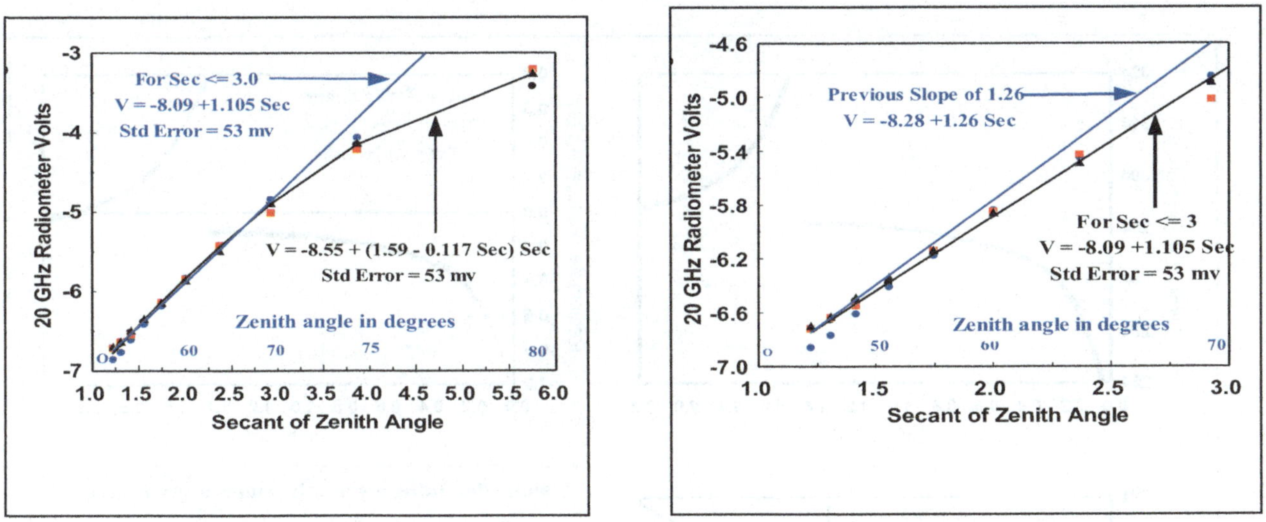

Figure 43 - Clear sky tipping curve measurements of the 20 GHz radiometer made August 26, 2018 using the iimproved setup in Figure 42. The left plot shows data for all zenith angles ($\theta = 35^0$ to 80^0 in 5^0 steps) while the expanded plot on the right is only for angles up to 70^0. Radiometer voltage is plotted as a function of Sec θ in both plots where the measurements were repeated four times. The full range of angles in the left plot displays a nonlinear increase in voltage for Sec $\theta > 3$ due to obscurations described in the text. The Figure also shows a best fit quadratic equation in addition to the linear tipping curve $V = -8.09 + 1.105$ Sec θ for Sec $\theta < 3$. The right-most Figure only shows data for Sec $\theta < 3$ along with the linear tipping curve equation. For comparison, it also shows the previous tipping curve equation of Figure 39 having a slope 1.26. This larger slope of 1.26 was obtained using the setup in Figure 38 which results in measurement errors (see text).

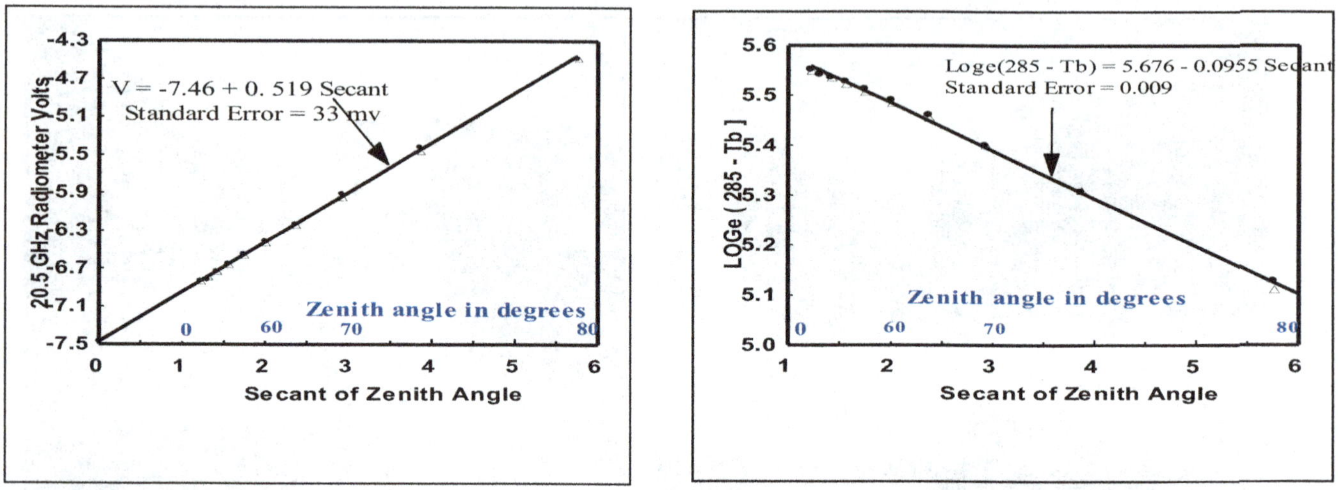

Figure 44 - Clear sky tipping curve measurements of the 20 GHz radiometer obtained on April 3, 2019. At this time the trees were bare and the water vapor is less than that for the summer as exhibited in the measurements of Figure 43. As such, compared to Figure 43, the left-most plot now displays data for zenith angles between 35^0 up to 80^0 in 5^0 steps. The radiometer voltage is again plotted against Sec θ where the measurements are repeated twice. The tipping curve equation now becomes $V =- 7.46 + 0.519$ Sec θ so that the slope of 0.519 is much smaller than the 1.105 slope in Figure 43. Also shown on the right is a plot of the natural logarithm in (34) plotted against Sec θ. The negative slope of 0.0955 is the atmospheric opacity for this day which is much smaller than 0.229, as displayed in Figure 43. As a result, the *TPW* is now only 22 mm whereas it was 48 mm on August 26, 2018.

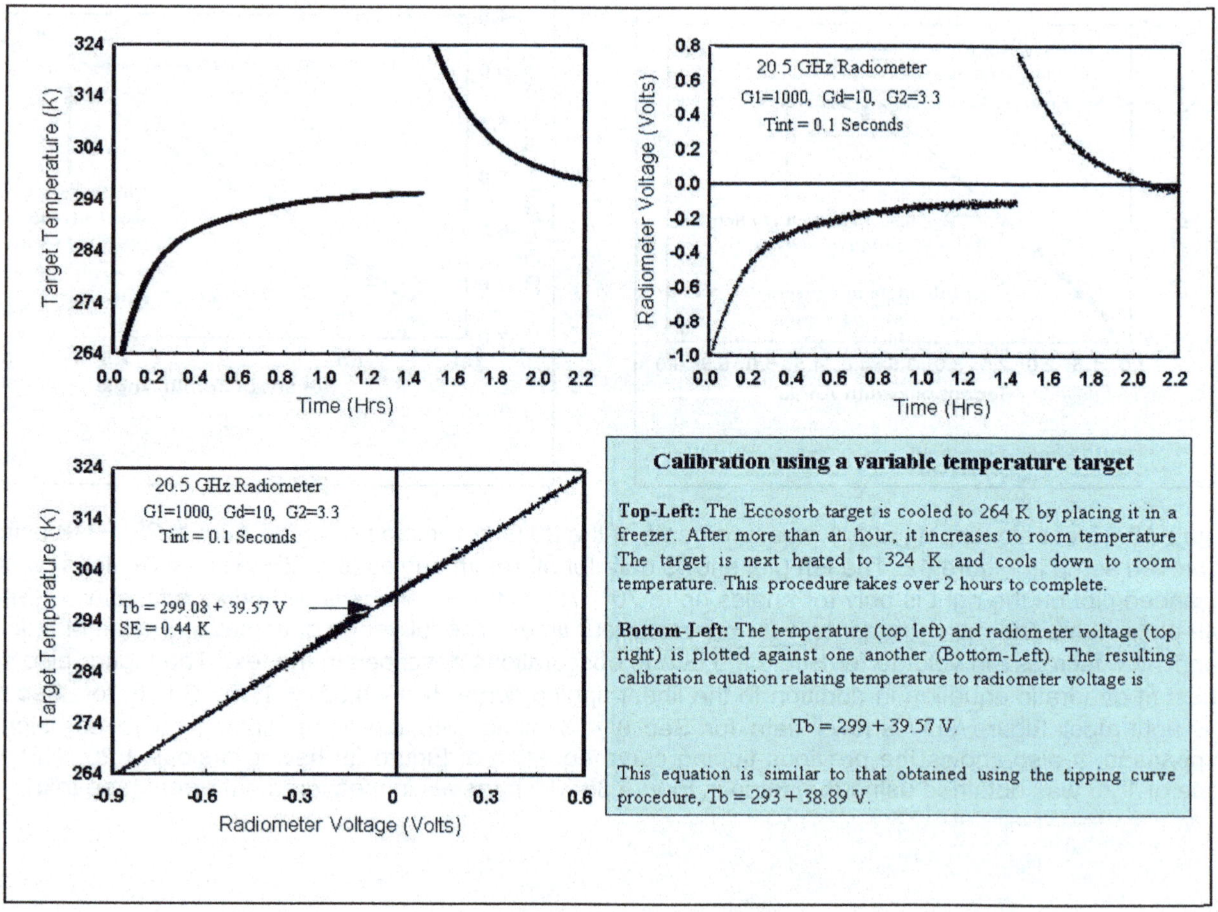

Figure 45 – Calibration of the 20.5 GHz radiometer using the near-field, variable target temperature procedure summarized in the insert. The calibration equation in the bottom left $T_b = 299+39.57$ V is similar to that obtained using the tipping curve procedure, i.e., $T_b = 293 + 38.89$ V.

8.5 Water Vapor and Cloud Water Simulations

Having calibrated the radiometer using tipping curve measurements the radiometer can be used to determine the water vapor by combining (27) and (37). The algorithm is obtained by considering $TPW << W(\nu) = 270$ mm in (37) and neglecting the small cosmic radiation term in (27), *viz.*,

$$T_b \cong (1-\tau^{Sec\theta})T_M \cong [\alpha_{O2}(\nu)+TPW/W(\nu)]T_M\ Sec\theta \qquad (43a)$$

$$\text{so}\quad TPW \cong -W(\nu)\alpha_{O2}(\nu)+[W(\nu)/T_M]T_b\ Cos\theta \qquad (43b)$$

where T_b is given by the calibration equation (41).

Substituting the α and W parameters from Table 5 into (42b) with $T_M \equiv 285\ K$,

$$TPW \cong -3.78+0.947\ T_b\ Cos\theta. \qquad (44)$$

However, as discussed next, equation (44) only provides accurate *TPW* measurements under clear sky conditions. For cloudy skies, dual frequency radiometer measurements are needed to account for cloud liquid water absorption. The second radiometer frequency is generally chosen to be greater than 22 GHz so that it is more responsive to clouds and less sensitive to water vapor than at 20.5 GHz. Fortunately, however, the 12 GHz radiometer is shown next to have sufficient sensitivity to clouds to correct the 20 GHz measurements. As such, water vapor measurements can be obtained for clear as well as cloudy skies by combining the 12 and 20 GHz radiometer measurements. An example showing actual radiometer measurements is given in Section 8.6.

Algorithms for determining *TPW* and Q are obtained using (27) with the transmittance given as the product of the clear transmittance (37) and cloud transmittance (16a), *i.e.*,

$$\tau(\nu) = \tau_{O2}\tau_{H2O}\tau_{Liq} = e^{-\alpha_{O2}(\nu)}\ e^{-TPW/W(\nu)}\ e^{-Q/Q(\nu)} \qquad (45)$$

with the transmittance parameters listed in Table 5 for the 12, 20 and 22 GHz radiometers having center frequencies of 11.7, 20.5 and 22.2 GHz.

Table 5: Atmospheric Transmittance Parameters, $\alpha_{O2}(\nu)$, $W(\nu)$ and $Q(\nu)$

Radiometer	**Center Frequency**	**Oxygen**	**Water Vapor**	**Cloud ($T_{CLD} = 275\ K$)**
22 GHz	ν_1= 22.2 GHz	$\alpha_{O2}(\nu_1) = 0.016$	$W(\nu_1) = 154$ mm	$Q(\nu_1) = 10.3$ mm
20 GHz	ν_1= 20.5 GHz	$\alpha_{O2}(\nu_1) = 0.014$	$W(\nu_1) = 270$ mm	$Q(\nu_1) = 11.9$ mm
12 GHz	ν_2= 11.7 GHz	$\alpha_{O2}(\nu_2) = 0.010$	$W(\nu_2) = 3989$ mm	$Q(\nu_2)$= 35.8 mm

Substituting (45) into (27) and neglecting the small cosmic radiation term,

$$\frac{TPW}{W(\nu)} + \frac{Q}{Q(\nu)} + \alpha_{O2}(\nu) = -\psi(\nu) \tag{46a}$$

where $$\psi(\nu) = Cos\theta \ln[1 - T_b(\nu)/T_M] \,. \tag{46b}$$

Solving (46a) for *TPW* and *Q* using dual frequency (ν_1=20.5 GHz, ν_2=11.7 GHz) brightness temperatures,

$$TPW = \frac{W(\nu_2)}{\eta - \beta}\left[\psi(\nu_1) - \eta\,\psi(\nu_2) + (\rho - \eta)\,\alpha_{O2}(\nu_2)\right] \tag{47a}$$

and $$Q = -\frac{Q(\nu_2)}{\eta - \beta}\left[\psi(\nu_1) - \beta\psi(\nu_2) + (\rho - \beta)\,\alpha_{O2}(\nu_2)\right] \tag{47b}$$

where $$\beta = \frac{W(\nu_2)}{W(\nu_1)}, \quad \eta = \frac{Q(\nu_2)}{Q(\nu_1)}, \quad \rho = \frac{\alpha_{O2}(\nu_1)}{\alpha_{O2}(\nu_2)} \,. \tag{47c}$$

Except for very large amounts of water vapor and cloud liquid water, $T_b(\nu) << T_M$ so that $\psi(\nu_1) \cong -T_b(\nu_1)\,Cos\theta / T_M$ and $\psi(\nu_2) \cong -T_b(\nu_2)\,Cos\theta / T_M$. Equations (47a) and (47b) then become linearized as

$$TPW \cong \frac{\eta - \rho}{\beta - \eta} W(\nu_2)\,\alpha_{O2}(\nu_2) + \frac{W(\nu_2)\,Cos\theta}{(\beta - \eta)\,T_M}\left[T_b(\nu_1) - \eta\,T_b(\nu_2)\right] \tag{48a}$$

$$Q \cong -\frac{\beta - \rho}{\beta - \eta} Q(\nu_2)\,\alpha_{O2}(\nu_2) - \frac{Q(\nu_2)\,Cos\theta}{(\beta - \eta)T_M}\left[T_b(\nu_1) - \beta\,T_b(\nu_2)\right]. \tag{48b}$$

Substituting the transmittance parameters from Table 5 into (48a, b) we obtain

$$TPW \cong 7.59 + Cos\theta\left[1.19\ T_b(\nu_1) - 3.57\ T_b(\nu_2)\right] \tag{49a}$$

$$Q \cong -0.40 - Cos\theta\left[0.010\ T_b(\nu_1) - 0.157\ T_b(\nu_2)\right]. \tag{49b}$$

Equations (49a, b) use weighted brightness temperature differences to obtain *TPW* and *Q*. Note that *TPW* is positively correlated to the 20 GHz channel, ν_1, with the 12 GHz channel, ν_2, providing cloud corrections. Conversely, *Q* is positively correlated to 12 GHz measurements with the 20 GHz measurements providing small water vapor corrections. Furthermore, by plotting $T_b(\nu_1)$ against $T_b(\nu_2)$ as in Figure 35 one can determine which quantity, *Q* or *TPW* has the larger temporal variability. For example, if the slope is closer to β or 15.7, than water vapor is more variable. Alternatively, if the slope is closer to η or 3.0, then cloud liquid water is more variable.

The fact that the slope in Figure 35 is 2.0, which becomes 3.0 by including a glass reflectivity factor of 0.69 in equation (19) suggests that clouds are the more variable parameter.

More accurate coefficients than those in equations (49a) and (49b) are obtained using statistical regression analysis of simulated brightness temperature measurements (predictors) against the water vapor and cloud liquid water (predictands) in the data. The simulated result uses the radiation transfer equation (5) with the latest atmospheric absorption models to calculate the brightness temperature and provide the standard error of the retrieved parameters. As part of the calculations, the mean atmospheric temperature in (6a) is determined using a global sample of temperature, water vapor and cloud liquid water profiles. The vertical distribution of temperature and water vapor in (6a) was obtained using an historical sample of RAOB data with the total precipitable water increasing from 2 mm to 60 mm as the surface temperature increases from 245 *K* to 303 *K*.

Since cloud liquid water is not available from RAOB data, clouds are artificially introduced at different heights and thickness into each atmospheric profile. The liquid water amount is varied between 0 mm to a maximum of 1 mm, with the smallest liquid water applied to clouds having temperatures below freezing. For reference it should be noted that actual satellite and ground-based radiometer measurements have shown that liquid water greater than about 0.3 mm is generally associated with rain clouds. In fact, it has been customary to identify rain from satellites using such a liquid water threshold.

Upon applying a least squares regression analysis to the simulated data, the resulting dual frequency algorithms for water vapor and cloud liquid water are similar in form to equations (49a) and (49b) but with slightly different coefficients, *i.e.*,

$$TPW = 10.39 + 1.33\ T_b(20) - 3.68\ T_b(12) \quad \text{with} \quad SE = 0.86\ mm \qquad (50a)$$

$$Q = -0.80 - 0.010\ T_b(20) + 0.159\ T_b(12) \quad \text{with} \quad SE = 0.07\ mm. \qquad (50b)$$

For comparison, the optimal single frequency algorithms for water vapor and cloud liquid water are

$$TPW = -6.54 + 0.86\ T_b(20) \quad \text{with} \quad SE = 4.49\ mm \qquad (51a)$$

$$Q = -0.59 + 0.098 T_b(12) \quad \text{with} \quad SE = 0.11\ mm. \qquad (51b)$$

The above equations give the dual frequency and optimal single frequency algorithms for zenith viewing ($\theta = 0^0$) along with their standard errors. For visualization, Figure 46 plots the corresponding retrieved water vapor obtained from the algorithms against the actual data set values. Similarly, Figure 46 shows the results for cloud liquid water. Each Figure plots the results obtained using dual frequency brightness temperatures (Left) as well as that obtained using single frequencies (Right). I should also mention that all of the simulations are for zero bandwidth so that the *TPW* error is minimal.

Comparing (51a) with (50a) we see that the 12 GHz radiometer measurements provides cloud corrections of *TPW*, reducing the standard error (SE) from 4.49 mm to 0.86 mm. These errors, particularly for the single frequency algorithms, would be substantially less for non-precipitating atmospheres where Q is less than 0.3 mm. Also, upon comparing (51b) with (50b) we see that the 20 GHz radiometer provides small water vapor corrections of Q, reducing the error from 0.11 mm to 0.07 mm. Although this appears to be a small improvement it is important for measuring the small amount of liquid water for non-precipitating clouds. However, to better measure Q a higher frequency than 12 GHz is needed. For example, simulations obtained using 20.5 GHz and 31 GHz measurements result in a liquid water error of 0.06 mm compared to 0.07 mm when using the 20.5 GHz and 12 GHz measurements. Also, the *TPW* error was found to be reduced from 0.86 mm to 0.61 mm when substituting the 31 GHz for the 12 GHz measurements. Besides these frequencies, the next section discusses issues regarding the use of other frequencies for *TPW* measurements.

A unique dual frequency Dicke radiometer providing measurements at 20.6 and 31.6 GHz was constructed in 1979 by the late Dr. David Hogg [13]. To obtain the same beamwidth at both frequencies its antenna consisted of a wideband hybrid-mode corrugated horn and offset parabolic reflector [14]. This highly reliable instrument was used in an unattended continuous mode of operation at airports in Denver, Colorado and Washington, D.C. to provide real time data of water vapor and cloud liquid water under all weather conditions to the National Weather Service (NWS) forecast office. As a result of its high performance, similar dual frequency radiometers were developed by other organizations for research and operational use. The only notable difference I found was that instead of a 20.6 GHz, radiometers have begun using 23.8 GHz to reduce *RFI.* This frequency was originally allocated by the FCC to be in a protected region, having no active terrestrial sources. Also, for better protection the 31.6 GHz frequency was reduced slightly to 31.4 GHz. However, I understand the protection near 23.8 GHz has unfortunately been recently relaxed by the FCC.

Lastly, while the coefficients in equations (49a) and (49b) are similar to those in equations (50a) and (50b) these final coefficients are considered more accurate since they are based on a larger data base that is more representative. Also, the single frequency water vapor coefficients in (51a) are also considered more accurate than those in equation (44) due to the larger more representative data base used in their derivation. I should also mention that the use of the logarithmic predictors as in equation (46b) were also analyzed and found to decrease the *TPW* error from 0.86 mm to 0.48 mm. This reduction in error is mainly due to the saturation effect seen in Figure 46 (Top-Left) for $TPW > 50$ mm when using linear predictors. However, the error in Q was not reduced using the logarithmic predictors. This is due to the variations seen in the plots do to parameters such as cloud temperature, which can not be accounted for using linear or logarithmic predictors. Lastly, it should be observed that the algorithms (50a) and (50b) are primarily physically based, with minimal use of statistical correlation between water vapor and temperature. Although such correlation is contained in the RAOB measurements and simulated brightness temperatures, this conclusion is evident by the fact that the coefficients in (50a) and (50b) are similar to those in (49a) and (49b). This point is also discussed in the 2006 paper by Grody *et. al.* More will be said about the use of statistical correlation when discussing temperature retrievals in Appendix A19.

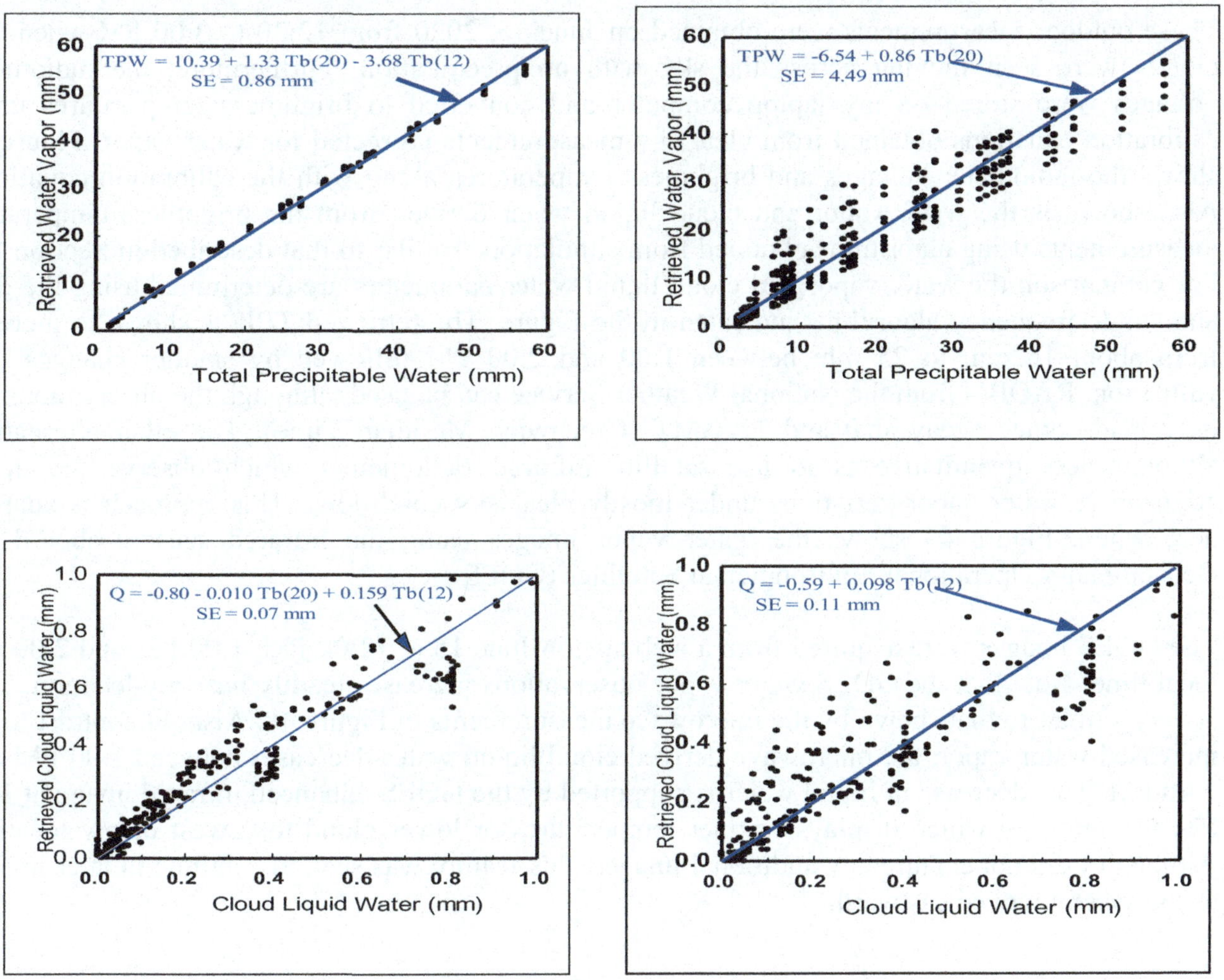

Figure 46 - Retrieval of *TPW* (Top) and Q (Bottom) are shown using simulations. The two top Figures show the *TPW* results while the two bottom Figures show the Q results. Both dual frequency (Left) and single frequency (Right) algorithms are shown in addition to their standard errors (SE).

8.6 Water Vapor and Cloud Water Measurements

As mentioned in the previous section, since the late 1970's, dual frequency ground-based microwave radiometers have measured the atmospheric water vapor, cloud liquid water and precipitation over land at various locations. While many radiometers operate at about 20.6 and 31.6 GHz, this section describes measurements obtained using the radiometers described in Section 8.2 and Chapter 2 which operate at 22.2 and 11.7 GHz, respectively. As discussed in Chapter 5, the 22 GHz radiometer is much less affected by *RFI* than at 20.5 GHz while the 12 GHz radiometer was chosen mainly due to its low cost and available components. The radiometers are also lightweight and small enough to be mounted on tripods. Also, instead of the 22 GHz radiometer shown in Figure 32, a smaller radiometer with lower noise was constructed using a higher frequency pin diode switch as well as the wideband Schottky diode detector described in Appendix A14. As such, this radiometer is the only one built using the commercially produced Schottky diode detector shown in Figure 16 rather than the homebuilt temperature compensated detector described in Appendix A6. Also, in contrast to most of the other observations reported here, Figure 47 shows the radiometers viewing the sky outside my house rather than through the glass patio door.

These outdoor measurements were obtained on June 15, 2020 from 12:30 to 3:00 PM when thin clouds were seen moving across the sky with no precipitation. Furthermore, the radiometer voltages were stored on my laptop computer and converted to brightness temperatures using calibration equations obtained from clear sky measurements corrected for water vapor. Figure 48 shows the radiometer voltages and brightness temperatures along with the calibration equations. Also shown is the water vapor and cloud liquid water derived from the brightness temperature measurements using algorithms obtained from simulations similar to that described in Section 8.5. For comparison the water vapor and cloud liquid water parameters are determined using the dual and single frequency algorithms indicated in the Figure. The retrieved *TPW* is shown to increase from about 15 mm to 25 mm between 1:00 and 2:00 PM followed by smaller changes. For validation, RAOB's from the National Weather Service can be used, although the observations are only made twice a day at 0 and 12 GMT (Greenwich Meridian Time). The other alternative, although less quantitative is to use satellite infrared radiometers which observe the upper atmosphere water vapor variations under mostly clear sky conditions. This approach is adapted here where Figure 49 shows the water vapor images using the infrared sensors aboard the Geostationary Operational Environmental Satellites (GOES).

The GOES images were acquired from a web site on June 15 at 12:00 PM, 1:00 PM and 2:00 PM local time. Note that the GOES water vapor observations increase steadily near my location. This is very similar to that shown by the microwave measurements in Figure 48. Also, in contrast to the increased water vapor, the microwave derived cloud liquid water decreases between 1:00 PM and 2:00 PM. This decrease in liquid water is supported by the GOES enhanced infrared image at 1:30 PM in Figure 50 which displays warmer temperatures or lower cloud tops west of my location. Although I did not capture any additional images, this feature was seen to continue in later images as the system moved eastward.

Figure 47. Top pictures show the radiometers mounted on tripods viewing the sky from my patio on June 15, 2020. The bottom-left shows the clouds seen while the bottom-right shows the setup.

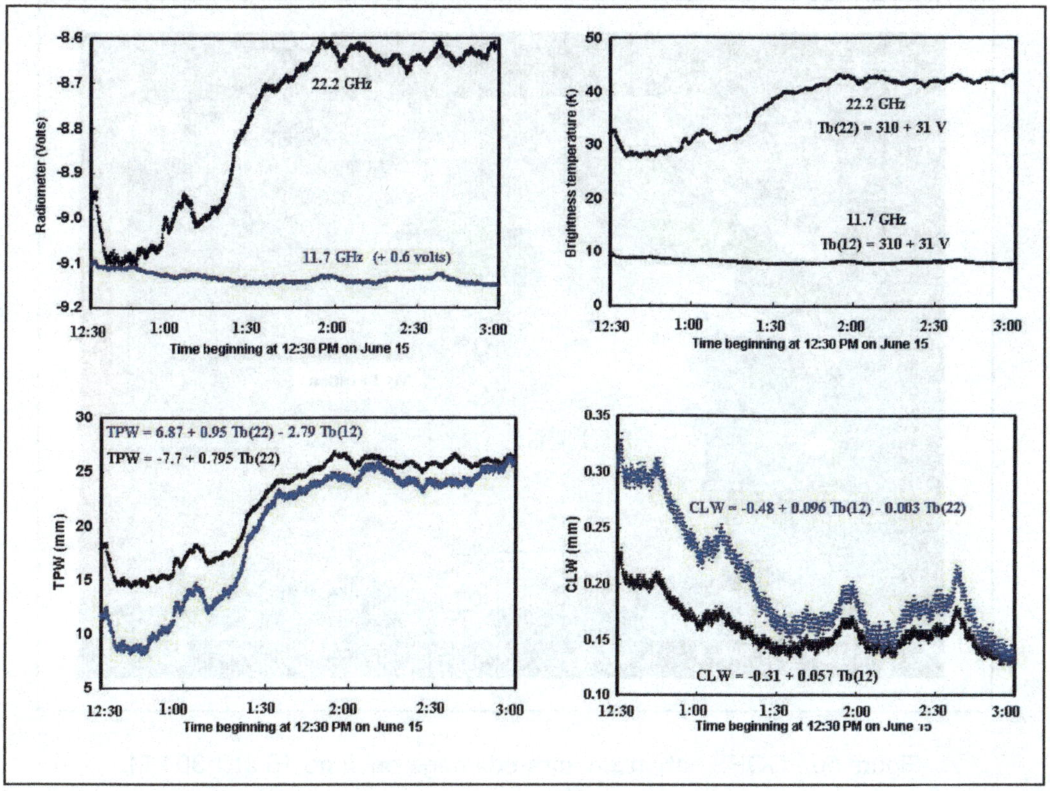

Figure 48. The top shows the 22 GHz and 12 GHz radiometer measurements (Left) and brightness temperatures (Right) on June 15 from 12:30 to 3:00 PM. The bottom shows the water vapor TPW (Left) and cloud liquid water CLW (Right) using the indicated single and dual frequency algorithms.

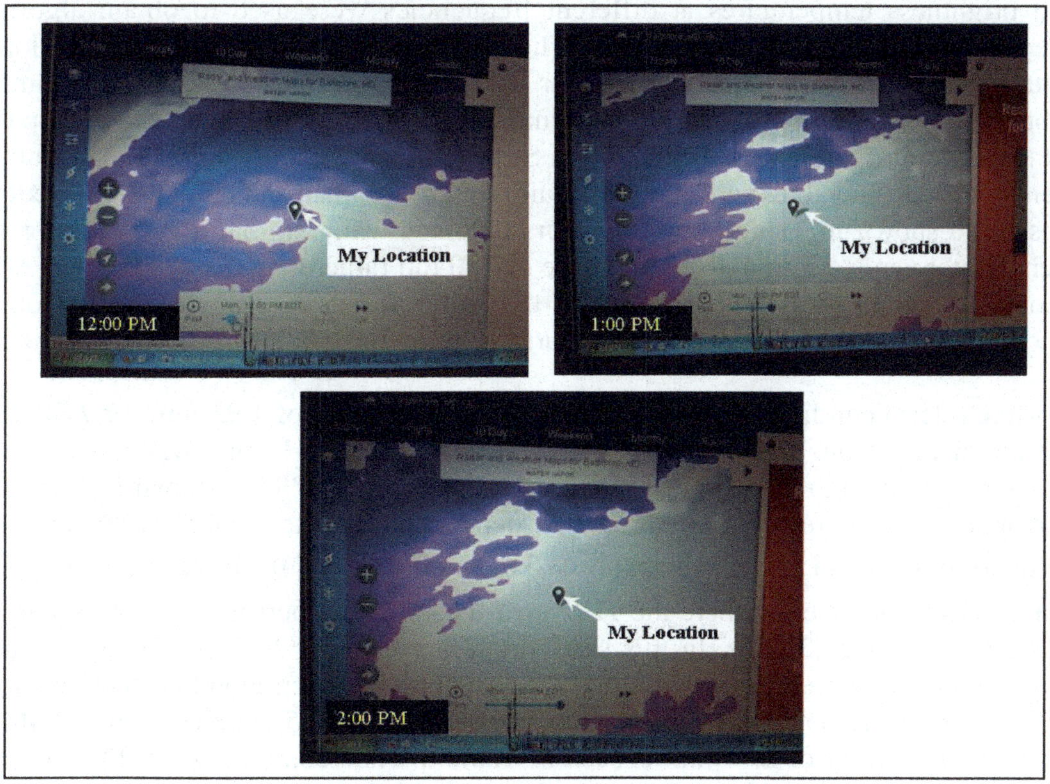

Figure 49. Shown are the water vapor images generated from the infrared sensors on the GOES satellite. The increase in water vapor at my location (Lat =39.49 ^{0}N, Lon = -76.31 ^{0}W) is consistent with the microwave derived *TPW* measurements between 12:30 PM to 3:00 PM in Figure 48.

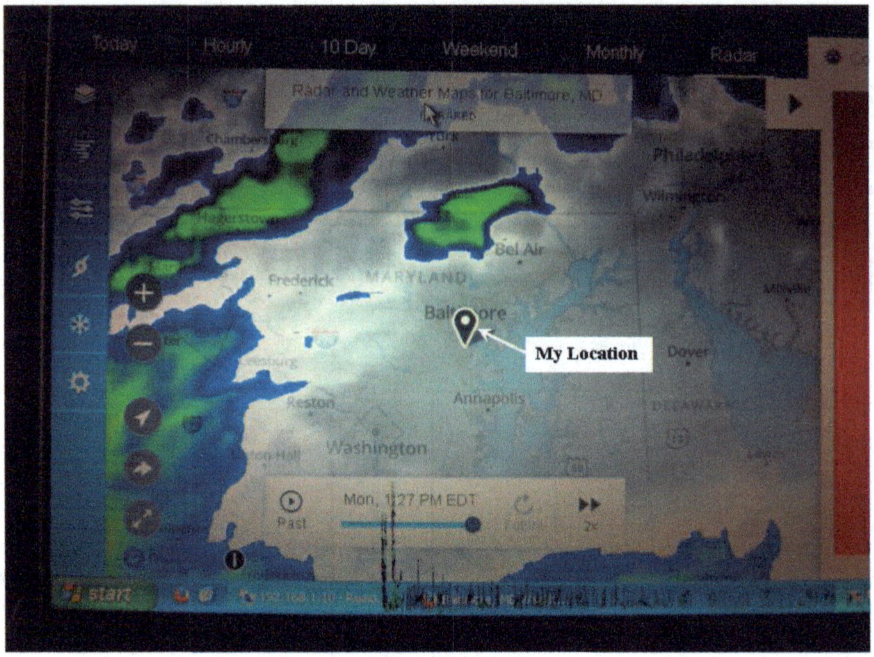

Figure 50. GOES enhanced infrared image on June 15 at 1:30 PM.

8.7 Water Vapor Retrieval Accuracy

Simulated brightness temperatures at different frequencies were used to obtain algorithms and determine the accuracy of the retrieved *TPW*. The procedure uses least squares regression analysis of the data resulting in linear equations similar to (50a) in Section 8.5. Table 6 summarizes the results obtained using different channel combinations. The results are also shown separately for clear and cloudy atmospheres. As mentioned in Section 8.5, the cloudy simulations contain clouds at different altitudes and thickness having liquid water less than 0.3 mm to represent rain-free conditions. Also shown in Table 6 are the errors obtained using similar algorithms for the upper atmospheric water vapor beginning at a pressure of 700 mb rather than the surface, *i.e.*, V700. This comparison is made to analyze the effects of the water vapor vertical distribution on different frequency measurements. The following summarizes the main findings of this part of the study.

For cloud-free (clear) conditions Table 6 shows a minimum error of 1.93 mm for *TPW* using 20.5 GHz measurements. Note that the *TPW* error increases to 2.51 mm when using 22.2 GHz measurements even though this frequency is more sensitive to *TPW* as shown in Figure 13. This smaller error at 20.5 GHz results from its smaller dependence on the vertical water vapor profile so τ_{H2O} in equations (37) and (45) can be expressed as Exp[$-TPW/W(\nu)$]. However, the opposite result occurs for cloudy atmospheres where the *TPW* error is shown to increase to 3.34 mm at 20.5 GHz while it is 3.03 mm at 22.2 GHz. In this case, the larger error at 20.5 GHz is due to its smaller water vapor signal compared to that of clouds. Consequently, when cloud corrections are applied using 11.7 GHz measurements, the 20.5 GHz error is reduced to 1.59 mm (**see large bold font**). In comparison, a larger error of 2.41 mm is seen in Table 6 when combining the 22.2 and 11.7 GHz measurements. To further analyze these results we examine the opacity function defined by equation (6b).

Using (6b), the water vapor opacity function can be expressed as,

$$\alpha_\nu(z) = \int_0^z \gamma_\nu(z')\,dz' = \int_0^z f_\nu(z')\,\rho_{H20}(z')\,dz' \tag{52a}$$

where
$$f_\nu(z) = \frac{S\,\nu^2\,\delta\nu(z)}{(\nu-\nu_0)^2 + \delta\nu(z)^2}\,. \tag{52b}$$

The opacity function is the vertically integrated absorption coefficient, which is the product of the water vapor density ρ_{H20} and approximate line shape function $f_\nu(z)$. While a more exact line shape containing an additional non-resonant term could be used, this single function referred to as the Lorentz line shape is adequate for our purposes. For water vapor, the line shape function (52b) is the response of a single resonant absorption line to thermal radiation following molecular collisions, which extends the spectral response beyond the resonant frequency ν_0 due to collisional broadening. It depends on the excitation frequency ν, line strength S and its resonant frequency ν_0 = 22.235 GHz. The vertical structure of $f_\nu(z)$ is defined by the line width $\delta\nu(z)$ which varies with altitude z. Assuming the line width is independent of altitude, the opacity throughout the atmosphere is $\alpha(\infty) = f_\nu \int_0^\infty \rho_{H20}\,dz = f_\nu \cdot TPW$ The transmittance then becomes $\tau_{H2O} = \mathrm{Exp}\left[-TPW/W(\nu)\right]$ which is the same as in equations (37) and (45). However, as shown next, the water vapor transmittance also depends on the vertical distribution of water vapor due to the line width dependence on altitude or pressure.

The line width in (52b) is proportional to pressure P due to molecular collisions. It can be approximated as $\delta\nu = \hat{\nu}(P/P_S)$ where P_S is the surface pressure and $\hat{\nu}$ is about 2.84 GHz. Therefore, when written in pressure coordinates using the hydrostatic equation $dp/dz = -\rho g$, where ρ is the air density and g is the gravitational constant, equation (52a) becomes

$$\alpha_\nu(p) = \frac{S\,\nu^2}{2(\nu-\nu_0)}\,\frac{1}{g}\int_p^{P_S} K_\nu(p')\,\omega(p')\,dp' \tag{53a}$$

where
$$K_\nu(p) = \frac{2\,x_\nu\,p}{x_\nu^2 + p^2} \quad \text{with} \quad x_\nu = \left[\frac{\nu-\nu_0}{\hat{\nu}}\right] P_S\,. \tag{53b}$$

The opacity function (53a) is the integral of the water vapor mixing ratio $\omega = \rho_{H20}/\rho$ weighted by the Kernel function $K_\nu(p)$. From equation (53b), the Kernel peaks at pressure $p = |x_\nu|$ with a width proportional to $|x_\nu|$. It is plotted in the left-most graph of Figure 51 as a function of pressure at frequencies from 19.0 to 22.2 GHz. Note that at 22.2 GHz it sharply peaks near 10 mb with a width and peak pressure decreasing as the frequency increases. At 20.5 GHz, the Kernel has a broad distribution from the surface to about 300 mb peaking at 610 mb. Because $K_\nu(p)$ weighs the mixing ratio nearly uniformly at 20.5 GHz, its opacity is proportional to TPW so $\tau_{H20} \approx \mathrm{Exp}[-TPW/W(\nu)]$. In contrast, $K_\nu(p)$ at 22.2 GHz has its strongest contribution at low pressures. However, the mixing ratio is largest near the surface and decreases rapidly with altitude so even at 22.2 GHz the opacity is proportional to TPW but with slightly less accuracy than at 20.5 GHz. Consequently the single frequency clear simulation results in Table 6 shows the 20.5 GHz channel more accurate in obtaining *TPW* than at 22.2 GHz. However, the opposite occurs for cloudy atmospheres since then the water vapor signal at 20.5 GHz can be less than the variations introduced by clouds so the stronger water vapor signal at 22.2 GHz is more accurate in determining *TPW*. This accuracy difference between channels is reversed in Table 6 when cloud corrections are obtained using a second channel. Also, as discussed next, besides *TPW* which is the dominant parameter, the different Kernels at 20.5 and 22.2 GHz produce a different response to the upper level atmospheric water vapor.

The right-most graph of Figure 51 shows the relationship between the 20.5 and 22.2 GHz measurement using the simulated clear and cloudy measurements. The plot also shows the regression equation relating the two measurements, *i.e.*, $T_b(20.5) = 2.33 + 0.623\ T_b(22.2)$. This equation has a standard error of 0.82 K due to the different Kernels at the two frequencies. As such, the 22.2 GHz measurement responds more to changes in the water vapor distribution. To further examine the effect on the water vapor distribution, Table 6 compares the water vapor burden for pressures less than 700 mb which is given by

$$V700 = 1/g\int_0^{700} \omega(p)\,dp \quad \text{while} \quad TPW = 1/g\int_0^{P_S} \omega(p)\,dp \qquad (54)$$

where ω is in units of *g/kg*, with *TPW* and V700 in units of mm and g = 980 cm/sec^2. Note from Table 6 that for cloudy atmospheres, the 22.2 GHz combined with 11.7 GHz measurements result in the smallest error of 1.15 mm for V700 (**large bold font**). This error increases to 1.28 mm when substituting the 20.5 GHz channel. However, as mentioned above, for *TPW* the 20.5 GHz results in a smaller error than that of 22.2 GHz. For this reason, the radiometer developed by Dr. David Hogg[5] used 20.6 rather than 22.2 GHz to measure *TPW*. This advantage of 20.6 GHz for deriving *TPW* was originally found from simulations performed by Dr. Ed. Westwater [15]. Also, instead of 20.6 GHz some later radiometers use 23.8 GHz since it is in a better protected region and provides the same water vapor response. Incidentally, I found the cloud liquid water accuracy is the same using 20.5 GHz or 22.2 GHz combined with 11.7 GHz measurements.

Table 6: Standard Error of *TPW* and V700 for Clear and Cloudy Atmospheres.

Radiometer Center Frequencies	**Atmosphere**	*TPW* (STD = 13.7 mm) ***TPW* Standard Error**	V700 (STD = 3.1 mm) **V700 Standard Error**
20.5 GHz	Clear	1.93 mm	1.30 mm
22.2 GHz	Clear	2.51 mm	1.16 mm
20.5 GHz	Cloudy	3.34 mm	1.37 mm
22.2 GHz	Cloudy	3.03 mm	1.22 mm
22.2, 11.7 GHz	Cloudy	2.41 mm	**1.15 mm**
20.5, 11.7 GHz	Cloudy	**1.59 mm**	1.28 mm
22.2, 20.5, 11.7 GHz	Cloudy	1.55 mm	1.04 mm

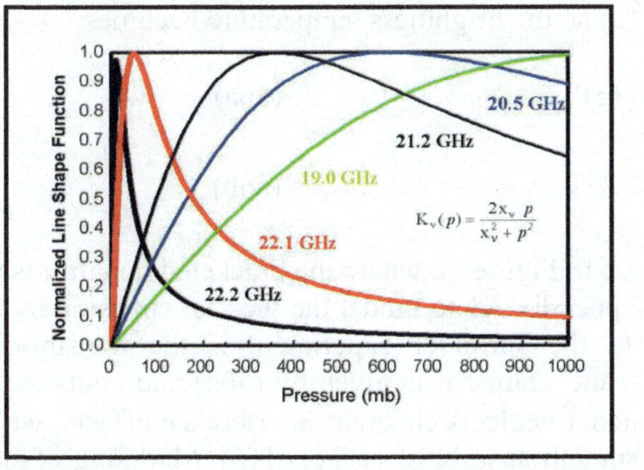

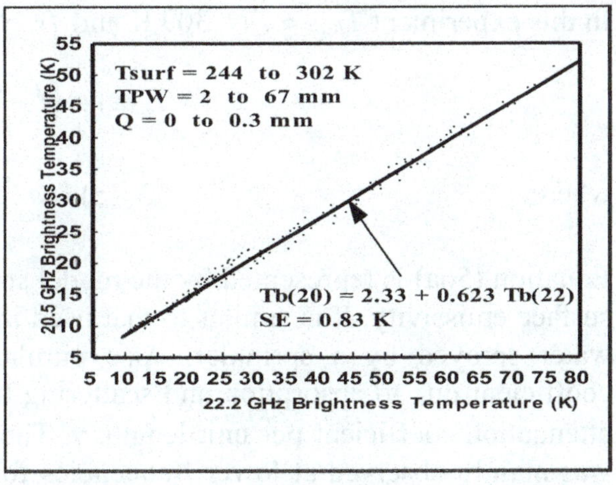

Figure 51. The left Figure shows the averaging Kernel (53b) as a function of pressure at different frequencies. The right graph plots the simulated 20.5 GHz against 22.2 GHz brightness temperatures for clear and cloudy atmospheres. Also plotted is the best fit regression equation and its standard error. The small slope of 0.623 results from the larger water vapor sensitivity at 22 than at 20 GHz. The 0.83 K standard error results from larger variations due to water vapor distribution at 22 than at 20.5 GHz.

9. Transmission Measurements of Sand

In general, active techniques employing signal generators together with transmission line, free space propagation or resonant cavity measurements are used to measure the transmission and reflection coefficient of materials as well as their dielectric constant. For very low loss dielectrics the resonant cavity approach must be used while the other techniques are used for more absorbing materials. In contrast to these active techniques, this chapter describes an alternative approach using radiometers to measure the transmission and reflection of surfaces.

The setup is shown in Figure 52 and consists of a radiometer and virtually transparent Styrofoam[3] (*i.e.,* polyurethane foam) container which is filled with granular material. The container is then placed over the horn antenna as shown in the Figure. In this experiment the material consists of fine grain sand used for landscapes. Its transmission coefficient is obtained using the 12 and 20 GHz radiometers by measuring the change in sky brightness temperature due to transmission by the sand. Although the particles reside close to the antenna aperture, near field distortion by the particles is considered negligible. As such, the schematic representation is similar to the sky radiation transfer through the glass door depicted in Figure 25. The representation is shown in Figure 53 where the brightness temperature measured by the radiometer is

$$T_b(z) = \left[1 - R_S(z) - \Im_S(z)\right] T_S + \Im_S(z) T_{Sky} + R_S(z) T_{Rad} \qquad (55)$$

where $\Im_S$ and R_S is the transmission and reflection coefficient of sand, whose depth z is varied during the experiment. The first term on the right side is the product of the sand emissivity in brackets and its temperature, T_S. The second term is the sand-attenuated sky radiation, T_{Sky}, which is given by equation (5) in Section 4.3. The last term is the radiometer emitted radiation T_{Rad} as it is scattered by the sand back into the antenna. As explained in Chapter 3, this thermally emitted radiation results from the LNB since the isolator blocks the coherent *LO* radiation from being transmitted out of the antenna.

In this experiment $T_{Rad} \approx T_S = 309$ K and $T_S >> T_{Sky}$ so the brightness temperature becomes

$$T_b(z) \cong [1 - \Im_S(z)]T_S \tag{56a}$$

where
$$\Im_S(z) = e^{-\gamma z}. \tag{56b}$$

Equation (56a) is represented by the model shown in Figure 52 where the bracketed quantity is the surface emissivity. It is similar to that used in Appendix A1 to model the thermal emission by the water sprayed by a sprinkler. Also similar to the sprinkler experiment is the transmission coefficient due to absorption and scattering by sand grains. It is given by (56b) and contains the attenuation coefficient per unit length, γ. The model neglects coherent interference effects, which was mainly observed at lower frequencies for smooth absorbing surfaces [16]. Modeling of such effects is described in Appendix A10 for stratified media, and applied to a patio door containing two glass panes separated by an air gap. Lastly, (56b) is also similar to the empirical atmospheric absorption equation A13-4 in Appendix A13, *i.e.*, $\alpha(z) \approx \alpha_0 Z/H$ when $Z/H << 1$ so $\tau(z) = e^{-\alpha_0 Z/H}$.

The attenuation coefficient of sand is obtained from (56a) by recognizing that $\gamma z << 1$ so that $\Im_S(z) \approx 1 - \gamma z$ and therefore

$$\gamma \cong \frac{1}{T_S}\frac{d\,T_b(z)}{dz} \tag{57}$$

where the penetration depth is $1/\gamma$. It corresponds to the distance at which $\Im_S = e^{-1}$.

Figure 54 shows the measured brightness temperature as a function of sand depth as obtained using the 12 GHz and 20 GHz radiometers. The 12 GHz attenuation coefficient is obtained from (57) using the best fit straight line slope of 7.1 *K*/cm so that $\gamma = 0.023$ cm-1 with $T_S = 309$ K. Nearly the same absorption coefficient is calculated from the 20 GHz radiometer measurements. Therefore, the corresponding penetration depth is 44 cm at both frequencies. However, measurements using ground-based [16] and satellite radiometers [17] found a decrease in penetration depth with increasing frequency for desert sand. Only cavity resonant measurements [18] of a small sample of sand found little change in the attenuation and penetration depth with frequency. These different measurements may be due to differences in grain size and impurities which affect the attenuation due to absorption and scattering [19].

In addition to measuring attenuation, radiometers can also measure the reflection coefficient by changing the sand temperature from T_S to $\hat{T}_S$ while keeping T_{Rad} and T_{Sky} unchanged. Using (56), we then obtain the following equations for R_S and γ,

$$R_S(z) \cong \frac{1}{T_{Rad}}\frac{T_b(z) - r\hat{T}_b(z)}{1 - r} \quad \text{where} \quad r = \frac{T_S}{\hat{T}_S} \tag{58a}$$

$$\text{and} \quad \gamma \cong \frac{1}{T_S}\frac{dT_b(z)}{dz}, \tag{58b}$$

with T_b and $\hat{T}_b$ being the corresponding brightness temperatures at temperature T_S and $\hat{T}_S$. This approach of measuring the reflection coefficient has not been done at this time since it requires a more precisely regulated temperature of sand than that of the attenuation coefficient, which only requires a single temperature measurement, T_S.

As an alternative means of measuring the reflection coefficient, the radiometer views the reflected radiation of sand from above as illustrated in Figure 55. The radiometer, which was originally below the sand, is now replaced by the high emissivity surface used for calibration at temperature T. Similar to equation (55), the brightness temperature is now given by

$$T_b(z) = \left[1 - R_S(z) - \Im_S(z)\right] T_S + \Im_S(z) T + R_S(z) T_{Sky} . \quad (59)$$

As in the previous discussion, $T \cong T_S$ and $T_S >> T_{Sky}$ so that the brightness temperature becomes

$$T_b(z) \cong \left[1 - R_S(z)\right] T_S \quad . \quad (60)$$

This equation is represented by the simple model shown in Figure 55 where the bracketed quantity in (60) is the surface emissivity. As such, the reflectance is given by

$$R_S(z) \cong 1 - \frac{T_b(z)}{T_S} \quad . \quad (61)$$

Equation (61) is much simpler to implement than the reflectance given by (58a) since it now only requires a single temperature measurement, T_S. However, the radiometer must be placed high above the sand at an oblique angle in order not to block the downward sky radiation from impinging on the surface. Also, a large enclosure must be constructed so that the antenna's field of view fully encompasses the sand [16].

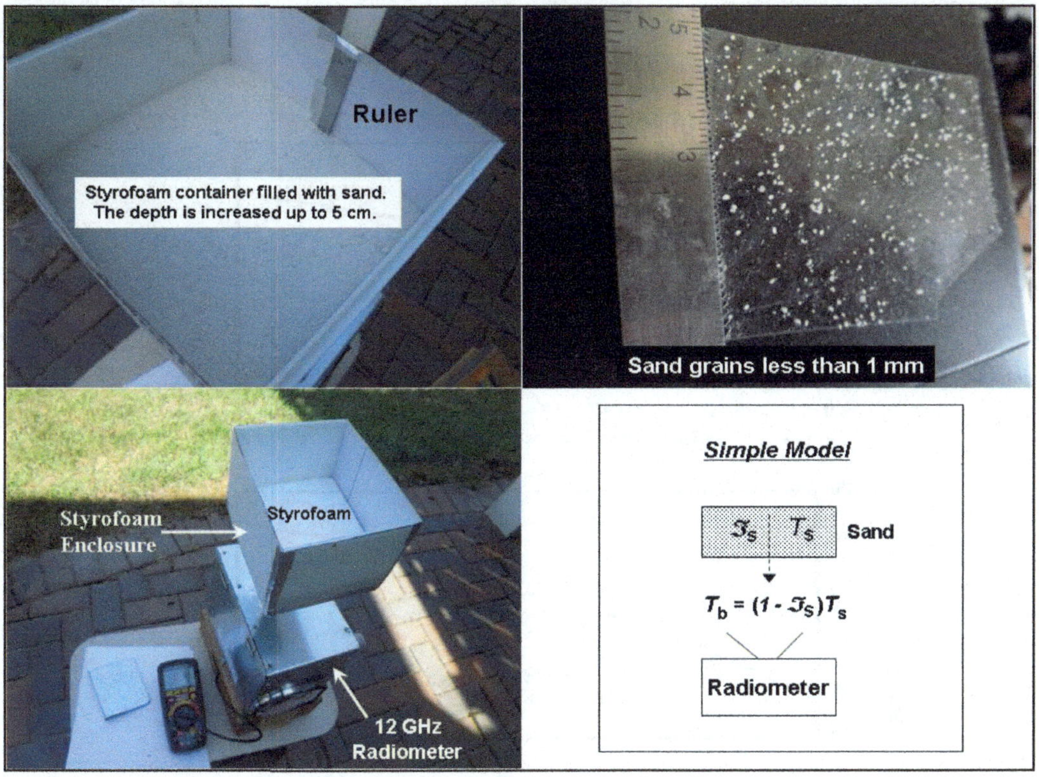

Figure 52- Bottom left shows the radiometer setup used to measure the attenuation and penetration depth of sand. A Styrofoam enclosure containing the sand is placed over the radiometer's horn antenna. The skyward viewing radiometer then primarily measures the attenuated thermal emission of the sand. A model describing the measurement is shown on the bottom right. The attenuation coefficient is obtained by varying the sand depth and plotting the radiometer measurement (see Text).

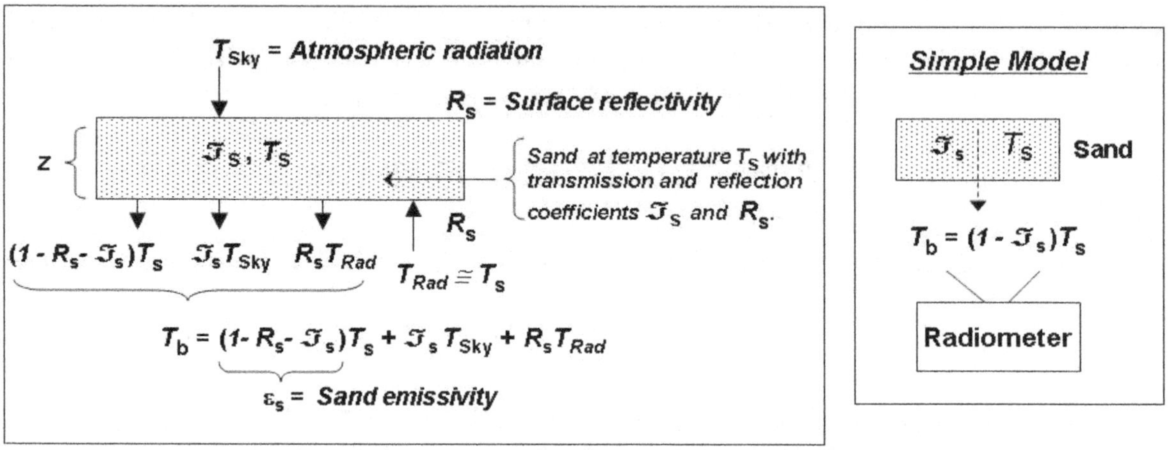

Figure 53 – The leftmost diagram is a schematic representation of the experimental setup shown in Figure 52. The test sample in this case is sand having depth z, transmission coefficient $\mathfrak{T}_S$, reflection coefficient R_S and temperature T_S. The diagram also shows the downwelling atmospheric radiation incident to the sand, T_{Sky}, as well as the three radiation components emanating from the bottom of the sample. The resulting brightness temperature from the combined components is indicated at the bottom of the diagram and given by equation (55) of the text. A simplified model representing the brightness temperature equation is also shown in the rightmost diagram and given by equation (56a) of the text.

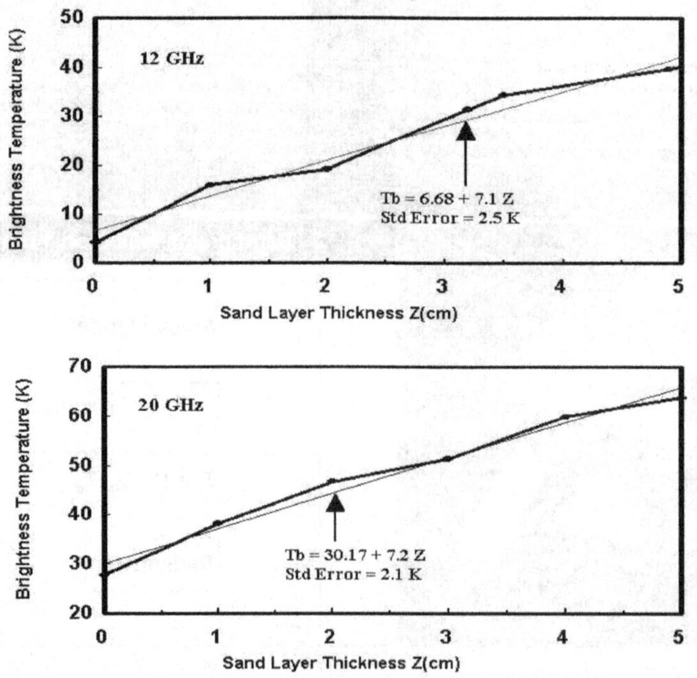

Figure 54 – Insertion loss measurements of sand using the 12 GHz (Top) and 20 GHz radiometer (bottom). These measurements were taken on a cloud free day on July 25, 2019 when the sand and outside temperature was 309 K.

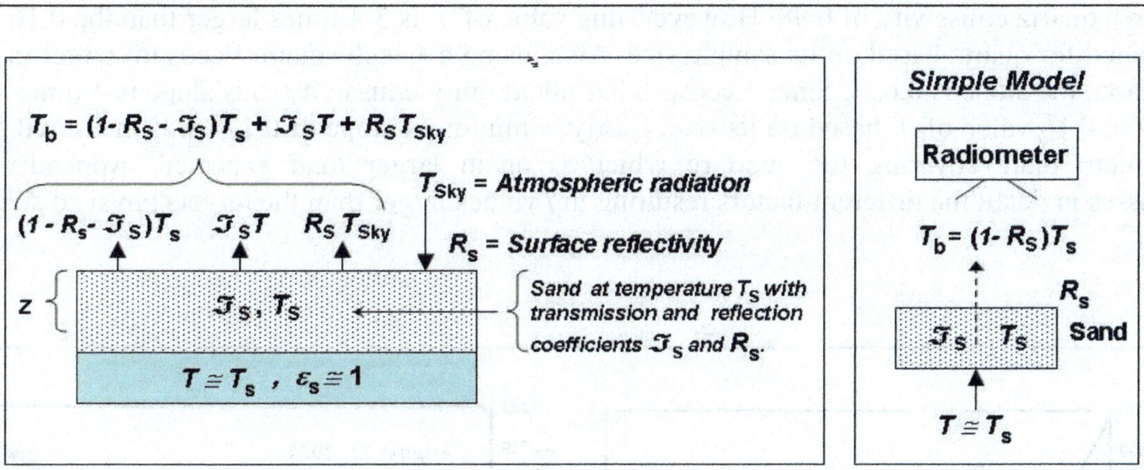

Figure 55 – The leftmost diagram is a schematic representation of a setup to measure the reflectance by viewing the sand from above. As in Figure 52, the test sample is sand having depth z, transmission coefficient $\mathcal{T}_S$, reflection coefficient R_S and temperature T_S. The diagram also shows the downwelling atmospheric radiation incident to the sand, T_{Sky}, as well as the three upwelling radiation components emanating from the top of the sample. Also included is a low reflectivity high emissivity target at temperature T below the sand. The resulting brightness temperature from the combined components is indicated at the top of the diagram and given by equation (59) of the text. A simplified model representing the brightness temperature equation is also shown in the rightmost diagram and given by equation (60) of the text.

9.1 Emissivity Measurement of Quartz

The next three chapters discuss the history of satellite radiometers and some measurements. Of interest here are the measurements over Saudi Arabia in Section 12.1. These measurements show quartz has the highest emissivity (~ 0.98) of all surfaces, including dense forests. This prompted using the 20 GHz radiometer to measure the quartz emissivity using a laboratory setup. Figure 56 shows the procedure, where a quartz sample is glued to a metal plate (Bot-Left). The unit is then placed over the antenna aperture (Bot-Right) after initially heating quartz to 334 K. The top-left displays the quartz temperature measured using a thermocouple, plotted against the corresponding radiometer brightness temperature T_B. Both temperatures are shown to decrease in time as they approach thermal equilibrium. For data analysis the top-right shows the brightness temperature T_B plotted against the quartz temperature T_Q. As discussed next, the measured slope of 0.536 is the quartz emissivity ε_Q multiplied by its fractional area f according to $\Delta T_B/\Delta T_Q = f\,\varepsilon_Q$.

As derived in Appendix A17, $T_B = [\varepsilon_Q T_Q + R_Q T_R]\,f + T_R\,(1 - f)$ where the 1st term in brackets contains the emissivity ε_Q and composite reflectivity R_Q of the quartz and metal backing. Also included is the quartz temperature T_Q and thermal radiation T_R emitted by the LNB. Both quantities are multiplied by the fractional area f viewed by the antenna. The 2nd term is the radiation reflected by the uncovered portion of the aluminum plate which has near unity reflectance. Based on energy conservation $R_Q = 1 - \varepsilon_Q$ so that $T_B = T_R + \varepsilon_Q\,(T_Q - T_R)\,f$. As such, $\Delta T_B / \Delta T_Q = f\,\varepsilon_Q$ where T_R is constant over the 24 minutes shown in Figure 56. This slope equation can be used to determine ε_Q given f. For example, the 0.536 slope shown in Figure 56 requires a fractional area of 0.55 to obtain a quartz emissivity of 0.98. However, this value of f is 3.4 times larger than the 0.16 value estimated for quartz based on its sample size. Also, using a 1 inch square Eccosorb target in place of quartz, the slope is 0.634. Since Eccosorb has about unity emissivity, this slope is 4 times larger than the 0.16 value of f based on its size. Lastly, a minimum slope of 0.174 was measured for an aluminum plate covering the aperture which is again larger than expected. Appendix A17 discusses in detail the different factors resulting in f values larger than the targets physical size.

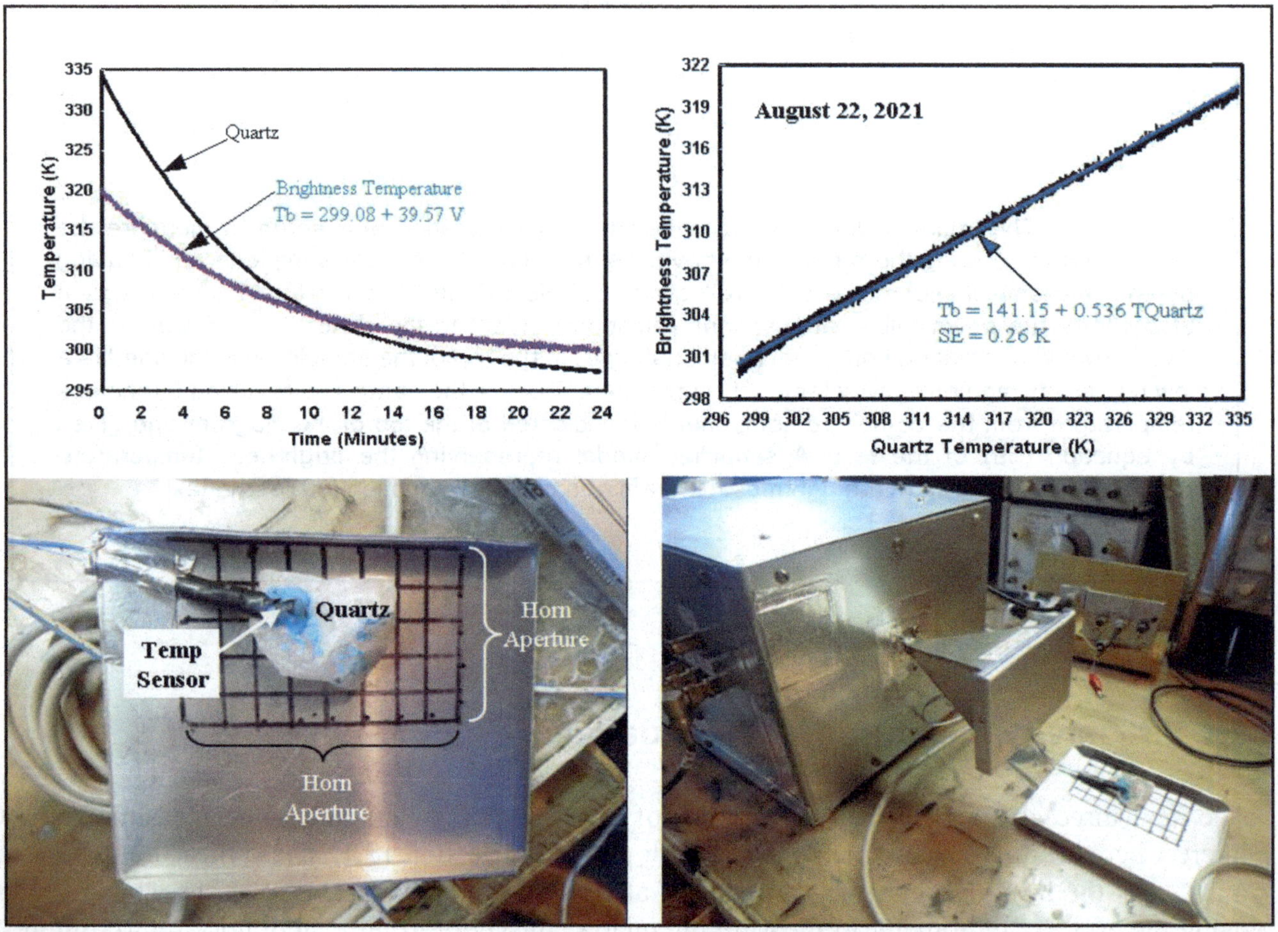

Figure 56 – Procedure to measure emissivity. A quartz sample attached to an aluminum plate (Bot-left) is heated to 334 K and placed over the horn antenna (Bot-Right). Top-left shows the quartz and radiometer temperatures while the top-right shows them plotted against each other.

10. Satellite - Microwave Radiometers

The previous chapters discussed the construction, measurement and analysis of ground-based radiometers. However, besides measuring the downward radiation, the upwelling radiation is measured by radiometers flown on satellites. Figure 57 shows the three main radiation components seen by satellite radiometers viewing Earth. Neglected is the cosmic radiation which is attenuated by the atmosphere, reflected by the ground, and further attenuated on its path to the satellite. Shown is the upwelling T_u and downwelling radiation T_d reflected by the ground and attenuated by the atmospheric transmittance $\tau^{Sec\theta}$. The Secθ exponent accounts for the increased attenuation along the slant path at zenith angle θ. Also, the ground reflectance is approximated by the specular[6] reflectivity (1-ε_S) where ε_S is the surface emissivity. The third component shown in the Figure is the surface emitted radiation at effective temperature T_{Eff} with is attenuated by the atmosphere. For most surfaces the effective temperature is the surface temperature. However, as discussed in Section 12.1, differences between temperatures occur for porous surfaces such as deserts.

Combining the three components shown at the top of Figure 57 we obtain

$$T_b = T_u + \tau^{\mathrm{Sec}\theta}\left[\varepsilon_S T_{\mathrm{Eff}} + (1-\varepsilon_S)T_d\right] \qquad (62)$$

where this equation can be simplified depending on the frequency. The largest simplification occurs for frequencies in the opaque portion of the oxygen band between 54 to 60 GHz. At these frequencies the transmittance is so small that the brightness temperature only depends on the upwelling radiation T_u which is given by equation (64a) in Chapter 11. Also, for frequencies in the more transparent window regions far from the oxygen band and strong water vapor line at 183.3 GHz, the upwelling and downwelling radiation components are nearly the same and can be approximated as $T_M(1-\tau^{\mathrm{Sec}\theta})$. In this case (62) reduces to

$$T_b = T_M\left[1-\tau^{2\mathrm{Sec}\theta}(1-\varepsilon_S)\right] + \tau^{\mathrm{Sec}\theta}\varepsilon_S\left[T_{\mathrm{Eff}} - T_M\right] \qquad (63)$$

where this equation is applied in Section 12.1 when discussing desert surface measurements.

Satellites provide the best platform for calibration since the radiometer can view cold space with no atmospheric absorption. Satellite radiometers also provide global measurements which is necessary to measure climate change. I therefore added three chapters to discuss satellite radiometers. A brief history of space-borne microwave radiometers is described next while the following two chapters give examples of atmospheric and surface measurements. For conciseness I have limited the examples to some of the most revealing observations. In particular, Chapter 11 describes the use of satellite radiometers to measure the warm core temperature structure of tropical storms and the globally averaged temperature trend. These examples were chosen to demonstrate the cloud penetration property of microwaves as well the long term performance of the instruments and their accuracy. Chapter 12 also describes measurements over deserts and snow to reveal features that are not as well understood. Also discussed in Section 12.3, is the measurements of sea ice concentration. Along with temperature this variable has been found to be important for monitor climate change.

Historically, the first space-borne microwave radiometer was launched in 1962 on NASA's Mariner-2 satellite. This instrument was a dual frequency (15.8, 22.2 GHz) Dicke radiometer developed by the Jet Propulsion Laboratory (JPL) for NASA to view the planet Venus and measure its water vapor. However, besides observing our solar system the Russians were the first

---

[6] It is customary to use the specular approximation, although for very rough (i.e., Lambertian) surfaces the downwelling reflected radiation T_d is at a different angle than in the specular direction. In fact, it can be shown that only for viewing angles near 60 degrees is the reflectance the same for specular and Lambertian surfaces.

to view Earth using a satellite microwave radiometer. In 1968 they launched a nadir viewing four channel (3.5, 8.8, 22.2, 37 GHz) radiometer to view earth from their Cosmos-243 satellite. In fact, much of my early knowledge about microwave remote sensing from satellites came from reading the early Russian literature, much of which has been translated [20].

I joined NASA in 1971 just prior to when they began launching microwave radiometers to view Earth on their Nimbus series of polar orbiting satellites. Before that time the Nimbus satellites only carried visible cameras to image Earth and infrared radiometers to measure the atmospheric and surface temperature under cloud-free situations. Following 1971 NASA began placing microwave radiometers aboard their satellites to probe through clouds and measure the underlying atmospheric temperature and surface features. For reference, Figure 58 shows the time-line of radiometers developed by NASA as well as by NOAA and the U.S. Department of Defense (DOD) as part of their Defense Military Satellite Program (DMSP). This Figure shows the chronology from 1972 to 2005 while Table 7 lists the frequencies, spatial resolution and applications of each radiometer.

The first microwave radiometers on NASA's Nimbus series of satellites were the Nimbus-E (called Nimbus-5 after launch) Microwave Spectrometer (NEMS) and Electronically Scanning Microwave Radiometer (ESMR-1). As indicated in Table 7, these instruments were launched on Nimbus-5, but had few channels with poor spatial resolution. The NEMS was a non-scanning nadir viewing five channel radiometer with three channels in the oxygen band (53.65, 54.90, 58.80 GHz) with the other two at the 22.2 GHz water vapor line and 31.4 GHz window region. On the other hand the ESMR-1 only had a 19.35 GHz vertically polarized channel, but used an electronically scanned phased array antenna to scan Earth. These radiometers were soon replaced by more advanced instruments until the Nimbus program ended in 1978 with the launch of the dual polarized Scanning Multichannel Microwave Radiometer (SMMR) on Nimbus-7. Measurements from these experimental radiometers were used to advance our knowledge of instrument design and improve the use of microwaves for observing Earth from space. Some of these sensors will be discussed later.

Table 7 lists two types of radiometers, one called sounders and the other imagers. The sounders were primarily used to determine the vertical atmospheric temperature profiles and obtain water vapor information. Temperature sounders have channels within the 50 to 60 GHz portion of the oxygen band to derive the vertical profile of atmospheric temperature while water vapor information is obtained from channels near the weak and strong water vapor absorption lines at 22.2 GHz and 183.3 GHz respectively. All radiometers use heterodyne receivers similar to the LNB's discussed in Chapter 3. While digital SDR receivers were not considered, they are most useful for the high spectral resolution needed for temperature and water vapor sounders. As shown in Figure 59, these sounders use cross-track scanning antennas to obtain the widest possible swath width to view Earth. This type of antenna also provides the most accurate means of viewing cold space for calibration by simply rotating its reflector away from Earth. The reflector is also rotated to view a high emissivity target mounted on the instrument at its ambient temperature for the second calibration point as explained in Section 4.3.

In contrast to sounders, imaging radiometers have channels in the more transparent window portion of the electromagnetic spectrum which lies outside the oxygen and water vapor regions. As shown in Figure 59, the radiometers reflector scans conically to minimize changes in the footprint or FOV and slant angle as it rotates to scan Earth. Also, the radiometer's front end which contains the feed horns is also shown to rotate with the reflector to minimize the rotation of polarization. This approach was used for SSM/I but not for SSMR whose radiometer was fixed so that analysis was needed to separate the V and H polarization components. The need to maintain a constant polarization for all scan positions is particularly important when viewing surfaces such as oceans whose emissivity is highly polarized. In fact, when viewing surfaces and precipitation the constant FOV, viewing angle and polarization obtained with imagers greatly improves the observations and simplifies the data analysis.

Although not constant, cross-track scanners can account for the changes in FOV by averaging all measurements to a fixed resolution. Also, as mentioned in the next chapter, the effect due to changes in slant angle can be accounted for by normalizing the measurements to nadir viewing using a limb correction procedure. However, sounders are primarily used to measure large scale features such as atmospheric temperature and water vapor, so that the changes in polarization and FOV are not as important compared to the calibration accuracy, which can be made more accurate for cross-track scanners. This is evident by noting in Figure 67 that for calibration purposes a conical scanning radiometer such as the SSM/I require a sub-reflector to direct the cold space radiation into the feed horn. This additional structure complicates the design of imagers and can lead to reduced accuracy.

The next chapter discusses the application of satellite microwave radiometers to measure atmospheric temperature, which is followed by another chapter on surface measurements including precipitation. To minimize the size of these final chapters, surface measurements are mainly limited to those features observed over land. Therefore, oceanographic features such as sea surface winds based on emissivity measurements are not described except in regard to its effect on sea ice measurements. Due to its importance in monitoring climate change, the use of microwave radiometers to measure sea ice concentration is described in some detail in Section 12.3. Also not discussed here is the subject of soil moisture, which requires low frequencies (< 10 GHz) to penetrate the vegetation cover and probe within the underling soils. However, as discussed in Chapter 9, and in Section 12.1, high frequencies can be used in desert areas to measure the various minerals and penetrate below its sandy surface.

As discussed previously, while surface wetness, water vapor, cloud liquid water and rain measurements are easily obtained using ground-based radiometers, they can also be measured from satellites but at poorer spatial resolution. For a given antenna size the highest resolution occurs at higher frequencies but with reduced visibility due to atmospheric absorption by water vapor and clouds. This results in a tradeoff although there is generally no problem observing surface features even at frequencies of 85 GHz and higher. It is only when one has to be more quantitative that atmospheric corrections are needed. Also important is the limited spatial and temporal sampling from polar orbiting satellites, which affect some parameters more than others.

Except for the use of geostationary satellites, the measurements from a single polar orbiting satellite only occurs twice a day over a given region and the spatial coverage from successive orbits is contiguous only at high latitudes. This limited temporal and spatial coverage is an important issue for small scale features such as rain while it is less important for larger scale parameters such as temperature and water vapor which are more continuous in space and time. Since the antenna footprint on Earth from microwave radiometers would be very large from geostationary altitudes, the practical means of reducing the spatial and temporal sampling issue is to use multiple polar orbiting satellites. Each satellite's orbit is then spaced apart to obtain overlapping coverage, and the launch time of each satellite is chosen to provide multiple observations over a given area each day.

Table 7: Early Satellite Microwave Radiometers (1968 - 2003)

Satellite	Sensor	Center Frequencies (GHz) V = vertical polarization H = horizontal polarization	*IFOV* *(km) X-Scan=Cross Track C – Scan = Conical	Applications (*see Legend*)
Cosmos-243 1968	- - - -	3.5, 8.8, 22.2, 37.0	13 (Nadir)	I, V, Q, T
Nimbus-5-1972	ESMR	19.35	25 (X - Scan)	I, M, P
Nimbus-5 1972	NEMS	22.23, 31.40, 53.65, 54.90, 58.80	200 (Nadir)	t, I, S, V, Q, P
Nimbus-6 1975	SCAMS	22.23, 31.65, 52.85, 53.85, 55.45	150 (X - Scan)	t, I, S, V, Q, P
Nimbus-6 1975	ESMR	37.00 (V + H)	25 (C - Scan)	I, M, P
Nimbus-7 1978	SMMR	6.6, 10.7, 18.0, 21.0, 37.0 (V + H)	25 - 100 (C - Scan)	I, M, S, V, Q, P, W, T
NOAA 1978-1994	MSU	50.30, 53.74, 54.96, 57.95	110 (X - Scan)	t, Q, P
DMSP 1979 -1997	SSM/T	50.50, 53.20, 54.35, 54.90, 58.40, 58.82, 59.40	175 (X – Scan)	t, Q, P
DMSP 1987- 1999	SSM/I	19.35, 37.0, 85.5 (V+H), 22.23 (V)	15 – 60 (C – Scan)	I, M, S, V, Q, P, R, W
DMSP 1991- 1997	SSM/T2	90.0, 150.0, 183± 7, 183± 3, 183± 1	50 (X-Scan)	v, V, P
TRMM 1997	TMI	10.7, 19.4, 37.0, 85.5 (V+H), 21.3 (V)	6 – 50 (C – Scan)	I, M, S, V, Q, P, R, W, T
NOAA 1998 - - -	AMSU/A2	23.8, 31.4	50 (X - Scan)	V, I, S, Q, P
	AMSU/A1	89.0, 50.3, 52.8, 53.6, 54.4, 54.9, 55.0, 57.29, F=57.29 ± .217, F±..048, F±..022, F± .010, F± .0045	50 (X - Scan)	t, S, P, R
	AMSU/B	89.0, 150.7, 183±7, 183± 3, 183± 1	15 (X - Scan)	v, P, R
AQUA 2002	AMSR-E	6.9, 10.7, 18.7, 23.8, 36.5, 89.0 (V+H)	5.4 – 56 (C - Scan)	I, M, S, V, Q, P, R, W, T
DMSP 2003 - - -	SSMI/S	19.35, 22.2, 37.0, 91.6, 50.3, 52.80, 53.60, 54.40, 55.50, 57.29, 59.40, 62.283 ±. 285 , F = 60.79 ± .357, F ± .050, F ± .016, F ± .006, F ± .002 91.0, 150.0, 183±7, 183± 3, 183± 1	15 – 75 (C - Scan)	t, v, I, M, S, V, Q, P, R, W, T

* *The X - Scan IFOV is listed at nadir viewing while the C - Scan IFOV varies inversely with frequency.*

LEGEND

ATMOSPHERIC PARAMETERS	*SURFACE PARAMETERS*
V = Precipitable Water (Oceans)	I = Sea Ice Concentration
Q = Cloud Liquid Water (Oceans)	M = Soil Moisture
P = Precipitation (Oceans)	S = Snow Cover and Depth
R = Precipitation (Land)	T = Sea Surface Temperature
v = Water Vapor Sounding	W = Sea Surface Wind Speed
t = Temperature Sounding	

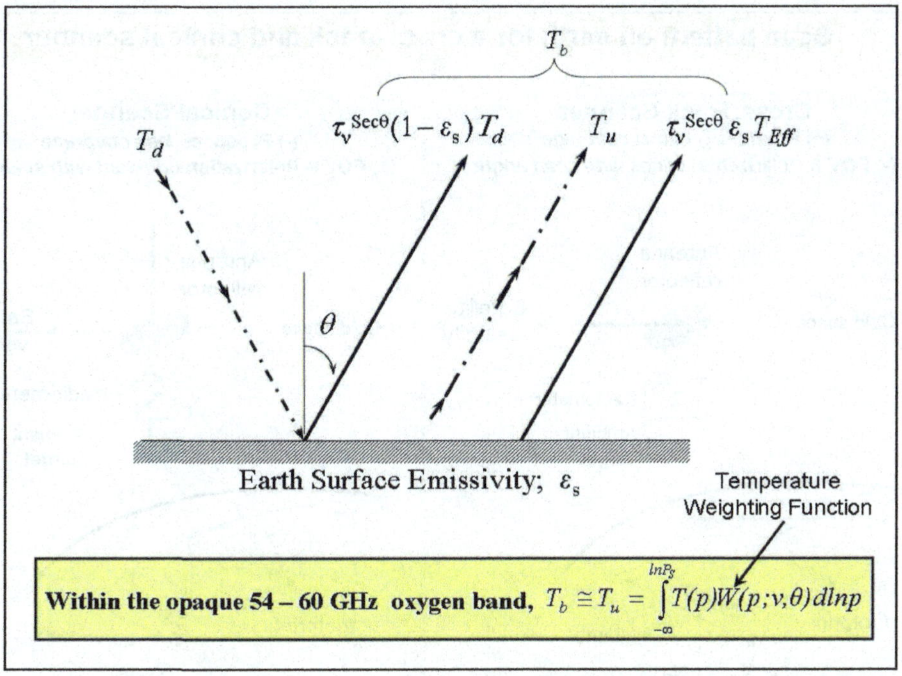

Figure 57 – The brightness temperature T_b measured by satellite radiometers contains the three terms shown in the top of the Figure. This includes the upwelling radiation T_u, downwelling radiation T_d reflected by the ground and attenuated by the atmosphere, and the surface emitted radiation at effective temperature T_{Eff} attenuated by the atmosphere. Neglected is the 2.7 K cosmic radiation which is attenuated by the atmosphere and reflected by the ground.

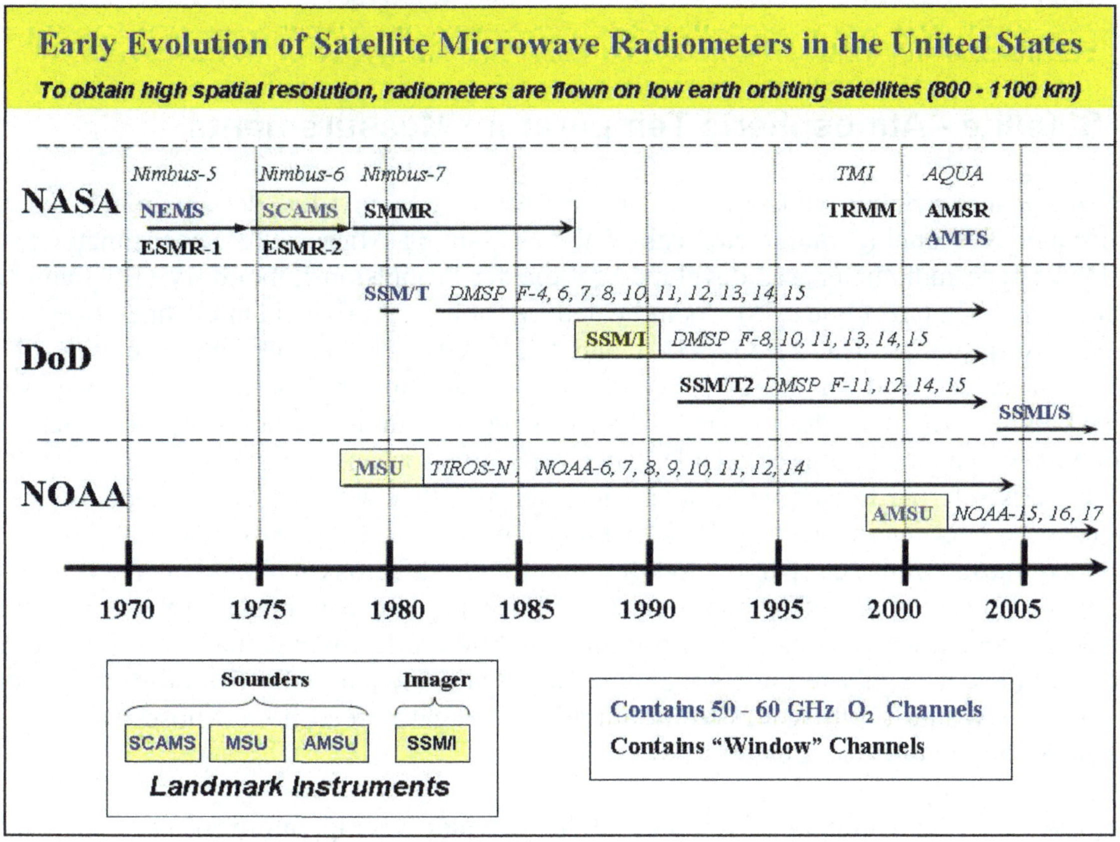

Figure 58 – Satellite microwave radiometer development in the U.S. from 1972 to 2005.

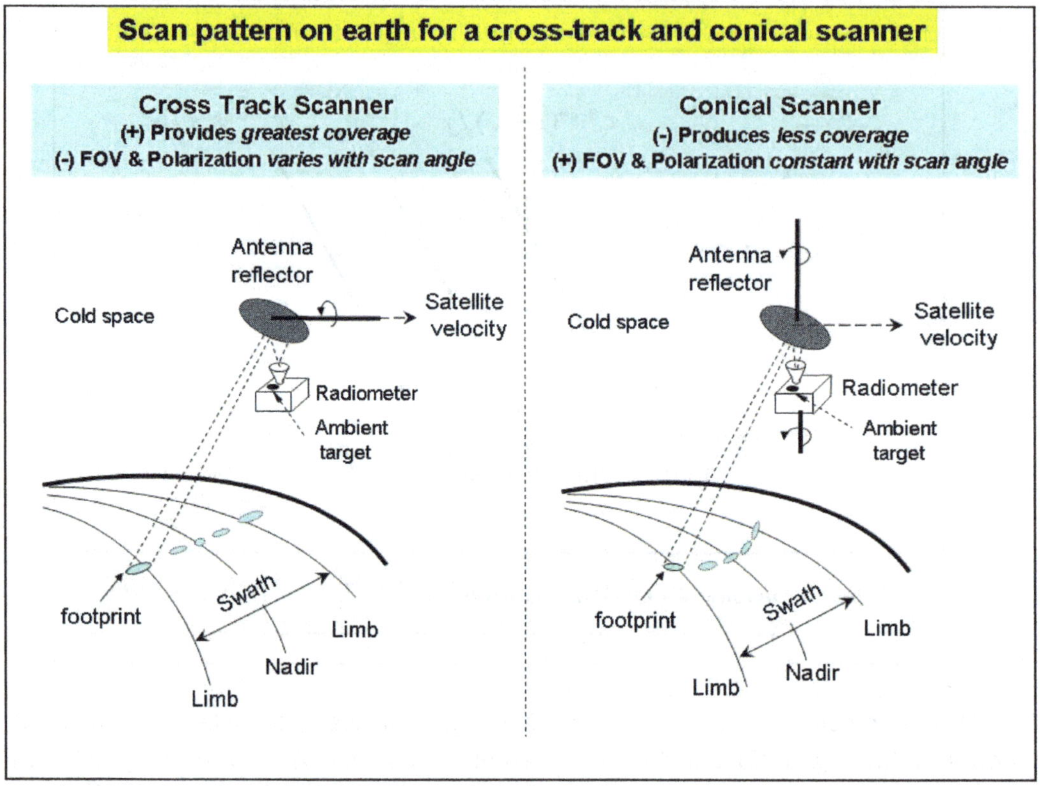

Figure 59 – Scan geometry of antennas for satellite radiometers. The left displays the scan pattern for a cross-track antenna while the right is for a conical scanning antenna. The cross-track scanner provides the widest coverage and is used in temperature sounders, while the conical scanner is used in imagers to view surfaces since the spatial resolution, slant angle and polarization are fixed.

11. Satellite - Atmospheric Temperature Measurements

The nearly all weather capability of microwaves was vividly demonstrated by the first multifrequency scanning radiometer called the SCAnning Microwave Spectrometer (SCAMS). This Dicke-type radiometer was developed by the Jet Propulsion Laboratory (JPL) and flown on Nimbus-6 in 1975 (see Figure 58). As indicated in Table 7, its five channels have three within the oxygen absorption band at 52.85, 53.85 and 55.45 GHz for temperature sounding. The other channels were centered on the 22.23 GHz water vapor line and between the water vapor and oxygen band at 31.65 GHz for viewing the surface and obtaining the liquid water content in clouds and rain as well as the total precipitable water vapor over oceans. The most transparent channel at 31.65 GHz is referred to as a window channel and together with the other channels is similar to those used on later microwave sounders. A picture of this pre-flight instrument is shown in Figure 60. It is composed of three separate modules, one for the three oxygen band channels and the other two for the water vapor and window channel. Each module contains its own warm calibration target and antenna, which consists of a rotating parabolic reflector and stationary feed horn. As the reflector rotates, it rapidly re-directs the incident radiation sequentially from the calibration target, cold space view and Earth scenes to the circular corrugated feed horn whose waveguide output connects to the radiometer receivers.

As mentioned in the beginning of Chapter 10, the brightness temperature (62) only depends on the upwelling radiation for SCAMS most opaque oxygen channel at 55.45 GHz. The equation has the same form as the mean temperature obtained for a ground-based radiometer (6a) and becomes

$$T_B(\nu) = T_u(\nu) = \frac{\int_0^\infty T(z) \frac{d\hat{\tau}_\nu(z)^{Sec\theta}}{dz} dz}{\int_0^\infty \frac{d\hat{\tau}_\nu(z)^{Sec\theta}}{dz} dz} \qquad (64a)$$

$$\text{with } \hat{\tau}_\nu(z) \equiv \tau_\nu(\infty)/\tau_\nu(z) = e^{-\int_z^\infty \gamma_\nu(z')\,dz'}, \qquad (64b)$$

where the upwelling transmittance $\hat{\tau}_\nu(0)$ is virtually zero at frequency ν.

Since the satellite measurement at 55.45 GHz is opaque to the surface the denominator in (64a) is unity so the brightness temperature is the atmospheric temperature vertically weighted by $-d\hat{\tau}_\nu(z)^{Sec\theta}/dz$. This weighting function is different than ground based radiometer which is $-d\tau_\nu(z)^{Sec\theta}/dz$. Unlike ground-based radiometers whose weighting function decreases exponentially with height (see Figure A13-1), the weighting function for satellite radiometers increase with height, reaching a maximum when $\hat{\tau}_\nu(z)$ becomes e^{-1}. The weighting function when expressed in pressure coordinates is Gaussian-shaped for uniformly mixed gases such as oxygen (also carbon dioxide). As shown in Appendix A20 the weighting function in (64a) has the approximate form given by equation (A20-8), *i.e.*,

$$-\frac{d\hat{\tau}_\nu(p)^{Sec\theta}}{d\ln p} = 2X_\nu\,\varepsilon^{-X_\nu} \quad \text{with} \quad X_\nu = \left(\frac{p}{P_\nu\sqrt{Cos\theta}}\right)^2 \qquad (65a)$$

$$\text{where } T_B(\nu) = -\int_{-\infty}^{\ln Ps} T(p)\frac{d\hat{\tau}_\nu(p)^{Sec\theta}}{d\ln p}\, d\ln p\,. \qquad (65b)$$

This weighting function peaks when $X_\nu = 1$ or at pressure $P_\nu\sqrt{Cos\theta}$. Furthermore, its width at half maximum is also proportional to $P_\nu\sqrt{Cos\theta}$. Equation A20-7 in the Appendix also shows that P_ν is proportional to the frequency displacement from the line center and inversely proportional to the line strength. Therefore, the weighting function peaks highest at frequencies near the most intense lines near the 60 GHz band center. All of these features are discussed next.

To begin the discussion, Figure 61 (Left) shows the exact SCAMS weighting function for its 55.45 GHz channel at nadir viewing, calculated using the oxygen absorption model mentioned in Section 4.3. For comparison the Right-most Figure plots examples of temperature profiles for three different atmospheres, with both plots using the same vertical scale. As such, this channel's response peaks at 200 mb, which is just below the tropopause for tropical atmospheres. Therefore, its brightness temperature (65b) is the vertically weighted temperature centered near 200 mb. Figure 61 (Left) also shows a similar weighting function for an infrared radiometer which has a channel in the carbon dioxide region at a wavenumber of 695 cm^{-1} or a wavelength of 14.5 μm corresponding to a frequency of 21,000 GHz. However, water drops and ice crystals formed in clouds are highly absorbing in the infrared so it's weighting function and brightness temperature only represent the measurement under clear sky conditions. This difference between infrared and microwave measurements was observed soon after SCAMS was launched when it viewed Typhoon June over the Pacific Ocean on November 21, 1975. This storm had the highest intensity resulting in the lowest central pressure (875 mb) of any previous Typhoon. Because of its large size its temperature structure could be observed by SCAMS resolution of 150 km at nadir viewing (see Table 7).

Figure 62 displays the Typhoon's warm core temperature structure near 200 mb as measured by SCAMS 55.45 GHz channel. This type of measurement with its wide spatial coverage was previously only possible by dropping radiosondes from specialized aircraft as they fly over and through storms, such as by the National Hurricane Center (NHC) of NOAA. Furthermore, as explained in Appendix A20, the brightness temperature gradient obtained from the measurement can be equated with the winds at a pressure near 500 mb [21]. This temperature and its gradient information can only be obtained from a radiometer whose measurements are unaffected by clouds, or more specifically ice clouds in the case of convective storms. In comparison to the microwave observations, Figure 62 also shows corresponding measurements obtained by an infrared sounder. The infrared measurement was obtained at nearly the same time as the SCAMS from the Vertical Temperature Profile Radiometer (VTPR) on a NOAA satellite. It's measurement in the carbon dioxide region at 14.5 μm is seen to be completely contaminated by cloud cover except in the eye of the Typhoon which is cloud free. In fact, the infrared measurement at 14.5 μm displays nearly the same features as the 11.5 μm infrared image which is in the most transparent region of the infrared spectrum. As a result of such demonstrations, the importance of microwave sounders to probe through clouds was well recognized by NOAA, who subsequently developed its first microwave temperature sounder. Previously their satellites only carried infrared temperature sounders.

The first operational microwave sounder developed by NOAA was the Microwave Sounding Unit (MSU), which had four channels in the 50 to 60 GHz portion of the oxygen absorption band with a FOV of 110 km at nadir viewing (see Table 7). The weighting functions are shown in Figure 63 along with a picture of a spare flight model that I took some time ago at the Smithsonian Institution, Air and Space Museum. Note how the weighting functions peak higher as the frequency approaches the center of the oxygen band at 60 GHz. The peak height is also increased as the antenna scans from nadir to the limb viewing angle of 56.6^0 in accordance with X_ν in (65a). As discussed in Appendix A20, the lowest frequency channel at 50.3 GHz mainly serves to measure surface emissivity and identify precipitation. Even with only three temperature sounding channels at 53.74, 54.96 and 57.95 GHz the MSU was able to derive relatively smooth temperature soundings from the surface to about 50 mb. The temperature profile T(p) is obtained using least squares linear regression analysis [22] using

$$T(p) = \mathrm{a}_0(p,\theta) + \sum_{\mathrm{n}=1}^{3} \mathrm{a}_\mathrm{n}(p,\theta)\, T_\mathrm{b}(\nu_\mathrm{n},\theta) \qquad (66)$$

where the retrieval coefficients $\mathrm{a_n}$ are obtained by correlating T_b against T using simulated or actual measurements as discussed in Appendix A19. Much more details on the analysis and error characteristics of satellite temperature retrievals is given in the Appendix.

To aid in the display and analysis of brightness temperature measurements, a limb correction procedure is used to reduce the angular change of weighting functions indicated by equation (65a) and shown in Figure 63. The algorithm to obtain the angle corrections is

$$T_\mathrm{b}(\nu_\mathrm{n},0^0) = \mathrm{b}_{\mathrm{n},0}(\theta) + \sum_{\mathrm{m}=1}^{3} \mathrm{b}_{\mathrm{n,m}}(\theta)\, T_\mathrm{b}(\nu_\mathrm{m},\theta) \qquad (67)$$

where the limb-correction coefficients $b_{n,m}$ are obtained using simulated or actual data. For example, the SCAMS measurements of Typhoon June shown in Figure 62 are limb corrected. As such, the temperature profiles obtained using equation (66) need only apply nadir viewing coefficients $a_n(p, 0^0)$ when using limb corrected measurements $T_b(\nu_n, 0^0)$.

This first microwave sounder developed by NOAA was designed based on the SCAMS instrument which as mentioned above was flown on Nimbus-6 in 1975. The first MSU was launched aboard the NASA TIROS-N polar orbiting satellite in 1978. Interestingly, in those early days of the satellite program it was possible to design, construct and launch instruments aboard satellites in record time. This was not the case for the next more advanced NOAA microwave sounder that is described at the end of this chapter. Following the first launch on TIROS-N, a number of identical MSU instruments were flown on NOAA satellites over a span of more than 25 years, each having exceptionally high stability and precision (see Figure 58). This sequencing of MSU's was done as a safety measure to minimize any possible instrument degradation due to changes in critical components such as the warm calibration target and the radiometer detector.

The high performance of the MSU's has made it possible to monitor very small climatic changes in atmospheric temperature with accuracy less than 0.1 K. As a striking example of the instruments capability, Figure 64 shows the global averaged nadir viewing brightness temperature measurements between 1978 and 2005 using its lowest sounding channel at 53.74 GHz who's weighting function at nadir peaks at 700 mb (see Figure 63). This time series was generated using all MSU instruments flown on satellites, beginning with the first MSU on TIROS-N to the one 27 years later flown on NOAA-17 (indicated as N-17 in Figure 64). The analysis used to calibrate the different MSU's and adjust for the different observing times (*i.e.*, diurnal variations) seen by each instrument was performed by Dr. Konstantin Vinnikov [23] who I collaborated with on this project. In addition to this adjustment, the analysis includes the small but important calibration corrections due to the detector's nonlinearity as discussed in Appendix A18.

The analyzed time series of the MSU global brightness temperature is represented by $T_b = a(t) + t\,b(t) + \delta(t)$ where a(t) is the periodic annual variation modulated by short time diurnal changes which is also periodic. The second term b(t) also contains annual and diurnal temperature variations but is multiplied by t, the long term trend. The last term contains non-periodic residual changes or anomalies $\delta(t)$ from volcanic eruptions due to Pinatubo and El Chichon in addition to climatic events from El Nino and La Nina. Figure 64 shows all of these features together with the global average trend of 0.17 K/decade. Interestingly, due to vertical convection between the surface and lower atmosphere this same climatic trend was found from surface temperature measurements [23]. Furthermore, and most disconcerting, this same trend continues up to 2024 based on additional satellite measurements from later microwave sounders. Incidentally, in recognition of MSU's unique ability to measure the global climate trend from space, it was displayed at the Smithsonian Air and Space Museum in 2004. None of these measurements could be achieved without the cloud penetration capability of microwaves.

This chapter concludes with the largest achievement by NOAA regarding its satellite program. It was the development of the 20 channel Advanced Microwave Sounding Unit (AMSU) which I had the good fortune to work on and witness its first launch in 1998 on the NOAA-15 satellite. The instrument development was led by the late Dr. David Staelin who was principal investigator for NEMS and SCAMS, and by many other individuals who advocated AMSU's development once they observed the nearly all weather capability of microwave radiometers to derive accurate and reliable temperature soundings. However, unlike the 4 channel MSU which took three years to develop and first launched on TIROS-N in 1978, AMSU was flown twenty years later due to various program delays involving many design changes to meet the required instrumental noise and retrieval accuracy.

Table 7 lists the AMSU channels and its instrumental parameters while Figure 65 shows a picture of the instrument and weighting functions for its 12 oxygen sounding channels. In addition, Figure 66 shows all 20 channels (including their bandwidth) on the overall brightness temperature spectra seen from space (top graph). An expanded portion for the 50-60 GHz oxygen band is shown in the bottom graph. This Figure was generated using clear atmosphere simulations over land ($\varepsilon_S = 0.95$) and ocean ($\varepsilon_S = 0.50$) based on radiation transfer calculations for a standard temperature profile that decreases with altitude in the troposphere and increase in the stratosphere as shown in Figure 61. As such, the sharp increases in brightness temperature near the line centers is caused by the stratospheric contribution while the weaker absorbing regions between lines results in the relatively colder temperatures near the tropopause which forms valleys in the spectrum. The least absorption occurs at frequencies below 53 GHz where the radiation results from temperatures near the warmer surface. As such, this spectral plot is a representation of the temperature profile in the frequency domain rather than pressure coordinates. Additional analysis of the spectral plot in Figure 66 is discussed in Appendix A20. Lastly, Appendix A19 describes an inversion technique similar to equation (66), which reconstructs the temperature profile from spectral measurements.

Unlike the MSU which was a Dicke radiometer, AMSU was designed as a total power radiometer to reduce the channels NEΔT by about a factor of two as discussed in Chapter 5. It's FOV at nadir was also reduced to 50 km for its sounding channels compared to 110 km for MSU. Also, compared to MSU, the AMSU was shown to provide much more accurate atmospheric temperature, wind speed and central pressure measurements for tropical cyclones by Dr. Stan Kidder *et. al.* [24]. As an example from his paper, Figure 67 shows the vertical temperature structure of Hurricane Bonnie on August 25, 1998 obtained from AMSU measurements. The structure is also compared on the right with an example of measurements for Hurricane Hilda in 1968, which as mentioned before was obtained by the National Hurricane Center of NOAA by dropping radiosondes from specialized reconnaissance aircraft as they fly over and through storms.

The AMSU development was a collaborative effort involving many organizations such as the National Weather Service (NWS) of NOAA. The NWS in fact was one of the main advocates as well as the primary user of its data for input to their short-range numerical weather prediction model. Previous to AMSU the NWS had gained much experience in using MSU data and realized that this primitive four channel sounder needed to be improved by adding additional channels to obtain more accurate temperature soundings with better vertical and horizontal resolution. Of coarse no one expected it to take twenty years to develop. I should also mention that the British were also among the strong advocates for the AMSU development. They witnessed first hand the use of microwave derived temperature soundings during the Falkland war in 1982. At that time, because of the war effort the only readily available temperature and wind information for the region came from satellites. They found that the MSU provided them reliable temperature measurements, including accurate gradient information to derive atmospheric winds using for example, the geostrophic relationship between thermal gradients and atmospheric winds. Only due to the cloud penetration capability of microwaves was such accurate spatial gradient information possible. More detailed analysis of the use of gradient information is presented in Appendix 20.

To further display the advantage of microwave soundings over infrared, NOAA performed an extensive study [25] comparing SSM/T soundings with that obtained from an infrared radiometer. As indicated in Table 7, the SSM/T was the first microwave sounder developed by the U.S. Air Force. It was launched aboard DMSP satellites beginning in 1979 and contained seven channels in the oxygen band to provide temperature soundings from the surface to the lower mesosphere. Until that time such temperature measurements were only obtained by the military using infrared sounders. As shown in Figure 68, this NOAA study compared the SSM/T temperature accuracy with that produced by the TIROS Operational Vertical Sounder (TOVS) system. While this system mainly relied on its infrared sounder, it also used MSU measurements to improve its accuracy under cloudy conditions. However, the SSM/T alone is seen to provide much greater accuracy than TOVS, particularly for cloudy atmospheres. Furthermore, since the "cloudy" TOVS results are mainly from MSU, this comparison also shows much greater accuracy of the 7-channel SSM/T soundings compared with the 4-channel MSU. As such, this study provided conclusive evidence of the need for NOAA to develop an improved microwave sounder with more channels than MSU, namely that of AMSU.

In addition to defining some of the AMSU channels my main responsibility was in developing algorithms for the non-sounding products and evaluating the instruments overall performance following its launch. Besides temperature soundings the other AMSU products consisted of rain rate, cloud liquid water, total precipitable water, surface wetness, snow cover and sea ice concentration. Most of these non-sounding products were never developed before by NOAA, particularly using microwave measurements. In fact, many of these products were not even envisioned at the time when AMSU was being developed. For that reason it only included four un-polarized non-sounding channels at 23.8, 31.4, 89 and 150 GHz, where the two highest frequency channels were not even flown in space at the time when AMSU was being designed in the early 1980's. In an unprecedented milestone, these products were generated within a year of its satellite launch in 1998. However, the main reason it was possible to accomplish this was because of knowledge gained previously by analyzing data acquired from the Special Sensor Microwave Imager (SSM/I) as part of a coordinated shared processing agreement between NOAA and DOD.

As indicated in Figure 58 the SSM/I was developed by the Navy and first launched in 1987 on a DMSP satellite. Table 7 shows that it contained six dual polarized channels at 19.35, 37.0 and 85.5 GHz and one vertically polarized channel at 22.23 GHz. Until the development of the SSM/I all other instruments at that time contained frequencies much less than 85 GHz. The 85 GHz channel also had the highest spatial resolution (15 km) of any previous microwave radiometer. As discussed in the next chapter, a number of important discoveries were made using this channel. While some of these measurements have already been reported in the literature, other observations have not been discussed in any great detail due to limitations in our understanding and physical models to fully explain the results. However, I feel it is important to discuss the measurements and provide some understanding based on our current knowledge. In this context, the most significant examples are SSM/I and AMSU measurements over deserts and snow covered surfaces, as shown in Chapter 12. These anomalous surface measurements and their analysis are discussed in detail in this last chapter.

Finally, I must mention that an even more advanced instrument than AMSU, the SSMI/S was developed and launched in 2003 by DOD (see Figure 58). It is a conical Imager and Sounder developed by the Air Force and Navy to combine the attributes of the SSM/I, SSM/T and SSM/T2 into a single instrument whose frequencies are listed in Table 7. As mentioned previously, the NOAA sounders such as ASMU were designed using a cross-track scanning geometry to minimize calibration errors while the DOD instrument uses conical scanning to obtain a nearly constant footprint, slant angle and polarization when viewing Earth (see Figure 59). In conclusion, after more than 20 years of development from SCAMS to MSU and more recently AMSU and SSMI/S, microwave sensors have matured so they are now the *primary instruments for measuring surface and atmospheric observations remotely*. Because of their nearly all weather capability, they routinely provide data to numerical models for analysis and prediction of both short-term and long-range weather forecasting.

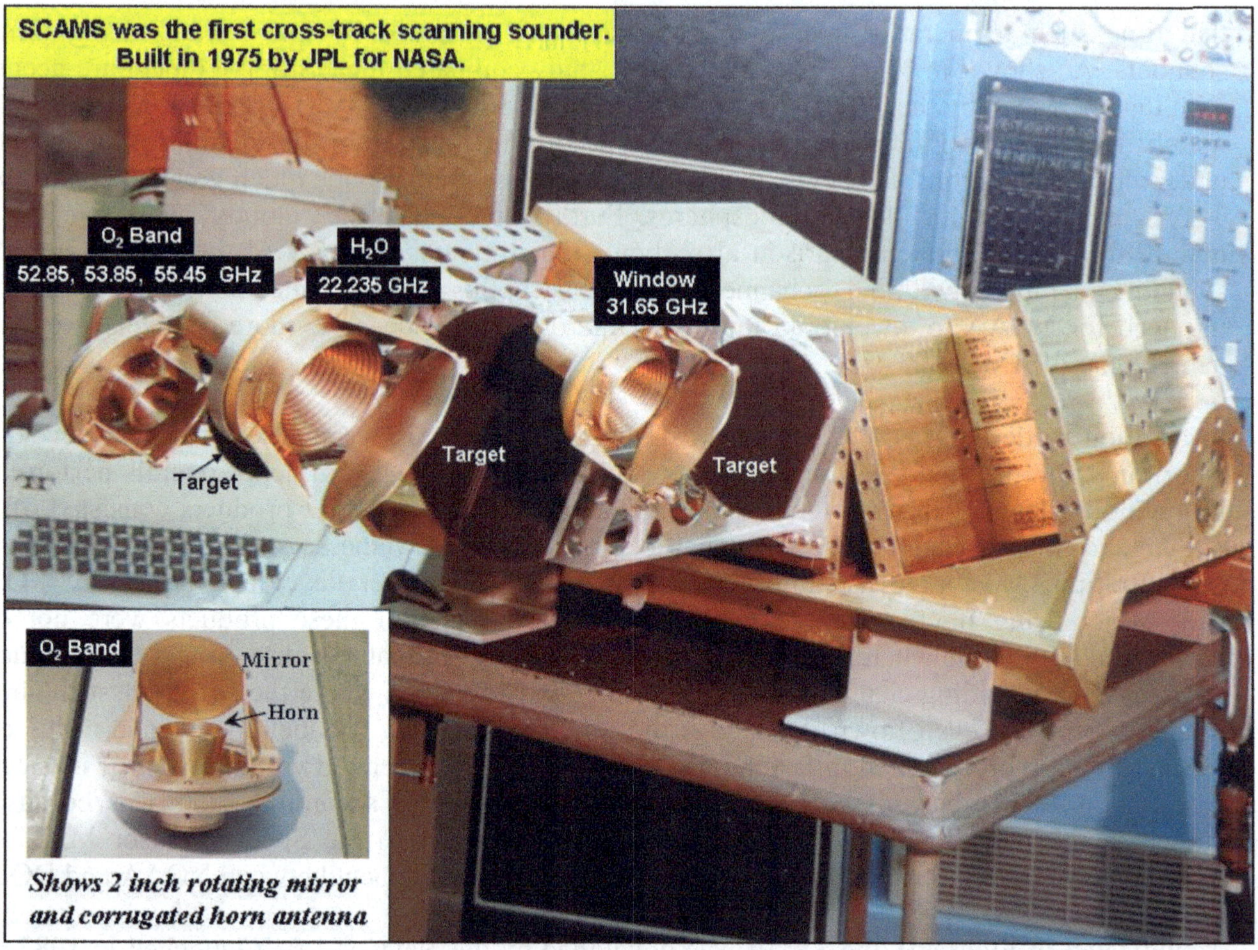

Figure 60 – First scanning microwave radiometer to provide global temperature soundings under nearly all weather conditions. This instrument called SCAMS was flown on NASA's Nimus-6 satellite in 1975.

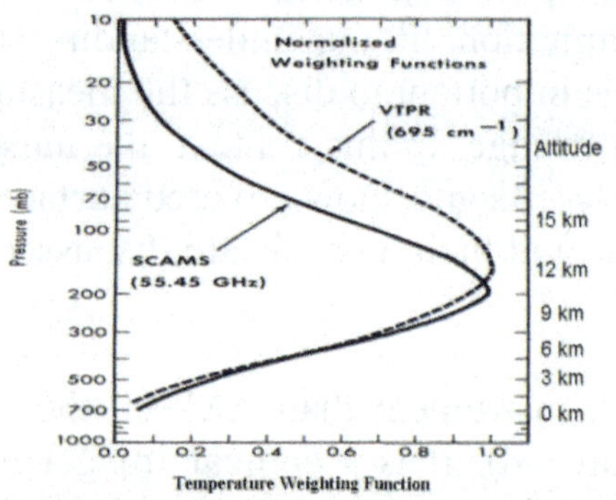

Clear atmosphere weighting functions at 55.45 GHz (O_2 band) and 695 cm^{-1} (CO_2 band). The 55.45 GHz channel is one of five measured by the SCAMS on Nimbus-6. Also shown is the weighting function of the infrared channel at 695 cm^{-1} which is one of six measured by the Vertical Temperature Profile Radiometer (VTPR) on NOAA-4.

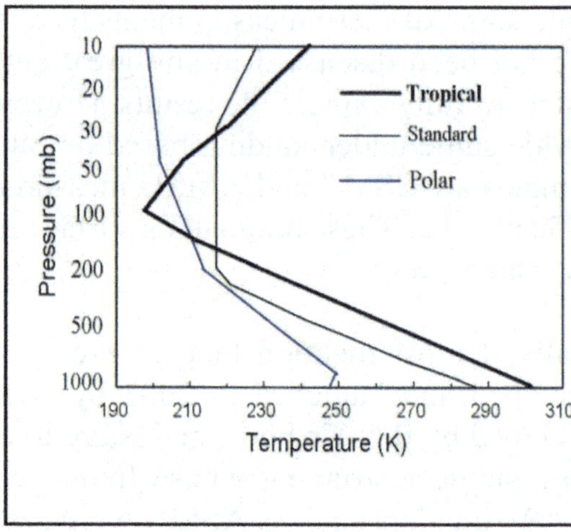

Figure 61 – (Left) Weighting function for SCAMS highest frequency channel at 55.45 GHz. Its brightness temperature senses the atmospheric temperature near 200 mb where the weighting function peaks. Also shown is the weighting function for an infrared radiometer VTPR which was flown on the NOAA-4 satellite. This infrared CO_2 channel at wavenumber 695 cm-1 has a similar weighting function to that of SCAMS. For reference the Right-Figure shows temperature profiles for tropical, polar and standard, or mid-latitude atmospheres. The brightness temperature (65b) averages these profiles over the weighting functions.

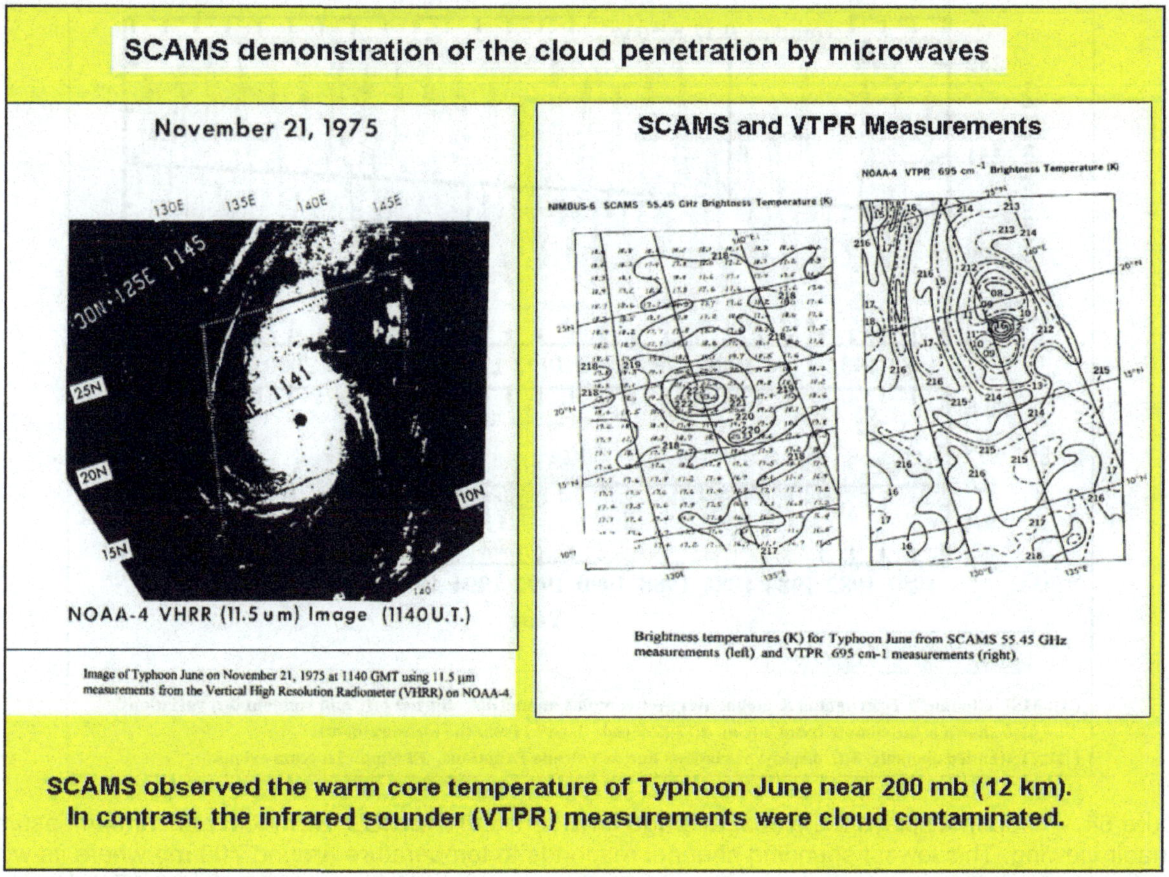

Figure 62 – The right-most segment displays a contour map of the SCAMS limb-corrected brightness temperature measurements at 55.45 GHz. It is compared to the VTPR infrared measurements at 695 cm^{-1} (14.5 μm) when viewing Typhoon June on November 21, 1975. Also shown on the left side is an image generated by the infrared window channel measurements at 11.5 μm.

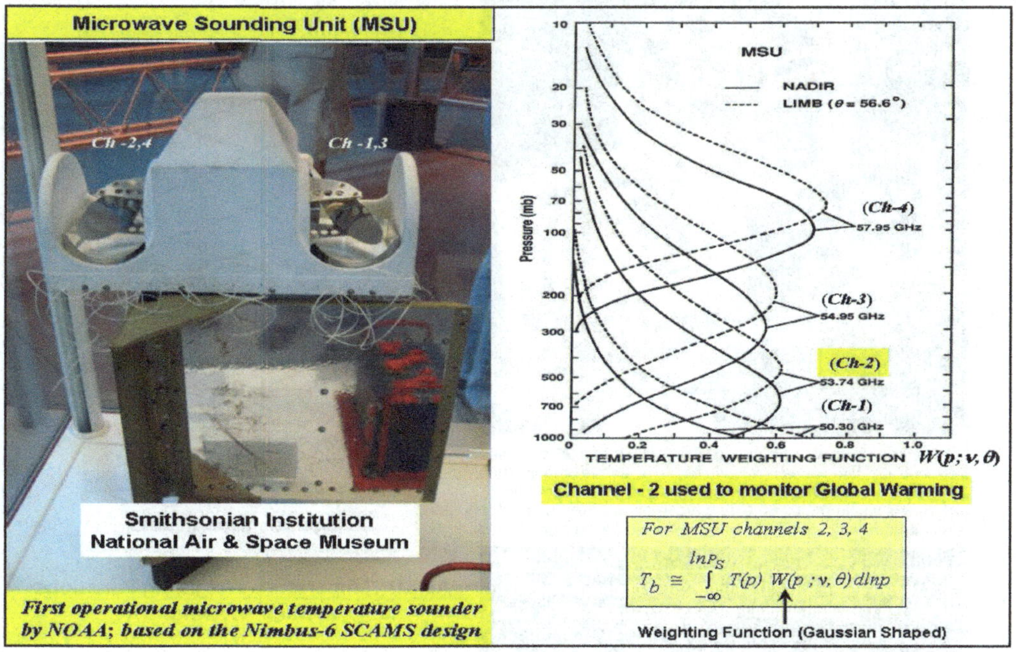

Figure 63 – The left-most picture shows the Microwave Sounding Unit (MSU) at the Smithsonian Air and Space Museum while the right segment shows the weighting functions of its four channels.

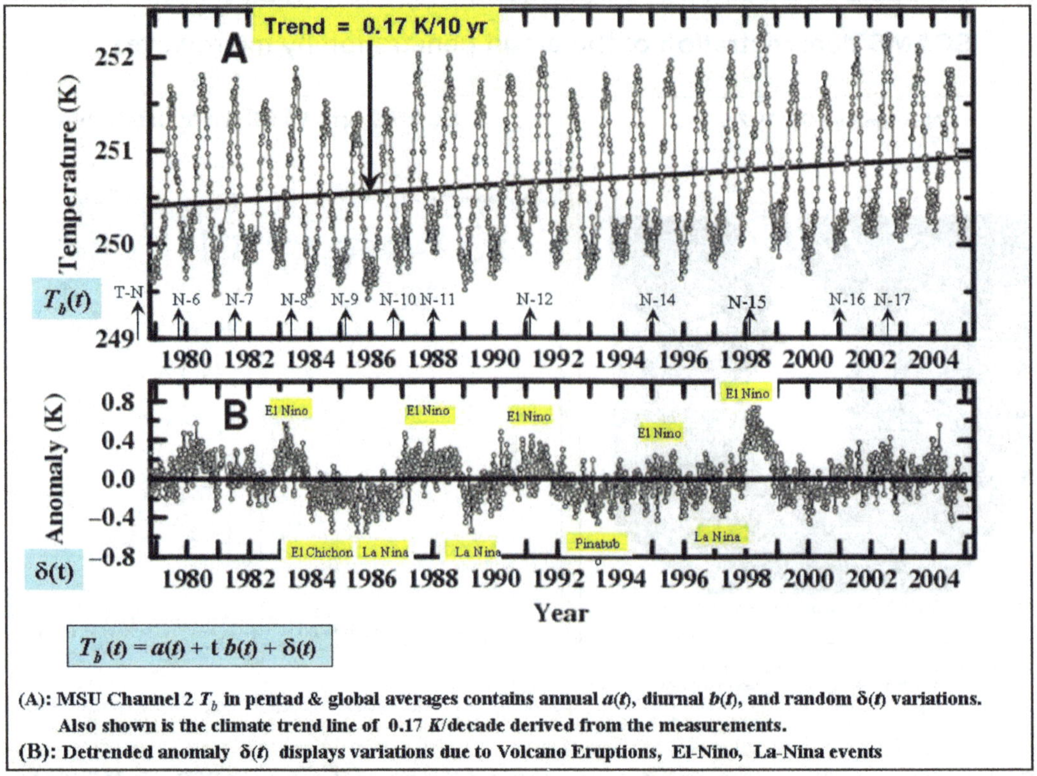

Figure 64 – The top displays the global average time series of the MSU 53.74 GHz channel measurements at nadir viewing. This lowest sounding channel responds to temperature around 700 mb where its weighting function peaks. It displays annual and diurnal variations in $a(t)$ and b(t) in addition to a 0.17 K/decade trend. The bottom displays the anomaly δ(t) due to Pinatubo and El Chichon volcano eruptions in addition to El Nino and La Nina climatic events.

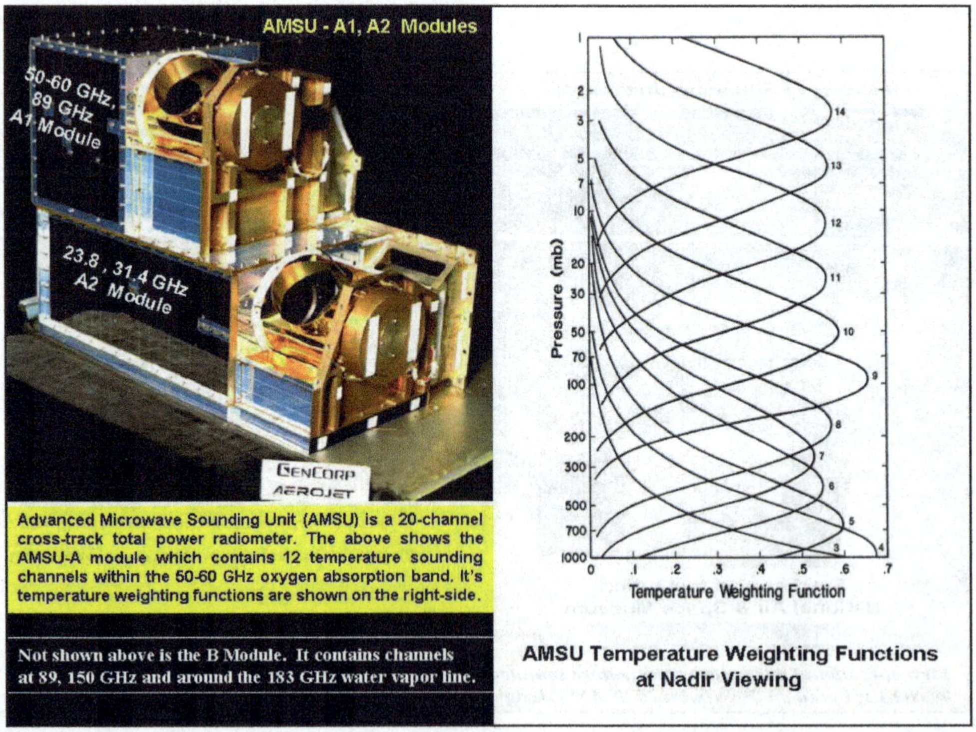

Figure 65 – The left picture shows two of the three modules of the AMSU instrument while the right is a plot of the weighting functions for the 12 oxygen channels whose frequencies are listed in Table 7. The AMSU-A modules were built by Aerojet while the AMSU-B module (not shown) was built by the British.

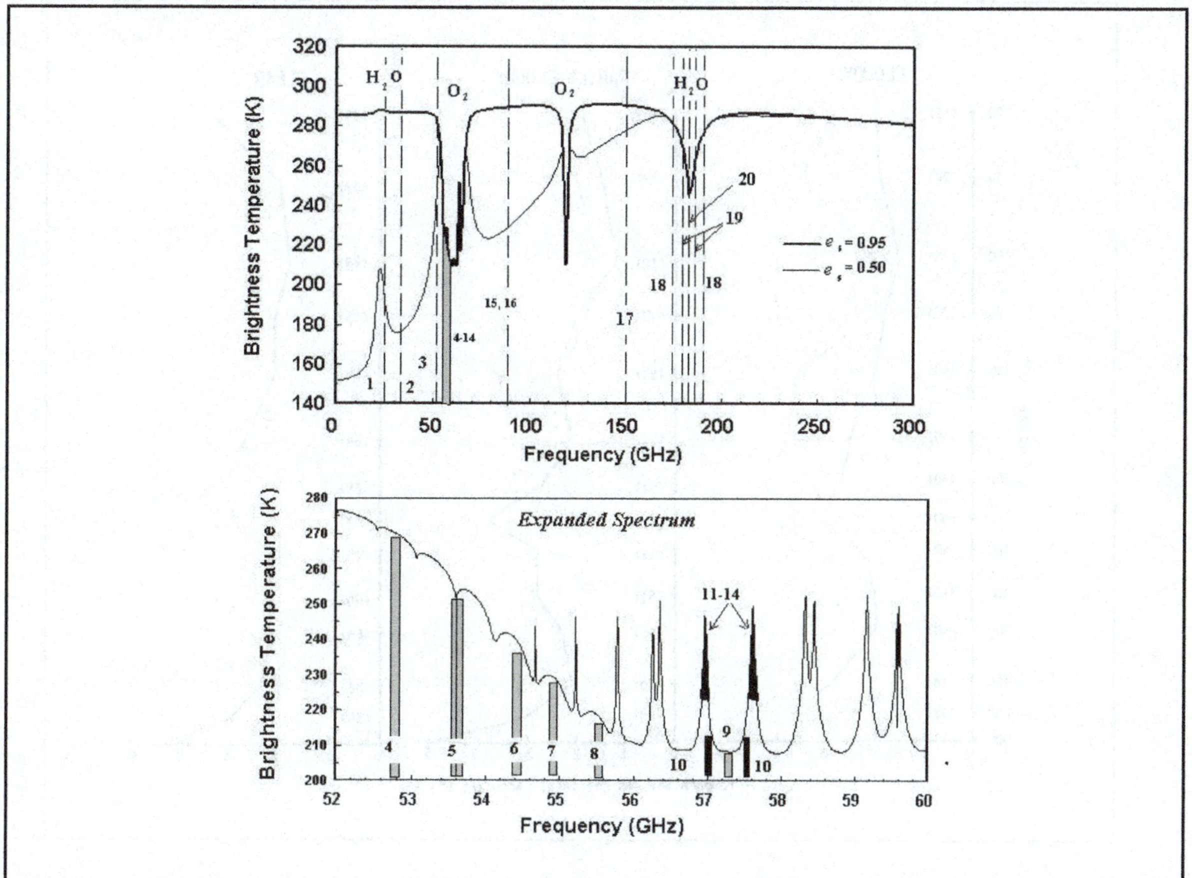

Figure 66 – Simulated brightness temperature spectra up to 300 GHz over land (εs = 0.95) and ocean (εs = 0.50) for a cloud-free standard atmosphere having 25 mm of *TPW* (Top). The bottom shows an expanded plot over the 50 - 60 GHz oxygen band. AMSU channel locations are indicated in both plots.

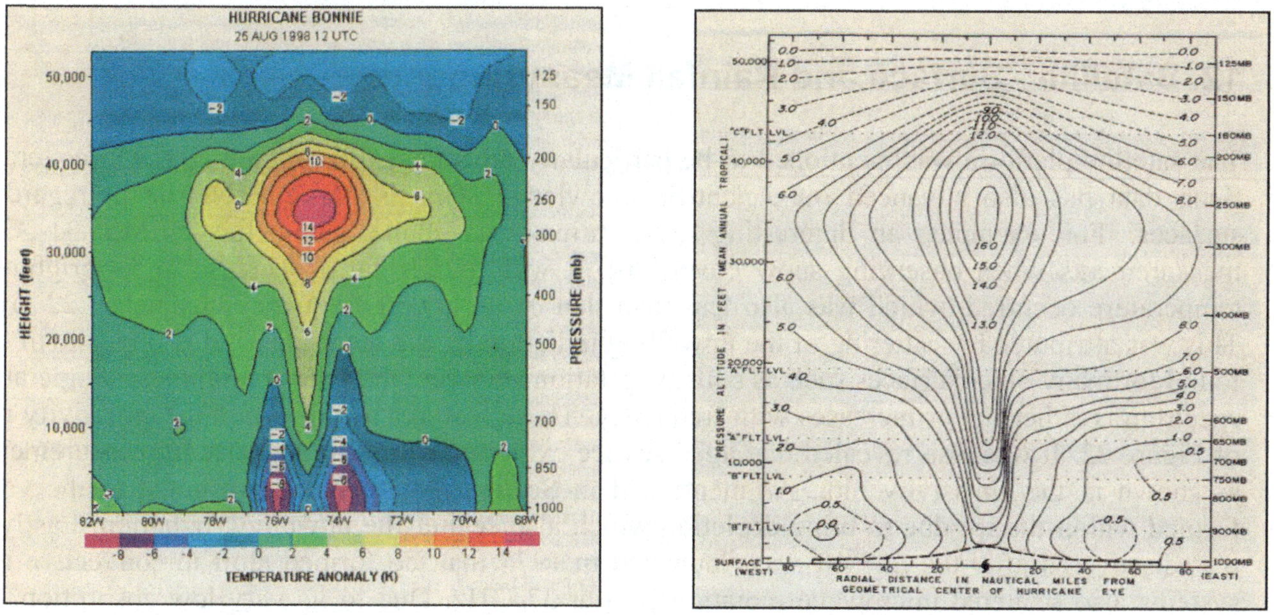

Figure 67 – The left-most Figure shows the cross section of temperature anomalies through Hurricane Bonnie at 1200 GMT on August 25, 1998 based on AMSU measurements. For comparison the right-most plot displays a similar cross-section of temperature anomalies through Hurricane Hilda in 1964 obtained using conventional dropsondes from aircraft. These examples are described by Dr. Stan Kidder [24].

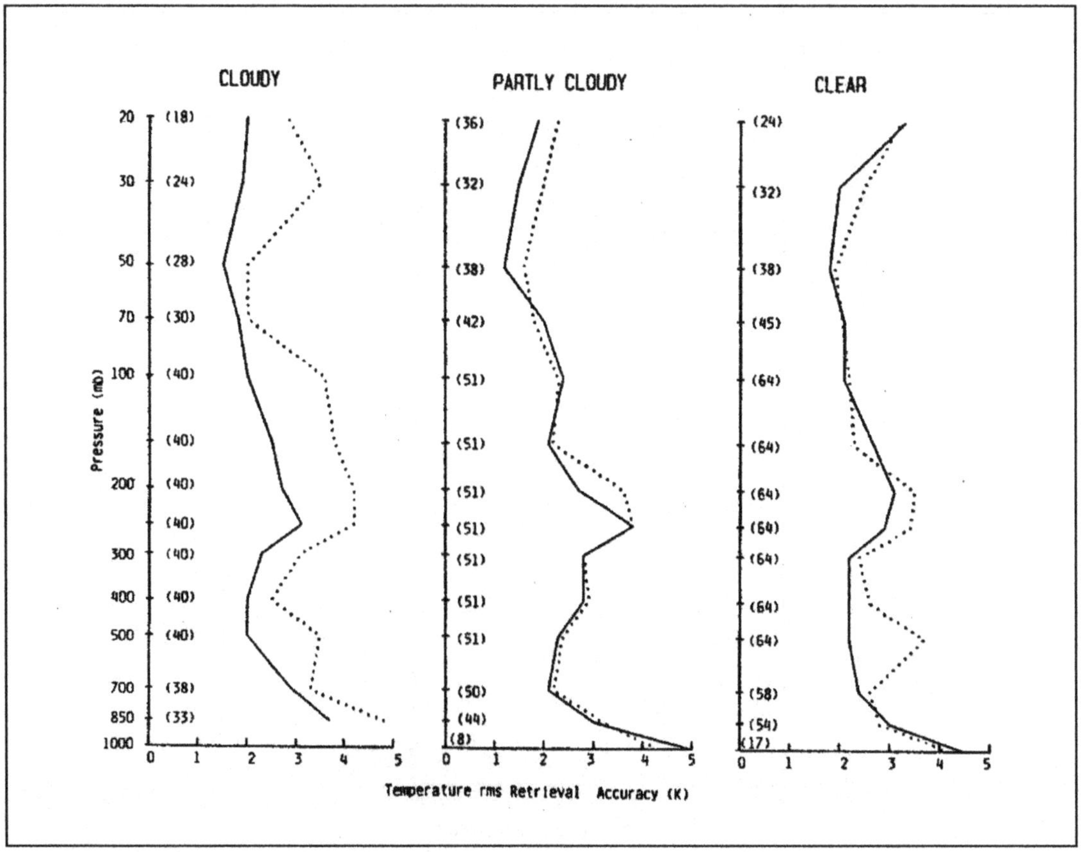

Figure 68 – Comparison between SSM/T (solid) and NOAA's TOVS (dashed) retrieval errors for November 14 through 17, 1983. Results are compared against radiosondes and stratified according to cloud condition. The sample size is indicated next to the pressure levels. Most importantly, as mentioned in the Text, the large error difference for cloudy atmospheres between SSM/T and TOVS also represents the error difference between the 7-channel SSM/T and 4-channel MSU soundings.

12. Satellite - Surface and Rainfall Measurements

The satellite platform was mentioned to be particularly advantages for obtaining global coverage. Such data has also advanced our scientific knowledge. This is particularly true in regard to surfaces. For example, an interesting observation was made using the NEMS 31 GHz measurements when observing snow cover.[26]. It was found that a decrease in its brightness temperature occurred, which was also less than that of its lowest frequency channel at 22 GHz. This was attributed to scattering of the upwelling radiation by the ice grains and is opposite to that found for many other surfaces such as soils, vegetation and wet land whose brightness temperature spectrum is either flat or increases with frequency. These features are shown in the emissivity plot of Figure 22. It was also revealed that aged sea ice exhibited a similar decrease in measurements, as shown in the emissivity plot. As mentioned in Section 7.1, this was attributed to the voids formed within the ice due to brine depletion which also scatter the upwelling radiation. Lastly, it was discovered after the SMMR was launched in 1978 that ice formed aloft in convective rain systems also scattered microwave radiation [27] at 37 GHz. Due to its very low absorption, ice mainly scatters radiation at high microwave frequencies, thereby decreasing the brightness temperature relative to the lower frequency channels. Previously, rain over land was mainly observed indirectly by monitoring the changes in surface wetness.

Scattering for sea ice and snow cover is well displayed by the emissivity spectra shown in Figure 22 even at frequencies less than 37 GHz. However, in the case of rain, it was only after higher frequency measurements at 85 GHz became available from the SSM/I in 1987 that the scattering approach to observe rainfall could be used reliably over land as well as oceans. Until that time precipitation could only be observed reliably over low emissivity oceans using the large contrast provided by the higher emissivity of liquid water drops in rain. Only until higher frequencies than 37 GHz became available could rain can be accurately detected over land as well as oceans using the scattering signature. In addition to detecting rain, algorithms were developed to estimate rain rate from the magnitude of the scattering signal. While less direct than the low frequency emission measurement obtained over oceans, it is statistically related to rain rate analogous to the scattering measurements obtained using radar. Lastly, it was found that in addition to aged sea ice, snow cover and rain, even deserts display a scattering signature at high frequencies. As discussed below, studies over deserts show that sand grains can also scatter microwave radiation but with a smaller magnitude due to their high density [19].

One of the best early examples of surface and atmospheric features observed from a microwave imager was that obtained using the SSM/I. This conically scanning radiometer developed by the Navy was first launched on the DMSP satellites in 1987. As mentioned in the previous section, and shown in Table 7, the radiometer contains four dual polarized channels at 19, 37, 85 GHz and a single vertically polarized channel at 22 GHz. An offset parabolic reflector is used to capture the Earth emitted radiation, which is then directed to separate feed horns for each channel. Figure 69 shows a picture of the reflector and other components of the flight model. It also shows a drawing of its scan geometry. To obtain the smallest antenna beamwidth the radiation pattern of each feed horn subtends the full reflector aperture. Due to diffraction limitations the antenna beamwidth at each frequency then increases linearly with wavelength so that the half power footprint or FOV seen on the ground increases from 15 km at 85 GHz to 60 km at 19.35 GHz.

As an example, Figure 70 shows a composite color image of the cloud liquid water, rain rate, snow cover and sea ice concentration determined using the SSM/I channels. The identification and separation of these parameters is based on a decision tree algorithm approach [10] Each of the daily products is monthly averaged to generate this global image for November 1987, the year the SSM/I was launched. Of particular importance was the use of the highest frequency channel at 85 GHz which as mentioned previously was particularly important to detect the volume scattering by millimeter size ice particles formed as part of the precipitation process. This channel enabled, for the first time, the most accurate detection of rain over land as well as oceans. As mentioned above, without this high frequency channel the earlier radiometers could only detect rain over low emissivity oceans. The 85 GHz channel also improved the detection of snow cover which at lower frequencies could only observe the scattering resulting from deep snow. Other surfaces such as sea ice and desert sand were also found to scatter at 85 GHz. The following begins with a discussion of the measurements obtained when viewing deserts and precipitation followed by snow cover measurements.

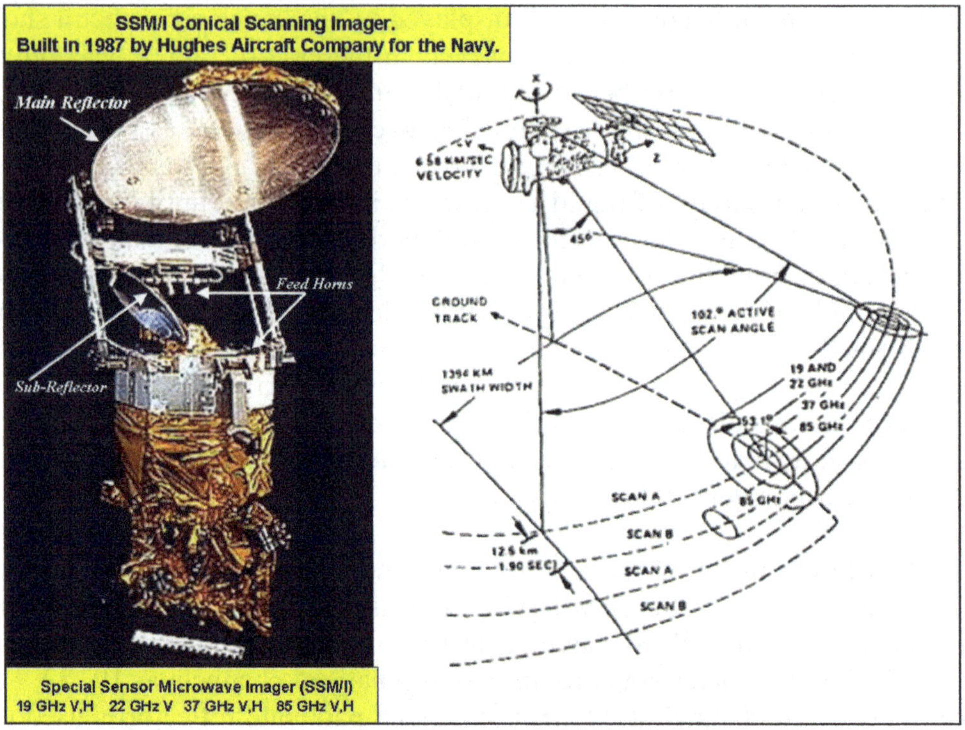

Figure 69 – The SSM/I was constructed by Hughes Aircraft Company for the U.S. Navy and flown on a sequence of polar orbiting satellites beginning in 1987. It's a 7-channel total power radiometer with dual polarization (V, H) at 19, 37 and 85 GHz and single polarization (V) at 22 GHz. It scans every 1.9 sec to view Earth and calibration sources (cold space and warm target).

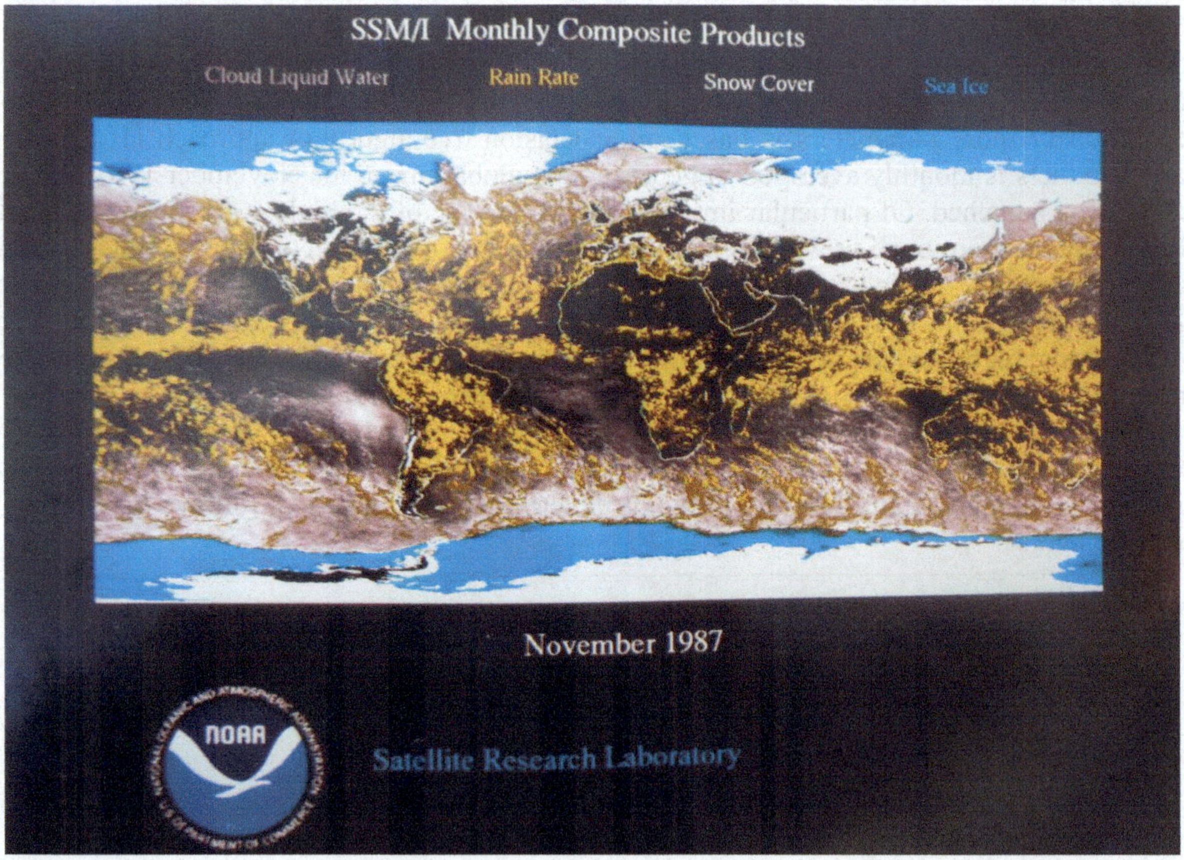

Figure 70 – Products such as shown above are determined daily using the Special Sensor Microwave Imager (SSM/I). The daily products are monthly averaged to obtain this composite image (see Text).

12.1 Desert Measurements

The high reflectivity over deserts has always been considered a unique target to vicariously help calibrate visible satellite instruments. Also, since the desert atmosphere has little moisture, there is little cloud cover and vegetation growth to obscure the different minerals within deserts. Furthermore, due to different dielectric constants the reflected and emitted radiation from desert minerals varies widely between the visible, infrared and microwave measurements. This radiation also varies depending on the size of the minerals. In particular, microwave frequencies are observed to scatter radiation in many ways similar to that of snow cover and precipitation. This of coarse can be a problem when using microwaves to identify precipitation and snow cover over arid regions which will be discussed here. On another feature seen over deserts, the lack of vegetation results in a large diurnal temperature variation at the surface, with much smaller variations occurring below the surface. This latter feature is most prominent at microwave frequencies due to its larger penetration depth and will also be discussed. In fact, all of these features have been measured using satellite microwave radiometers and analyzed using models [19].

Two of the most dominant minerals in deserts are quartz and limestone, which have widely different dielectric constants. Figure 71 (bottom-left) shows the calculated emissivity as a function of dielectric constant for the SSM/I viewing angle and polarizations as well as at nadir viewing. Of particular interest here is the large emissivity difference for quartz and limestone which is indicated in the plot. The quartz dielectric constant of 3.8 is a little larger than that of ice, while limestone has a dielectric constant between 6 and 9 which is slightly larger than glass. Also, the dielectric loss for ice and glass is probably less than desert minerals due to impurities occurring in nature.

The effect of the different microwave emissivities found in deserts is shown in the top image. This color composite image displays the monthly averaged vertically polarized 37 GHz channel measurements over Saudi Arabia for July, 1996. Also shown in the bottom right of Figure 71 is the area of limestone deposit over the Arabian peninsular by a surveyor map generated by geologists. Note how well the lowest brightness temperature measurements at 37 GHz correspond to the limestone in the map. A particular interesting feature is the crescent shaped region of limestone near the center of both figures, as well as being along the coast. The SSM/I image also shows a large southern region over Saudi Arabia having very high brightness temperatures, which is due to high emissivity minerals such as quartz. A very similar image is shown in Figure 72 using the MODIS infrared measurements in the 8.4 – 8.7 μm band. The difference being that the infrared emissivity is reversed, with quartz having the lowest emissivity while limestone has the highest emissivity.

Identification of high emissivity surfaces at microwave frequencies such as quartz can also be identified by measuring the difference between two closely spaced frequencies such as the 19 and 22 GHz SSM/I channels. For these high transmittance channels the brightness temperature equation (63) can be expanded in a Taylor series about $\tau^{\mathrm{Sec}\theta} = 1$ so that one can write

$$T_b \approx (2-\varepsilon_S)T_M + \tau^{\mathrm{Sec}\theta}(\varepsilon_S - \hat{\varepsilon})\,(T_M + T_{Eff}) \qquad (68a)$$

$$\text{where} \quad \hat{\varepsilon} = \frac{2T_M}{T_M + T_{Eff}} \qquad (68b)$$

The brightness temperature difference then becomes

$$T_b(19) - T_b(22) \approx \left[\tau_{19}^{\ \mathrm{Sec}\theta} - \tau_{22}^{\ \mathrm{Sec}\theta}\right] \left[\varepsilon_S - \hat{\varepsilon}\right] (T_M + T_{Eff}) \qquad (69)$$

where the emissivity ε_S and mean temperature T_M is about the same at these two frequencies. On the other hand, the atmospheric transmittance due to water vapor absorption (also clouds) is greater at 19 GHz than at 22 GHz so that the bracketed transmittance term in (69) is positive. The sign of the brightness temperature difference then depends on the second bracketed term which is positive for high emissivity surfaces where $\varepsilon_S > \hat{\varepsilon}$ and reverses sign for lower emissivities. The effective temperature T_{Eff} is close to the surface temperature while the mean radiating temperature T_M resides above the surface which is generally lower than T_{Eff}. For illustration, equation (63) is plotted as a function of transmittance in Figure 72 with the effective temperature set at 305 K and the mean temperature set at 285 K. The Figure shows the brightness temperature difference or slope to be positive for emissivities greater than 0.966 ($=\hat{\varepsilon}$), which for vertical polarization includes quartz whose calculated emissivity is 0.980 for a smooth surface based on a dielectric constant of 3.8. The slope becomes even smaller for the other emissivities indicated in Figure 71 (Bottom Left), with limestone producing a negative slope since its emissivity is less than 0.95. Also, as with equation (68a), the brightness temperature is seen to vary almost linearly with transmittance at 19 and 22 GHz (see dashed lines). For reference, the graph also shows the transmittance range at different frequencies due to water vapor, oxygen and cloud absorption.

The simulated result discussed above prompted the generation of the monthly average color images in Figure 74. The bottom image shows the 37 GHz vertically polarized measurements for the same time period in Figure 71 but for a larger area over Saudi Arabia as well as Africa. Of greater importance, the top image displays the difference between 19 and 22 GHz vertically polarized brightness temperature measurements. Note that only the area over central Saudi Arabia has brightness temperature differences greater than 5 K which is attributed to quartz while all other regions have a brightness temperature difference less than 5 K. The difference image also suggests that the regions of dense forest and vegetation cover over central Africa have a lower emissivity than quartz due to effects of surface roughness which depolarize and reduce the emitted radiation. Stated differently, the large quartz surface over Saudi Arabia provides a much higher emissivity target for calibration than all other natural surfaces including vegetated land and forests whose roughness depolarizes and decreases its emissivity. Lastly, I should mention that the difference measurements shown in the bottom of Figure 74 is less dependent on temperature and therefore provides a more accurate identification of high emissivity surfaces than the brightness temperature measurements shown in the top image.

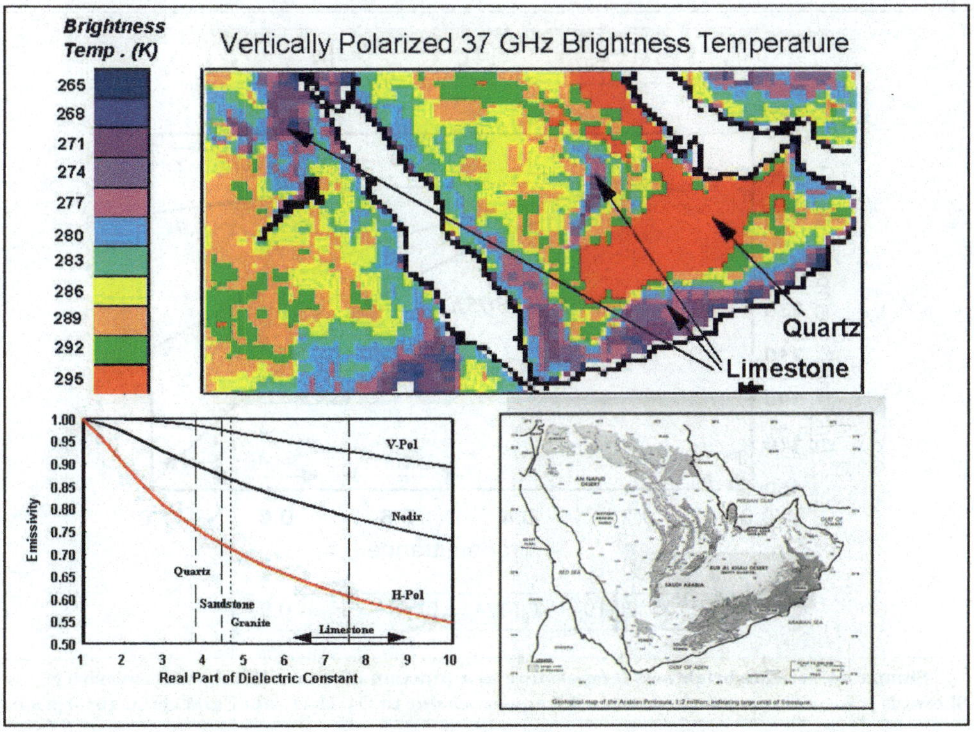

Figure 71 – The top image shows the SSM/I vertically polarized 37 GHz channel measurements for July, 1996 over Saudi Arabia. Note that quartz has the highest brightness temperature while limestone has the lowest measurements. The bottom-left shows the calculated emissivity at nadir and at vertical and horizontal polarization for the SSM/I viewing angle of 53.1^{0}. The bottom-right shows a map of limestone deposits over Saudi Arabia obtained from geologists in the oil Industry.

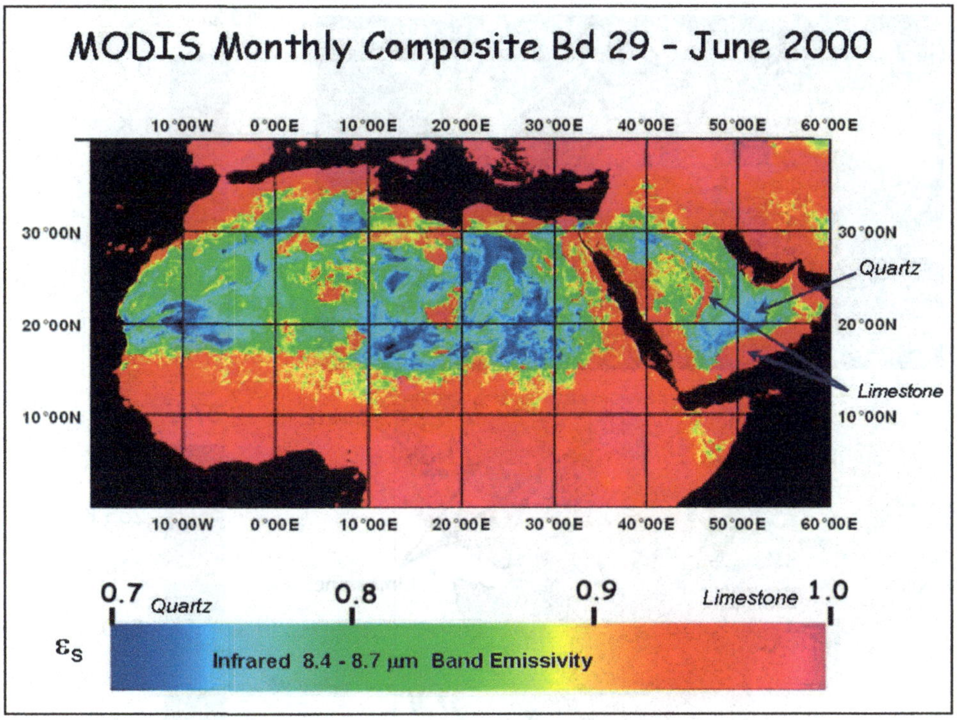

Figure 72 – Infrared image of North Africa and Saudi Arabia showing the surface emitted radiation in the 8.4 – 8.7 μm band. Note that quartz has the lowest emissivity while limestone has the highest emissivity. This is completely opposite to the emissivity at microwave frequencies.

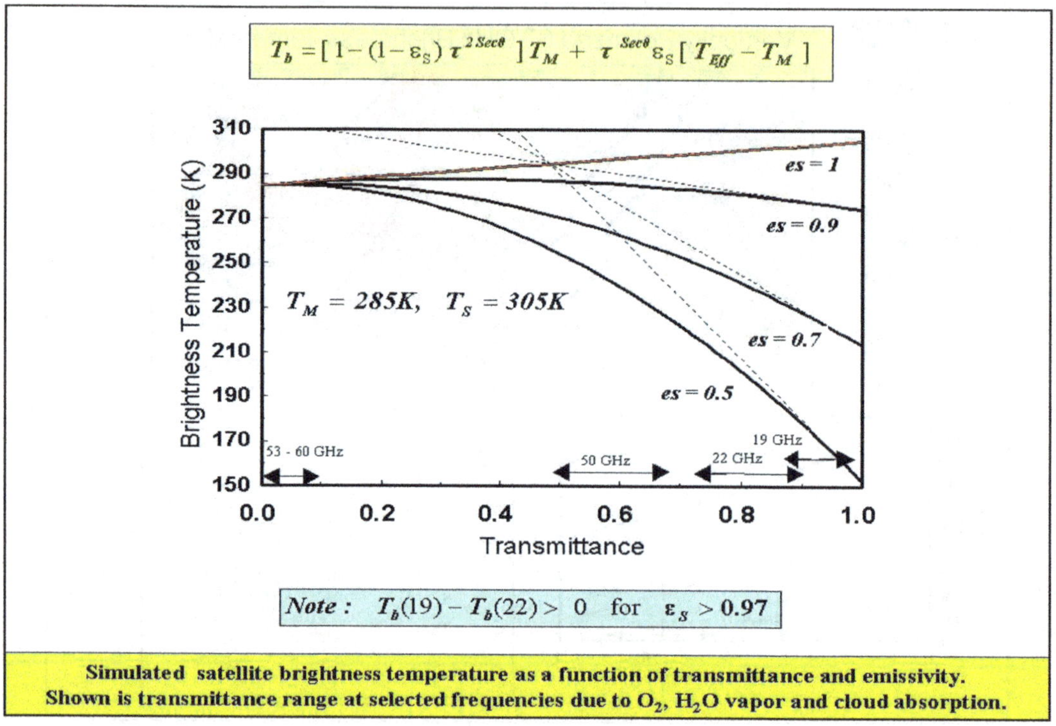

Figure 73 – Simulated brightness temperature T_b obtained using (63), which is plotted as a function of transmittance $\tau^{Sec\theta}$ with emissivity ε_S as a parameter. The dashed lines are obtained using the linear approximation (68a). The transmittance range shown for the different frequencies is due to water vapor, oxygen and cloud absorption. Note that $T_b(19) > T_b(22)$ for $\varepsilon_S > \hat{\varepsilon} = 0.966$ (see Text).

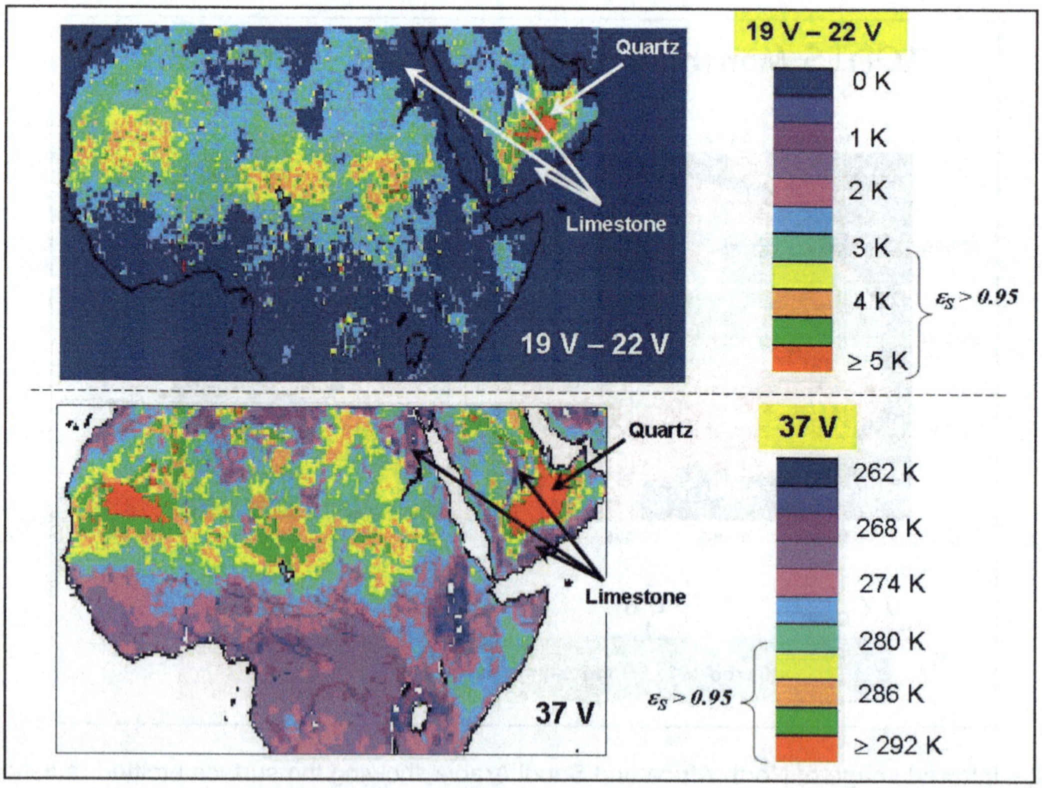

Figure 74 – The top image shows the difference between the SSM/I vertically polarized 19 and 22 GHz channel measurements for July, 1996. The bottom shows the 37 GHz measurement for the same area. Note that both are similar although the top image is the difference while the bottom is the amplitude.

12.1.1 Diurnal Effects

Another feature observed over deserts is the effect of diurnal temperature variations on the SSM/I channel measurements. As discussed in Chapter 9, the transmission coefficient in sand varies with frequency due to absorption and scattering by the sand grains. As a result, of the different penetration depths, the SSM/I channels were found to display different diurnal variations at each frequency. This is shown in Figure 75 by displaying the positive difference between the 19 and 37 GHz vertically polarized globally averaged measurements at 6:00 AM (left) and 9:30 AM (right) for May 1998. Note that the large desert areas over North Africa and Saudi Arabia only display a large positive difference between the two measurements at 6:00 A.M, whereas the difference becomes negative at 9:30 A.M. To aid in the interpretation, the bottom graph in Figure 75 shows the time series of temperature measured at different depths within a sand dune in the central Sahara in mid-August. Due to solar heating it shows large surface temperature variations at the satellite overpass times between 6:00 A.M. and 9:30 A.M. This diurnal variation in temperature is also shown to be absent at a depth of 30 cm. These insitu observations can be related to the SSM/I measurements since the 37 GHz channel responds to temperature near the surface while the 19 GHz radiation emanates from deeper layers where the effective temperature in equation (61) is less variable. These different diurnal variations at 19 and 37 GHz reverse the sign of the difference measurement in Figure 73 so that only positive values occur at 6:00 A.M.

To further examine the effect of diurnal temperature variations Figure 76 shows the difference between the 37 and 85 GHz measurements. However, unlike the previous measurements, the difference measurement in Figure 76 displays positive values at both observing times. This can result from two different physical mechanisms. It could be that the larger water vapor absorption at 85 GHz reduces its brightness temperature relative to 37 GHz, thereby producing a positive difference at all times. Alternatively, it can also result from larger scattering at 85 GHz than at 37 GHz, causing the difference to be positive even though both channels vary diurnally due to temperature. To examine this, simulations were performed using a dense media model to calculate the diurnal variation in brightness temperature at each frequency. The model assumes a frequency independent dielectric constant of $\varepsilon = \varepsilon_R + i\,\varepsilon_I = 4.0 + i\,0.08$ for the individual sand grains where the imaginary part represents the dielectric loss. The desert contains these randomly distributed spherical particles surrounded by air so that the effective dielectric constant $\varepsilon_{\mathrm{Eff}}$ of the mixture depends on the sand fractional volume, f, particle radius, r, and frequency (see Appendix 16). Using a simplified analytical result from dense media theory [29] the composite dielectric constant of the sand - air mixture can be written as,

$$\varepsilon_{\mathrm{Eff}} = \left(\frac{1+2f\,y_R}{1-f\,y_R}\right) + i\frac{f\,y_R}{(1-f\,y_R)^2}\left[\frac{2(kr)^3\,y_R(1-f)^4}{(1+2f)^2} + \frac{3y_I}{y_R}\right] \qquad \text{(70a)}$$

$$\text{where } y_R = \frac{\varepsilon_R - 1}{\varepsilon_R + 1}, \quad y_I = \frac{3\varepsilon_I}{(\varepsilon_R+1)^2} \quad \text{and } kr = \frac{2\pi r}{\lambda}. \qquad \text{(70b)}$$

In the low frequency limit (*i.e.*, $kr = 0$) the effective dielectric constant reduces to the Maxwell-Garnett mixing formula. The second term in brackets contains the dielectric loss and Rayleigh scattering loss which increases with frequency and particle size, *i.e.*, $(kr)^3$. This scattering term also contains the *form factor* $(1 - f)^4/(1+2f)^2$ to approximately account for dependent scattering as discussed in Appendix A16. However, although this dielectric model applies for all fractional volumes, it is only applicable in the Rayleigh limit where $kr < 1$. This limits the radius to less than 0.55 mm at 85 GHz. As such, the particle radius was varied up to 0.5 mm and the fractional volume was set at 1.0 for a non-scattering solid surface and 0.6 for more porous media such as deserts. Using this approximation, simulations[13] show that for non-scattering media where $f = 1$ the difference between the 37 and 85 GHz channels displays the same diurnal variation as that between the 19 and 37 GHz channels, *which is not observed*. On the other hand, when reducing the fractional volume to 0.6 to represent deserts, the difference between the 37 and 85 GHz measurements became positive at both observing times due to volume scattering, *which is observed* in Figure 76. While these early results are encouraging, additional studies using more advanced models requiring extensive numerical calculations are needed to more accurately demonstrate the effects of scattering for deserts at high frequencies.

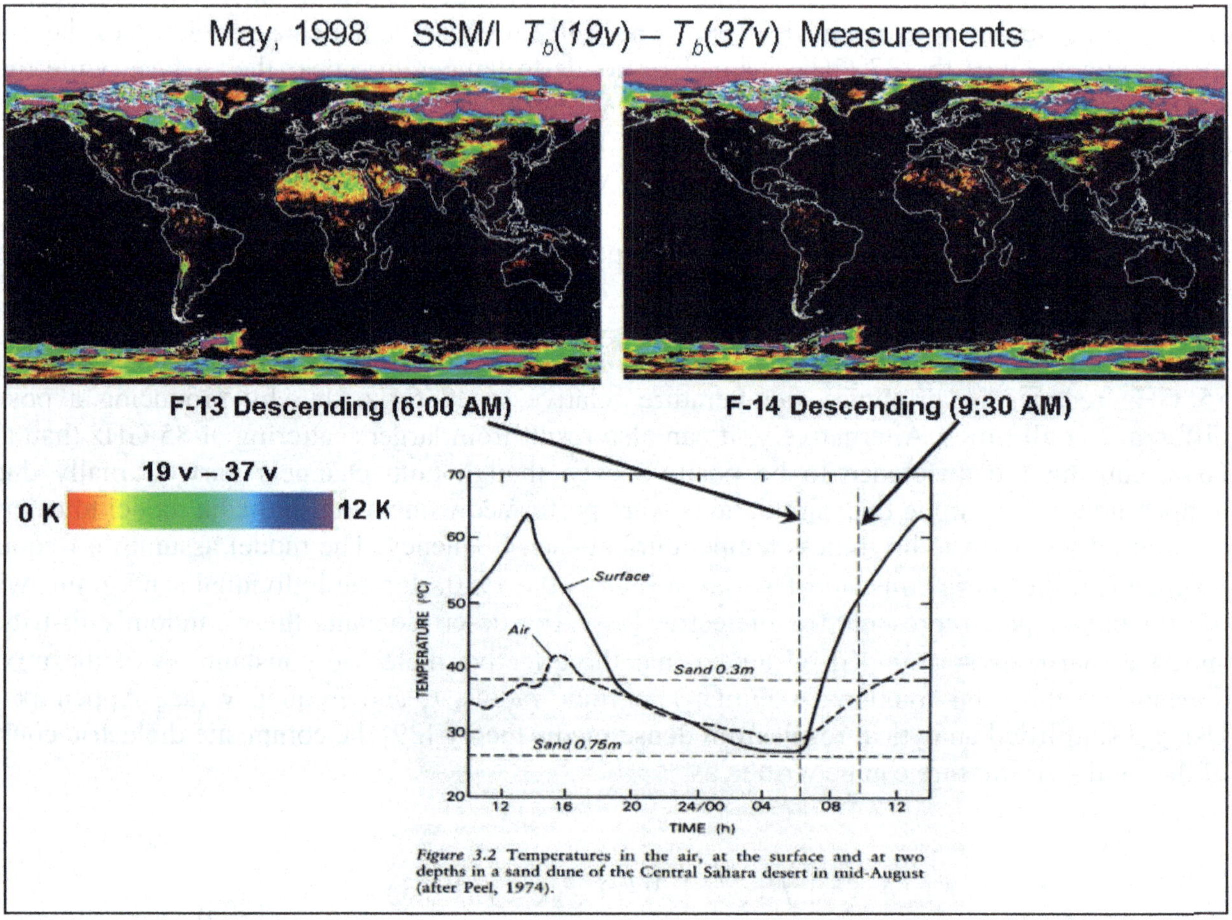

Figure 75 – Difference between the 19 and 37 GHz SSM/I Vertical polarized measurements at (left) 6:00 A.M and (right) 9:30 A.M for May 1998. For comparison the bottom image shows a time series of the temperature at different depths within a sand dune for the central Sahara in mid-August.

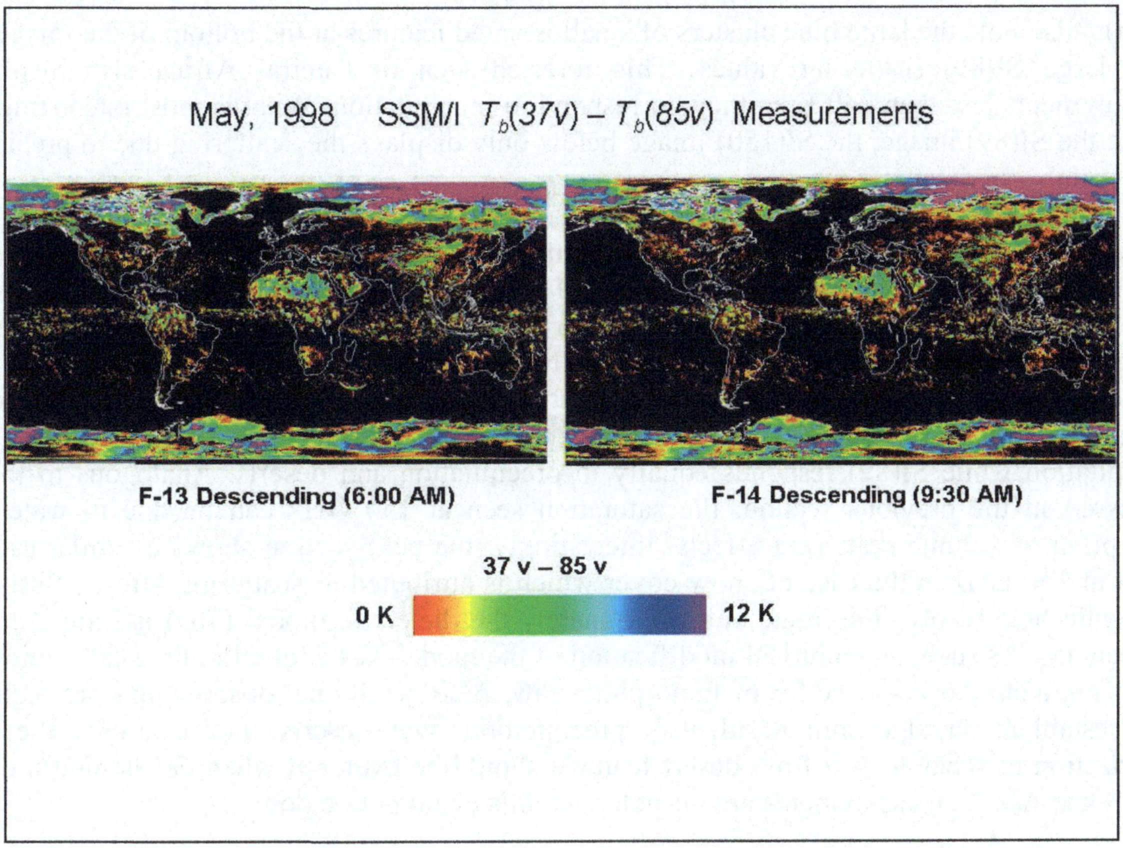

Figure 76 – Difference between the 37- and 85-GHz SSM/I V-Pol measurements at (left) 6:00 A.M and (right) 9:30 A.M for May 1998.

12.1.2 Scattering Effects

The effect of scattering in deserts was further examined at still higher frequencies than those of SSM/I using AMSU channels [19]. As listed in Table 7, the AMSU-A modules contain window channels at 23.8, 31.4 and 89 GHz channel, with a resolution of 50 km at nadir viewing. The AMSU-B module, which is not shown in Figure 65, contains a 150 GHz channel together with another 89 GHz channel, where both have 15 km resolution at nadir. Figure 75 shows the channel differences on January 3, 2000 at 7:30 P.M over Africa and Saudi Arabia. The top-left image is the scattering index SI(89) obtained by taking the difference between the 23.8 and 89 GHz channels while the bottom-left image is the scattering index SI(150) obtained by taking the difference between the 89 and 150 GHz measurements. Both images display differences greater than 5 K for the scattering indices. For comparison the Figure also shows an infrared (IR) image on the right obtained from the IR imager on the METEOSAT geostationary satellite. The IR imager identifies precipitation by its temporal variability and brightness temperatures less than 240 K, which is shown in yellow and green.

Identification of rain over deserts is difficult using SI(89) since similar signals are observed in the absence of precipitation. This is shown in the Top-Left image of Figure 77 which displays SI(89) greater than 5 K for the large irregular red patches throughout North Africa and Saudi Arabia. It has corresponding IR measurements greater than 280 K (purple and pink) so they are obviously not due to precipitation. Only the top-center blue diagonal line in the SI(89) image over North Central Africa centered at 22.3^0 N, 16.4^0 E has IR measurements much lower than 240 K (yellow and green). Also note the large blue clusters of small isolated features at the bottom of the image which have large SI(89) scattering values. This forested area of Central Africa also displays IR measurements less than 240 K so they correspond to precipitation. What is most interesting is that unlike the SI(89) image, the SI(150) image below only displays the scattering due to precipitation with no false signatures.

To better examine this, Figure 78 plots the brightness temperatures spectrum using AMSU-A and -B non-sounding channels between 23.8 and 150 GHz. The figure shows the measurements at the above mentioned heavy precipitation region over North Central Africa and at a rain-free westward location of 25.9^{0}N, 7.7^0 E (see triangle symbol). Note that the scattering due to precipitation results in a continuous decrease in brightness temperature while the spectrum in the rain-free location appears to saturate beyond 89 GHz. As such, only SI(150) displays large positive values due to precipitation while SI(89) responds equally to precipitation and deserts. Analogous to the issue discussed in the previous section, the saturation seen at 150 GHz can be due to water vapor absorption or volume scattering effects. Interestingly, the next section shows a similar saturation affect at 150 GHz in the case of snow cover which is attributed to scattering affects. Simulations can again help resolve this issue but unfortunately the dielectric model (70a) is limited to lower frequencies. As such, an empirical modification of the model was applied in the next section when observing a similar response for metamorphic snow. Also, additional observations are required to fully establish an algorithm to identify precipitation over deserts. For example, the use of polarization to separate rain from desert features should be explored when developing a decision tree. Since AMSU measurements are un-polarized this could not be done.

In summary, the *sand grains appear to not only emit microwave radiation but also scatter them.* This leads to false rain-scattering signatures over deserts particularly when using a lower frequency scattering index for identification, *i.e.* SI(89). Furthermore, unlike the strong scattering signals found at all frequencies due the ice formed aloft in many rain systems, the scattering amplitude for deserts appears less and saturates at very high frequencies. The reduced scattering for the high particle densities found in deserts is shown by calculating the single particle albedo, ω, using the effective dielectric constant (70a). The albedo is the ratio of the scattering to extinction coefficient and when using this model becomes [19],

$$\omega = \frac{(kr)^3 (1-f)^4 y_R^{\ 2}}{(kr)^3 (1-f)^4 y_R^{\ 2} + 1.5(1+2f)^2 y_I} \qquad (71)$$

Note that the albedo is highest for diffuse media having small fractional volumes such as precipitation and decreases for materials with higher fractional volumes such as snow cover and sand. The albedo is also reduced for lossy scatterers such as the water droplets in stratiform rain as opposed to the ice crystals formed in convective rain. All of these characteristic are shown in Figure 79 by plotting the equation as a function of fractional volume for different size parameters. In this example the individual particles either have a dielectric constant associated with ice (Left) or that of desert sand (Right). It's seen that even for large size parameters the albedo is small for the high fractional volume of 0.6 for sand. This figure also explains the much larger scattering signals observed for more diffuse media such as rain and snow surfaces which have smaller fractional volumes compared to desert sand. To summarize this, Figure 80 lists on the bottom the average values of the SSM/I scattering index, $T_b(22v) - T_b(85v)$, found for precipitation, snow cover and deserts. The figure also summarizes the many other features observed here for deserts based on SSM/I and AMSU measurements. Interestingly, the next section on snow cover reveals similar scattering features to that of deserts.

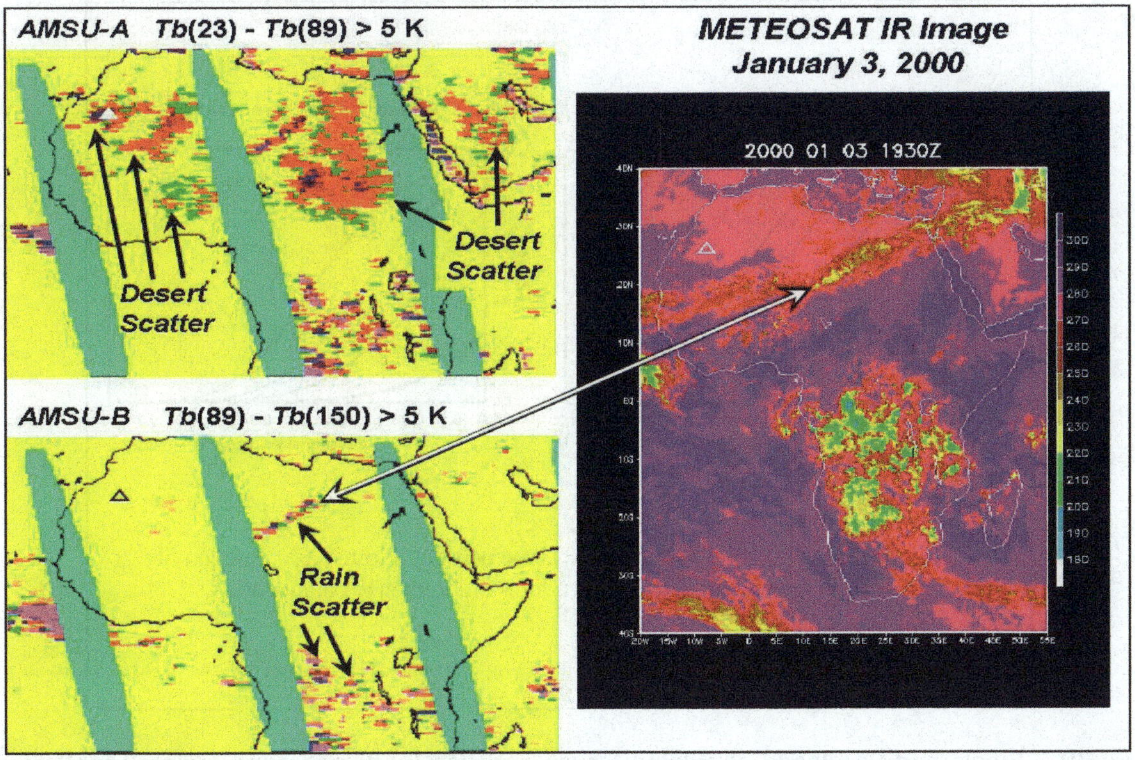

Figure 77 – Comparisons between NOAA-15 AMSU measurements on January 3, 2000 and the METEOSAT-IR image at the same time (7:30 P.M.) over Africa. The IR image on the right side shows regions having brightness temperatures of 240 K and lower (yellow and light green) associated with heavy precipitation. Of particular interest is the line of precipitation over North Africa centered around 22^0 N, 16^0 E. This region shows large positive values for both the AMSU-A 23 and 89 GHz channel difference (top, left) and the AMSU-B 89 and 150 GHz channel difference (bottom, left). However the AMSU-A image also displays a number of areas having IR brightness temperatures higher than 280 K (purple and pink) which is not due to precipitation but due to desert scattering.

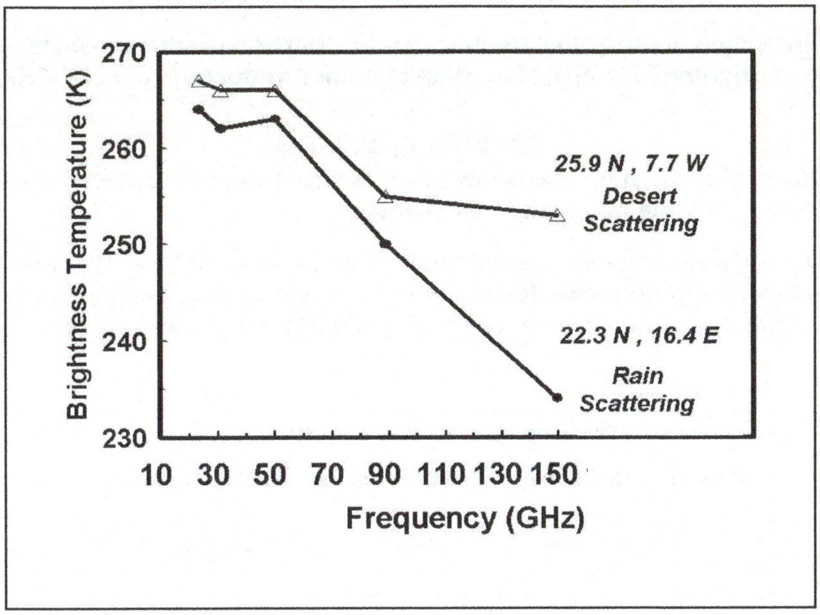

Figure 78 – Spectra of desert and rain scattering features. Brightness temperature spectrum from AMSU-A and AMSU-B measurements for the heavy precipitation region of north central Africa (22.3^0 N, 16.4^0 E) as well as the rain-free area shown in Figure 75 (left images) by the triangle symbol .

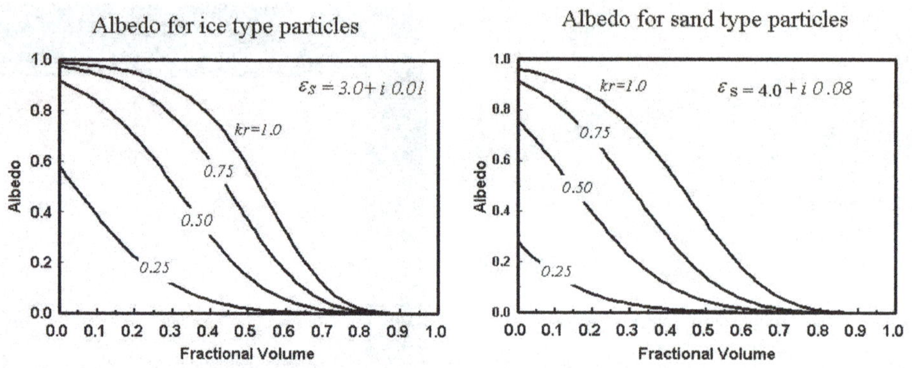

Albedo for ice (Left) and sand (Right) type particles plotted as a function of fractional volume. Each plot is shown for a different size parameter.

Figure 79 – Single particle albedo calculated for ice particles (Left) and sand grains (Right) plotted as a function of fractional volume for different size parameters, $k\,r = 2\pi\,r\,/\lambda$. Note the larger albedo for the ice formed in convective rain compared to the more lossy sand associated with deserts.

Summary of SSM/I and AMSU Measurements over Deserts

1. Emission Signals:

(a) **Dielectric constant produces a high emissivity for *Quartz* and low emissivity for *Limestone* at V-Polarization (*opposite 8.5 μm Infrared emissivity*).**

(b) **Large penetration depth at low frequency (*e.g.,* 19 GHz) produces smaller diurnal variation of brightness temperature than at higher frequency (*e.g.,* 37 GHz).**

2. Scattering Signals:

(a) **Scattering for precipitation and snow cover is larger than for deserts, due to its high fractional volume or density (see below).**

(b) **Unlike precipitation, deserts scatter similarly at 89 and 150 GHz. This provides a means of identifying rain over deserts. As shown in the next Section, a similar scattering effect occurs between new and metamorphic snow.**

Scattering Index for Different Materials

Materials	**Scatterers**	**Fractional Volume**	**$T_b(22v) - T_b(85v)$**
Precipitation	Ice Hydrometeors	< 0.01	0 - 100 K
Snow Cover	Ice Crystals	0.2 – 0.4	0 - 50 K
Deserts	Sand	> 0.5	0 - 20 K

Figure 80 – This figure summarizes what was covered in this Section regarding deserts. Also listed at the bottom insert are the scattering properties of different materials.

12.2 Snow Cover Measurements

This section discusses satellite measurements of snow cover. Beginning when the first radiometer NEMS was launched in space in 1972 it was found that snow cover can be detected using microwave radiometers. Just as in the case of precipitation, algorithms have been developed to not only detect different surfaces but to derive quantitative measures such as the rain rate for precipitation and the equivalent water content in the case of snow cover. However, since snow cover and precipitation both scatter microwave radiation, techniques were developed using a decision tree approach using most of the SSM/I channels to separate the variables [10]. This section however describes some unusual microwave characteristics found for snow. As with deserts, these measurements help provide a better understanding of the scattering properties of surfaces.

We begin by first discussing some ground-based observations taken of snow cover using microwave radiometers with frequencies at 10, 22, 35 and 94 GHz at an incident angle of 50^0. Figure 81 shows measurements taken over a year by Christian Matzler [30] for a site in Bern Switzerland. The surface conditions varied from bare soil, grass, frozen soil to various snow types. Each point is the average emissivity at vertical polarization for a particular frequency and surface type (categorized on the basis of ground truth). The first nine surfaces are snow free, followed by wet snow and dry snow types. All of the dry snow types excluding type 17 exhibit a monotonic decrease in emissivity with frequency due to volume scattering and represent one major snow class. Only the snow containing a bottom ice crust (type 17) exhibits an increase in emissivity at 94 GHz relative to 35 GHz, while the 35 GHz measurement is less than the lower frequencies. This "inverted spectrum" due to an ice crust provides a unique signature to characterize this snow type. Furthermore, one would think that this is a rare event since all other measurements show no such feature. However, as discussed below, this anomaly as well as others is frequently seen from satellite measurements.

This section begins by briefly describing a case study pertaining to metamorphic changes in the grain size of ice as snow ages [28]. The study is summarized by the images shown in Figure 82 whose left panel shows a time series in late November to early December, 2001 of the AMSU derived snow cover (white) based on the difference between the 23 and 89 GHz measurements. Also shown is the difference between the AMSU 89 and 150 GHz measurements using the same 5 K brightness temperature threshold used in Figure 77 when studying deserts. Note that the area formed using this threshold decreases in December even though the snow cover area shows no significant change during this time period. This is likely due to growth of ice grains due aging processes in snow and was observed for other snow events as well. It is important to understand that this same microwave feature at 150 GHz due to changes in grain size has been observed for many other snow events.

A simulation showing the effect of grain size growth is displayed in the right panel. It shows how an increase in ice size increases the difference between adjacent AMSU channels until the particle radius exceed 4 mm. At that point the difference between the 89 and 150 GHz brightness temperature reaches a maximum and then begins to decrease due to saturation of the 150 GHz measurement. A similar saturation effect was also observed for deserts in the previous section. This phenomenon appears similar to the geometric optics limit that causes the scattering cross section to approach a constant value for large size parameters, *kr*. However, the simulation in Figure 82 is approximate since as mentioned previously the effective dielectric constant (70a) is only valid for small size parameters (*i.e.*, $kr < 1$). In order to extend it to larger size parameters or for frequencies beyond 90 GHz a piecewise approach was used to bridge the gap between this dense media model and geometric optics theory which is limited to sparse media. However, while the analysis used in the simulation was not done rigorously it appears physically correct. Furthermore, I am unaware of any other model simulations describing the scattering properties of snow cover at frequencies up to 150 GHz. In fact most simulations only pertain to measurements up to 37 GHz with one discussed in Appendix A16 at 90 GHz.

A second study shows another effect rarely discussed. It is the effect of ice stratification in the form of ice layers or an ice crust on the microwave scattering signatures of snow. This type of snow is best identified using ground penetrating radar or altimeters and found responsible for the inverted spectra in Figure 81 and denoted as category 17. A very similar characteristic is seen in Figure 83, in addition to more pronounced anomalies. This Figure shows the SSM/I brightness temperatures for three different snow covered regions (Canada, Greenland, Russia) in February 1998. Two different locations are chosen in each region to display the normal spectrum as well as the abnormal spectra where the brightness temperature at 89 GHz is higher than the 37 GHz channel instead of being lower. However, the most unusual feature is the spectrum over Southern Greenland. This spectrum appears as a liquid water surface rather than snow cover in the middle of winter. Except for errors or malfunctions of the instrument, or super cooled liquid water, this unusual spectrum is best explained by stratified ice layers in the snow.

As reported by Grody and Basist [31], the attenuation of radiation by stratified ice layers causes the higher frequency channels to emit at higher brightness temperatures than the lower frequencies, thereby reversing the normal spectrum seen for snow cover. To show that this feature as well as other anomalies extends over large areas, Figure 84 displays latitudinal cross-sections of the SSM/I measurements for the three regions in Figure 83. The cross sectional transit used for each region is shown in the snow cover image at the bottom of Figure 83. Lastly, as with deserts, Figure 85 summarizes the many features observed regarding snow using the SSM/I and AMSU measurements. It also shows at the bottom a spectral diagram summarizing the measurements found in these studies. Note that many of the anomalous features such as inversions and saturation effects are only observed at frequencies beyond 85 GHz. As such, all of the algorithms developed for SSM/I did not have to account for these characteristics. It was only after studying AMSU measurements that these anomalies became apparent. These higher frequency signatures (> 85 GHz) depicted in the spectral diagram need to be further investigated and modeled to better understand their physical basis and see if they can lead to new products such as the classification of snow type.

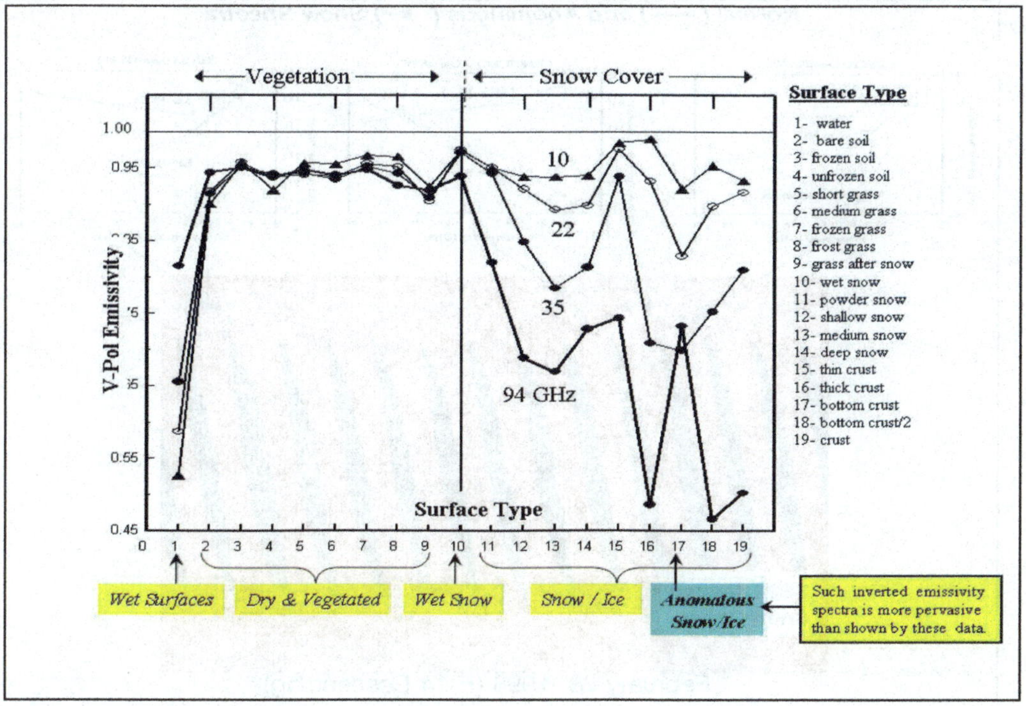

Figure 81 – Ground-based vertically polarized emissivity measurements at 10, 22, 35, and 94 GHz for different surfaces. Surface types greater than 10 correspond to different classes of dry snow whose emissivity generally decreases with increasing frequency.

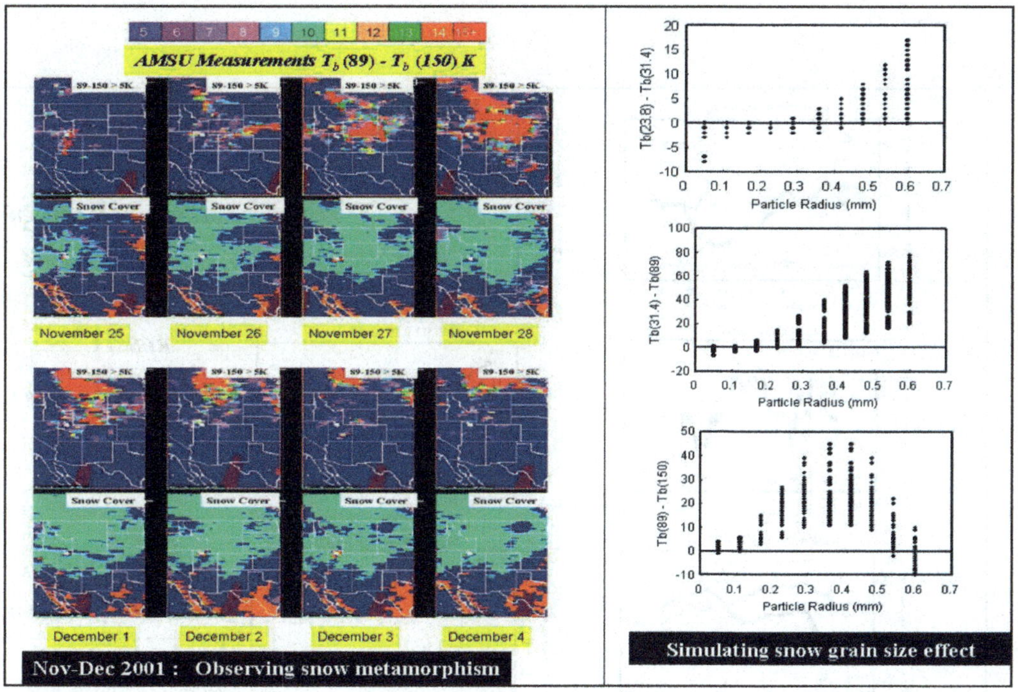

Figre 82 – The left panel shows AMSU measurements of snow cover in white during late November and early December 2001. Two separate sections are used to show the complete time series of snow cover based on the 23 and 89 GHz difference measurements. For comparison, the top of each section shows differences between the 89 and 150 GHz measurements using a 5 K threshold. Note that this difference drops beyond 5 K as time increases even though the snow cover shows no large changes. To aid in the explanation the right panel shows simulated differences between the 23 and 31 GHz, 31 and 89 GHz and 89 and 150 GHz measurements as a function of ice particle size (see Text).

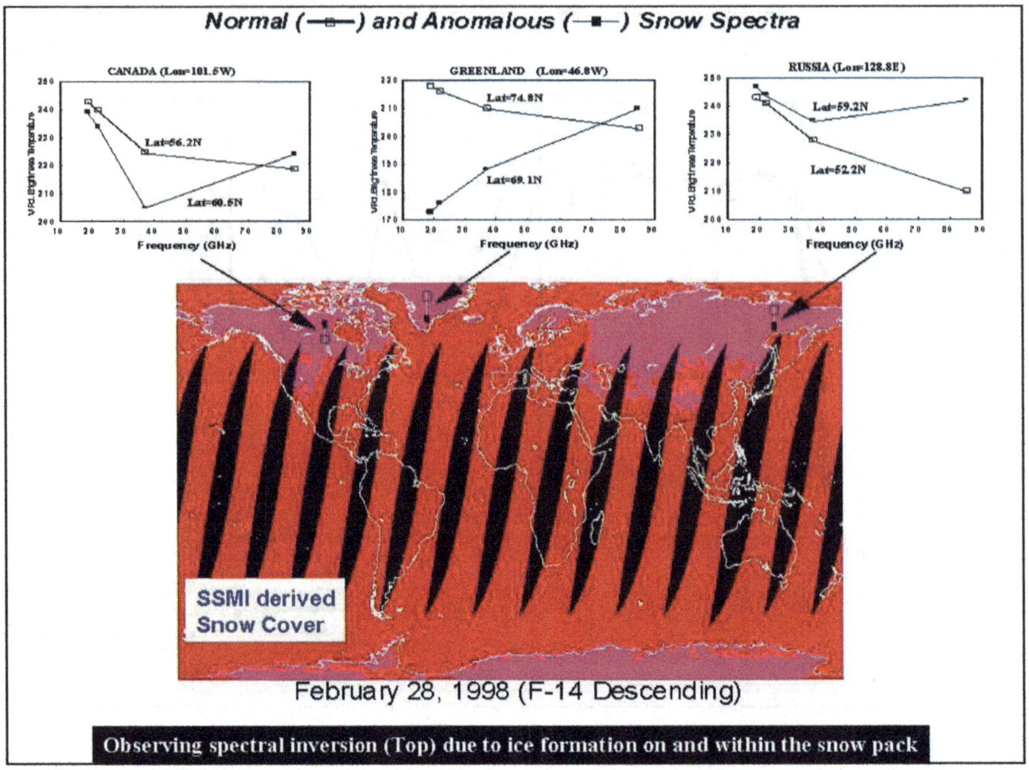

Figure 83 – The top show SSM/I brightness temperature measurements for three different snow covered regions (Canada, Greenland, Russia) in February 1998. Two different sets of measurements are shown for each region, one displaying the normal spectrum while the other is anomalous. The bottom shows the global snow cover derived from SSM/I measurements with each region depicted.

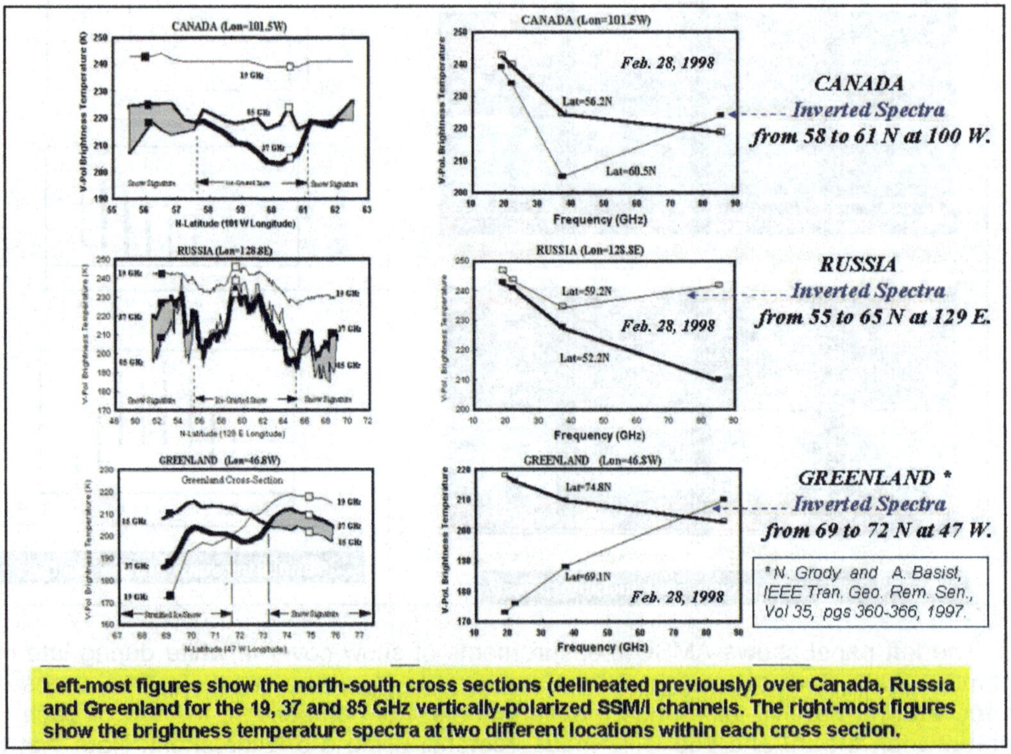

Figure 84 – The left panel shows cross sections of the SSM/I brightness temperature measurements at 19, 37 and 85 GHz for the three regions in the global snow cover display at the bottom of Figure 83. Also shown in the right panel are the three brightness spectra at two locations in each cross-section.

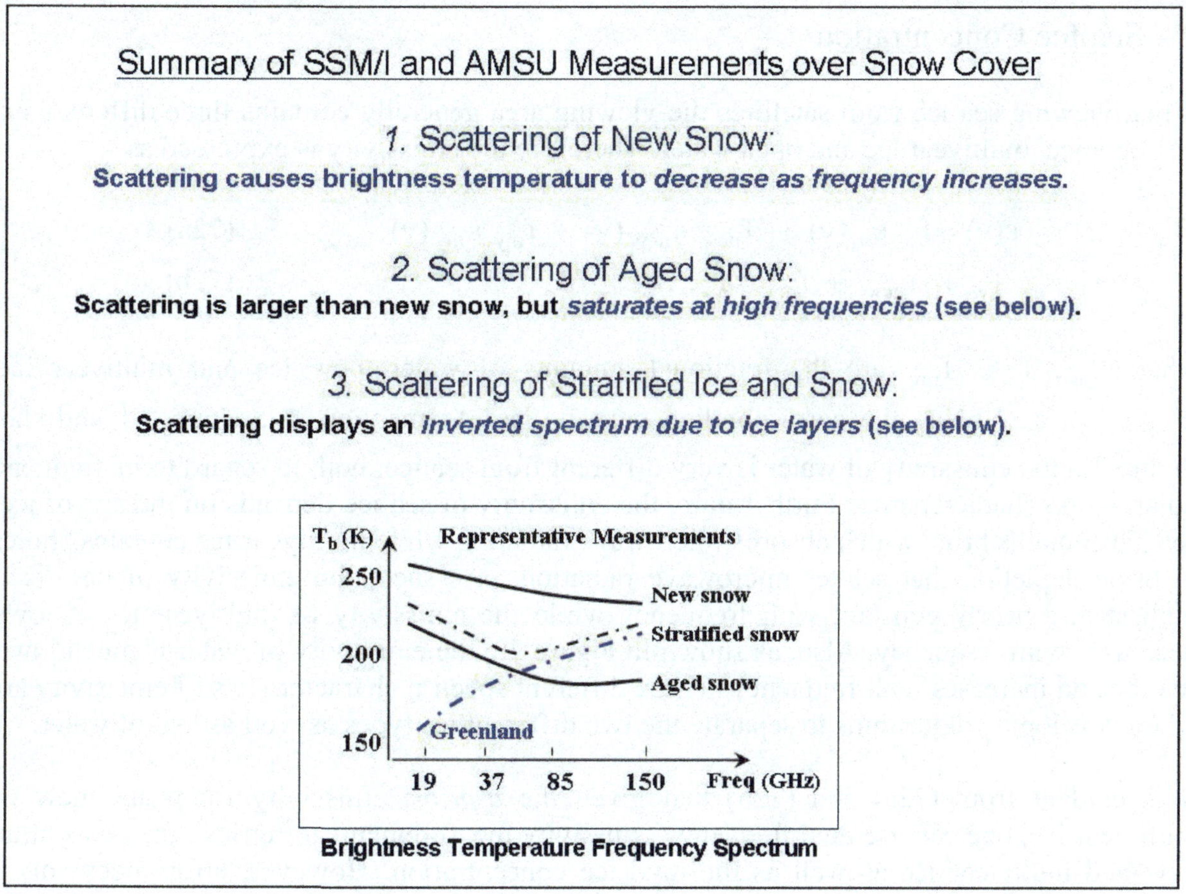

Figure 85 – This figure lists what was discussed in this section regarding snow. A spectral diagram at the bottom summarizes the anomalous findings reported here.

12.3 Sea Ice Measurements

This last section discusses the use of microwave radiometers to measure the sea ice concentration. Like many other geophysical parameters observed from passive microwave measurements the study of sea ice has a long history. However, this brief overview only describes an algorithm for determining the total ice concentration based on the ASMU channels. As explained below, this instrument provides a number of challenges for measuring sea ice concentration due its limited number of channels and lack of polarization. Also, as when using any microwave radiometer, one must account for the emissivity variations by different ice types in addition to the effects of melting ice, snow cover, ocean winds, precipitation and cloud cover. Most of these effects are discussed and accounted for in the following presentation.

Passive microwave measurements have been used since the early 1970's to determine the ice coverage or sea ice extent [32]. As with other surface measurements, during periods of extensive cloud cover the microwave observations of sea ice extent are often the only data available and have been used to infer global temperature change in polar regions [33]. Microwave measurements are also used to examine the percent of ice concentration within the sensor's viewing area or FOV. This section outlines the algorithm developed from AMSU measurements for deriving the total sea ice concentration and shows examples of its performance. This study was done in collaboration with Dr. Konstantin Vinnikov who I also worked with on using MSU data for climate studies, as reported in Chapter 11

1 – Sea Ice Concentration

When viewing sea ice from satellites the viewing area generally contains three different surfaces, *i.e.*, new ice, multiyear ice and open water. Therefore the emissivity is expressed as

$$\varepsilon(\nu) = f_w\, \varepsilon_w(\nu) + f_{new}\, \varepsilon_{new}(\nu) + f_{my}\, \varepsilon_{my}(\nu) \tag{72a}$$

$$f_w + f_{new} + f_{my} = 1 \tag{72b}$$

where f_w, f_{new}, f_{my} are the fractional amounts of water, new ice and multiyear ice with ε_w, ε_{new}, ε_{my} being the corresponding emissivities. As mention in Section 7.1 and shown in Figure 22, the emissivity of water is very different from sea ice, both in regard to its frequency and polarization characteristics. Furthermore, the emissivity of sea ice depends on the age of ice. New sea ice contains brine which absorbs microwave radiation while multiyear ice contains "holes" due to brine depletion that scatter microwave radiation. As such, the emissivity of new ice is the highest and nearly constant with frequency while the emissivity of multiyear ice is lower and decreases with frequency. Also, as shown in Figure 22, the emissivity of water is much lower than sea ice and increases with frequency. These different spectral characteristics of emissivity are used when developing algorithms to separate the two different ice types as well as that of water.

It is evident from (72a) and (72b) that given the *a-priori* emissivity for water, new ice and multiyear ice, one can use dual frequency emissivity measurements to retrieve the concentration of new and multiyear ice as well as the total ice concentration. However, errors occur due to the emissivity variations of multiyear ice and the effects of melting ice and snow cover [33, 34]. Since the total ice concentration is less effected by these errors, which affect the *a-priori* emissivity, the total ice concentration is more accurate than the separate concentrations.

By combining (72a) and (72b), the total ice concentration, $f = f_{new} + f_{my}$, is given by

$$f = \frac{\varepsilon(\nu) - \varepsilon_w(\nu)}{\varepsilon_{ice}(\nu) - \varepsilon_w(\nu)} \tag{73a}$$

where the composite emissivity of ice is

$$\varepsilon_{ice}(\nu) = \varepsilon_{new}(\nu) - f_{my}\left(\frac{\varepsilon_{new}(\nu) - \varepsilon_{my}(\nu)}{f_{new} + f_{my}}\right) \tag{73b}$$

which contains the emissivity and fractional amount of both ice types

The error in sea ice concentration is obtained by differentiating (73a) with respect to each emissivity, *i.e.*,

$$df = \left.\frac{\partial f}{\partial \varepsilon}\right)_{\substack{\varepsilon_w \\ \varepsilon_{ice}}} d\varepsilon + \left.\frac{\partial f}{\partial \varepsilon_{ice}}\right)_{\substack{\varepsilon_w \\ \varepsilon}} d\varepsilon_{ice} + \left.\frac{\partial f}{\partial \varepsilon_{water}}\right)_{\substack{\varepsilon \\ \varepsilon_{ice}}} d\varepsilon_w \tag{74a}$$

so from (73a),

$$\left.\frac{\partial f}{\partial \varepsilon}\right)_{\substack{\varepsilon_w \\ \varepsilon_{ice}}} = \frac{1}{\Delta\hat{\varepsilon}}, \quad \left.\frac{\partial f}{\partial \varepsilon_{ice}}\right)_{\substack{\varepsilon_w \\ \varepsilon}} = \frac{-f}{\Delta\hat{\varepsilon}}, \quad \left.\frac{\partial f}{\partial \varepsilon_w}\right)_{\substack{\varepsilon \\ \varepsilon_{ice}}} = \frac{-(1-f)}{\Delta\hat{\varepsilon}} \tag{74b}$$

with $\Delta\hat{\varepsilon} \equiv \varepsilon_{ice}(\nu) - \varepsilon_w(\nu)$.

The rms error of sea ice concentration is found by taking the ensemble average <> after squaring (74a) and using (74b), *i.e.*,

$$\sqrt{< df^2 >} = \frac{1}{\varepsilon_{ice} - \varepsilon_w} \sqrt{< d\varepsilon^2 > + f^2 < d\varepsilon_{ice}^2 > + (1-f)^2 < d\varepsilon_w^2 >} \qquad (75)$$

where $< d\varepsilon_x^2 >$ is the mean squared error of the different variables, x.

Equation (75) shows that the error in total ice concentration depends on four quantities,

1. Retrieved emissivity error, ε
2. *A-priori* emissivity error of sea ice, ε_{ice}
3. *A-priori* emissivity error of water, ε_w
4. Fractional amount of sea ice, f

The lower limit is obtained by only considering the error due the retrieved emissivity. As discussed next, the standard error of the emissivity error is 0.01 based on simulated data. There, the expected error of ice concentration obtained from (75) is only 2 percent. However this error is increased due to the effects of precipitation on the retrieved emissivity measurement and the uncertainties in the *a-priori* emissivity of water for $f = 0$ and ice for $f = 1$. These *a-priori* sources of error can be minimized by properly choosing the microwave frequencies and its polarization. For example, at vertical polarization the emissivity of sea surfaces has the smallest dependence on wind roughening of the ocean surface. At this polarization, the emissivity variation depends mainly on the changes due to sea surface temperature and sea state, due to wind generated foam. In addition, at low frequencies the multiyear and new emissivity is nearly the same so that from (73b) the composite emissivity is the new ice emissivity which is independent of ice-age. However, AMSU does not have very low frequencies or dual polarized channels so the composite emissivity and ocean emissivity can vary significantly, and must be accounted for.

2 – Emissivity Algorithms

Equation (73a) is used to obtain the total ice concentration. However, to use the equation, an algorithm must be developed to retrieve the emissivity $\varepsilon(\nu)$ and determine the *a-priori* emissivity of sea water $\varepsilon_w(\nu)$ and the composite emissivity of sea ice $\varepsilon_{ice}(\nu)$. As listed in Table 7, AMSU contains its lowest frequency window channels at 23.8 and 31.4 GHz in its A2 module, and 50.3 GHz in the A1 module. Although not optimum, these three channels are sufficient to derive the sea ice concentration. To develop the emissivity algorithm, $\varepsilon(\nu)$, brightness temperature were simulated using a radiation transfer program that includes atmospheric absorption by oxygen and water vapor in addition to that of clouds and precipitation. The model also contains surface emissivity models of the ocean and sea ice.

Brightness temperatures are then obtained for each channel and for each surface. These surfaces include new and multiyear sea ice as well as ocean surfaces with different wind speeds. To replicate the AMSU channel measurements the simulations also include the change in path length and variation of emissivity with scan angle. As explained below this angular variation is mainly due to the cross-track antenna used by AMSU. Regression analysis is then applied to the set of simulated brightness temperatures to obtain an algorithm for retrieving emissivity at the lowest AMSU frequency of 23.8 GHz. The resulting algorithm is given by

$$\varepsilon = A + BT_b(23) + C\,T_b(31) + DT_b(50), \qquad (76)$$

$$A = 1.84 - 0.723\ \text{Cos}\ Z,\ \ B = -0.00088,\ \ C = 0.0066 + 0.0029\ \text{Cos}\ Z,\ \ D = -0.00926.$$

where the coefficients for some channels depend on the cosine of local zenith angle Z. The following demonstrates the accuracy of this equation.

To begin the evaluation Figure 86 (Top) shows a scatter plot of the 50.3 GHz emissivity against the 23.8 GHz emissivity. The data set contains the sea surface, dry land, wet land, new sea ice, multiyear ice and shallow snow. Each surface is identified by the numbers shown while the sea surface is labeled separately. All surfaces have fixed emissivities except for sea surfaces whose emissivity varies due to temperature, wind speed and viewing angle. The accuracy of the retrieval algorithm is also displayed in Figure 86 (Bottom), which plots the retrieved emissivity using equation (76) against the actual emissivity in the simulated data set. Note that the standard error of the retrieval algorithm is 0.01 with an explained variance of 99% for the 6932 observations. Having established the accuracy we next determine the a-priori emissivity of the sea surface emissivity $\varepsilon_W(\nu)$ at 23.8 GHz, which is also used in equation (73a).

Just as in the case of the retrieval equation (76) the sea surface emissivity is obtained using regression analysis of the model used in the radiation transfer program. Although the emissivity varies with wind speed and surface temperature, for calm seas it is approximately given by

$$\varepsilon_w = 0.1824 + 0.9048\ \text{CosZ} - 0.6221\ \text{Cos}^2\text{Z}. \qquad (77)$$

Figure 87 shows the angular variation of (77) together with the emissivity obtained for sea surface winds of 10 /sec. The figure also shows an example of the actual emissivity measured by AMSU, using the retrieval equation (76). All three of these emissivities are represented by equations having the form of (77) but with different coefficients. The main angular variation of emissivity is due to polarization rotation by AMSU's cross-track scanning antenna. This rotation results in the equation $\varepsilon_w = \varepsilon_V - (\varepsilon_V - \varepsilon_H)\text{Cos}^2\text{Z}$, where the angular dependence of the individual emissivity components ε_V and ε_H produces the additional Cos Z term in equation (77).

Lastly, the *a-priori* composite emissivity of sea ice $\varepsilon_{ice}(\nu)$ used in (73a) is also obtained using the emissivity model contained in the radiation transfer program. As explained below, depending on the difference between the 23 GHz and 31 GHz brightness temperatures the composite emissivity changes from new to multiyear ice, *i.e.*,.

$$\varepsilon_{ice} = \left\{ \begin{array}{lll} \varepsilon_{new} = 0.93 & \text{where} & T_b(23) - T_b(31) < 5\text{ K} \\ \varepsilon_{my} = 0.87 & \text{where} & 10\text{ K} \geq T_b(23) - T_b(31) \geq 5\text{ K} \\ \varepsilon_{my} = 0.83 & \text{where} & T_b(23) - T_b(31) > 10\text{ K} \end{array} \right\} \qquad (78)$$

The channel difference $T_b(23) - T_b(31)$ in (78) represents a scattering index. It is used to identify if the surface acts more as a scatterer (*i.e.*, ε_{my}) or absorber (*i.e.*, ε_{new}). Unlike a similar approach used to identify precipitation and snow cover [10] this discrimination of ice-type uses a lower frequency scattering index to minimize the effect of precipitation on the measurements. As explained next, the brightness temperature thresholds and emissivity values in (78) are based on simulations followed by actual measurements.

Using simulated brightness temperature measurements, Figure 88 shows the scattering index $T_b(23) - T_b(31)$ plotted against the retrieved emissivity obtained using equation (76). The largest change in the index occurs for emissivities less than 0.6. This is due to water vapor variations (designated by V symbol) which mostly affect the 23.8 GHz channel measurement. The next highest scattering index (> 5 K) is due to multiyear sea ice which is labeled # 6 and has a retrieved emissivity about 0.75. All other surfaces are labeled 1 through 5, and have emissivities greater than about 0.90 with the smallest scattering index. Using these simulated measurements, it is evident that a scattering index threshold greater than 5 K can generally be used to identify multiyear sea ice, whose emissivity is between 0.7 and 0.8. However in order to identify multiyear ice from the scattering index one must restrict the measurements to regions having low amounts of water vapor ($V < 30$ mm). Actual AMSU measurements are used to obtain better values for the threshold and *a-priori* emissivity of multiyear ice.

Figure 89 shows the AMSU retrieved emissivity (using equation (76)) over sea ice, plotted against the scattering index. Note that the retrieved emissivity of multiyear ice is not constant but is shown to decrease as the scattering index increases. This variation may be due to the different emissivity characteristics combined with the spatial averaging of AMSU measurements. Rather than use a continuous function, the *a-priori* multiyear ice emissivity ε_{ice} uses a step-wise approach. The equation for ε_{ice} is given three different values in (78) depending on the scattering index. Also, regarding the water vapor issue it is evident that in order to use the scattering index to separate multiyear from new ice, the measurements are only applied at high latitudes (>50^0 N) where the water vapor is low.

3 – Case Study

The accuracy of ice concentration using (73a) depends on the AMSU retrieved emissivity (76) and *a-priori* emissivities (77) and (78). As mentioned before, errors occur when the AMSU field of view contains the more variable emissivity of multiyear sea ice. However, the relationship developed before for this *a-priori* emissivity using the scattering index and thresholds should reduce this error. One problem not accounted for is the presence of water on the ice due to summer-melt. Due to the shallow penetration depth of microwave radiation, water-coated ice has an emissivity similar to deep water, and thereby results in a reduction in the ice concentration. Fortunately, only a small fraction of the AMSU footprint contains melted iice, so the application of (73a) does not result in missing ice. Therefore, while the effects of *a-priori* emissivity and ice-melt are important concerns, they generally do not lead to false ice identification. Unfortunately, however, precipitation can result in false ice detection.

The 31 and 50 GHZ channels used in (76) minimize the dependence the effects of non-precipitating clouds whose cloud liquid water is less than 0.2 mm. They also minimize the temperature dependence on the 23 GHz measurements. However, precipitation greatly increases the 23 and 31 GHz brightness temperatures and results in overestimates of the retrieved emissivity and the derived sea ice concentration. Studies using actual AMSU measurements show that the error in sea ice concentration, SI, increases in proportion to the cloud liquid water, Q, content. An example of this effect is displayed in Figure 90.

Figure 90 shows the retrieved sea ice concentration (Top-Right) and cloud liquid water (Top-Left) around Greenland on January 15, 1999. The cloud liquid water over oceans is obtained from a separate algorithm using the 22 and 31 GHz channels. It is similar to that described Section 8.5 for ground-based radiometers. To display the problem areas the range of both parameters (Q and SI) begins at the lower limit of zero, where the scale for both images is shown on the right. Since both products use the same window channels it is not surprising to find *cross-talk* between them, where high cloud liquid water also results in large ice concentration. Stated differently, it is difficult to screen out the effects of precipitation where $Q > 0.2$ mm without removing some actual sea ice. Fortunately, we find that for latitudes greater than 50 degrees, virtually all precipitation is removed by only considering ice concentration greater than 40 percent. This is seen by further examining Figure 90.

The images in Figure 90 demonstrate a number of important issues regarding the effects of clouds, precipitation and even sea surface winds on the sea ice product. The areas of high ice concentration (> 30 %) shown in red appear correct. However the sea ice retrieved south of Iceland is misclassified due to precipitation arising from an occluded front. The front contains cloud liquid water amounts greatly exceeding 0.3 mm and results in sea ice concentrations approaching 40 percent. In addition, the large area near the southern tip of Greenland has ice concentrations up to 20 percent with liquid water amounts of 0.3 mm in some places. Although the false ice in this region was originally believed to be due to precipitation, further analysis suggests that it also due to sea surface wind effects.

Ocean winds roughen the seas and generate foam cover. Although rough surfaces have a higher emissivity than a smooth surface, foam produces the largest increase in emissivity the same way as sea ice does. In fact, (73a) can be used to estimate the fraction of foam cover by replacing ε_{ice} with the near unity emissivity of foam. Rather than compute the foam coverage from the emissivity retrievals, Figure (90) shows the foam coverage (Bottom-Left) from based on analysis by the National Environmental Prediction Centers (NCEP) of NOAA. from reanalysis of wind stress [36] for this day. While there are overlapping regions of large foam cover (>20 %), clod liquid water (> 0.2 mm) and sea ice concentration (.>20 %), there is also a region further West where only the foam coverage and sea ice concentration is large (see arrows). In this region, the sea ice concentration is misclassified due to sea state and not precipitation. The foam coverage in this region increases the emissivity from 0.46 (see surrounding area) to 0.56, *i.e.*, 20 percent.

The above example demonstrates the effects of precipitation and sea surface winds on the retrieved sea ice concentration. The simplest way to remove these effects is to increase the threshold of minimum sea ice concentration. This is shown in Figure 91 where the threshold has been increased from 10 percent to 40 percent. Note that the lowest threshold displays false ice due to precipitation and surface effects. In addition to this example, studies were also done during the warmer season when precipitation events were more intense. More intense precipitation can produce errors exceeding the 40 % ice concentration shown in Figure 91. Fortunately, however, these events only occur at latitudes below 60 degrees.

4 – AMSU versus SSM//I derived Sea Ice

An automated system for retrieving ice concentration from SSM/I measurements was developed by Robert Grumbine [37] which was based on the NASA team algorithm [38, 39]. The SSM/I measurements was validated using high-resolution sources of sea ice information and was estimated to have an accuracy of about 3 to 5 %. This high accuracy is presumably due to its improved cloud editing using spatial consistency steps checks for "weather filtering" and is why it is used as ground truth for validation of the AMSU sea ice product. As an example, Figure 92 displays the SSMI sea ice concentration map in the Northern and Southern hemispheres for January 15, April 15, July 15 and October 15. 1999. Also, Figure 93 shows the AMSU sea ice concentration map for the same periods. Noticeable differences in the maps are seen between the

SSM/I and AMSU maps for concentrations exceeding 90 %. However much less difference is seen for concentrations greater than 50 % (shown in gray). To better compare the two, Table 8 lists the calculated sea ice extent in million km^2 measured by each satellite instrument. In both cases the calculations are done for latitudes greater than 50 degrees North and South. The results are shown separately for the Northern and Southern hemispheres and for the four different times. Since the SSM/I is expected to be the more accurate it the AMSU sea ice is used operationally to fill the gaps, particularly near the North Pole.

Table 8: Sea Ice Extent (million km^2) in Northern and Southern Hemispheres for selected Dates from SSM/I and AMSU algorithms

Date	SSM/I N - Hemisphere	AMSU N - Hemisphere	SSM/I S - Hemisphere	AMSU S - Hemisphere
01/ 15/ 1999	12.52	12.51	5.58	4.11
04/ 15/ 1999	13.28	13.05	7.18	6.66
07/ 15/ 1999	9.13	8.63	16.58	15.73
10/ 15/ 1999	8.28	8.67	18.83	16.38

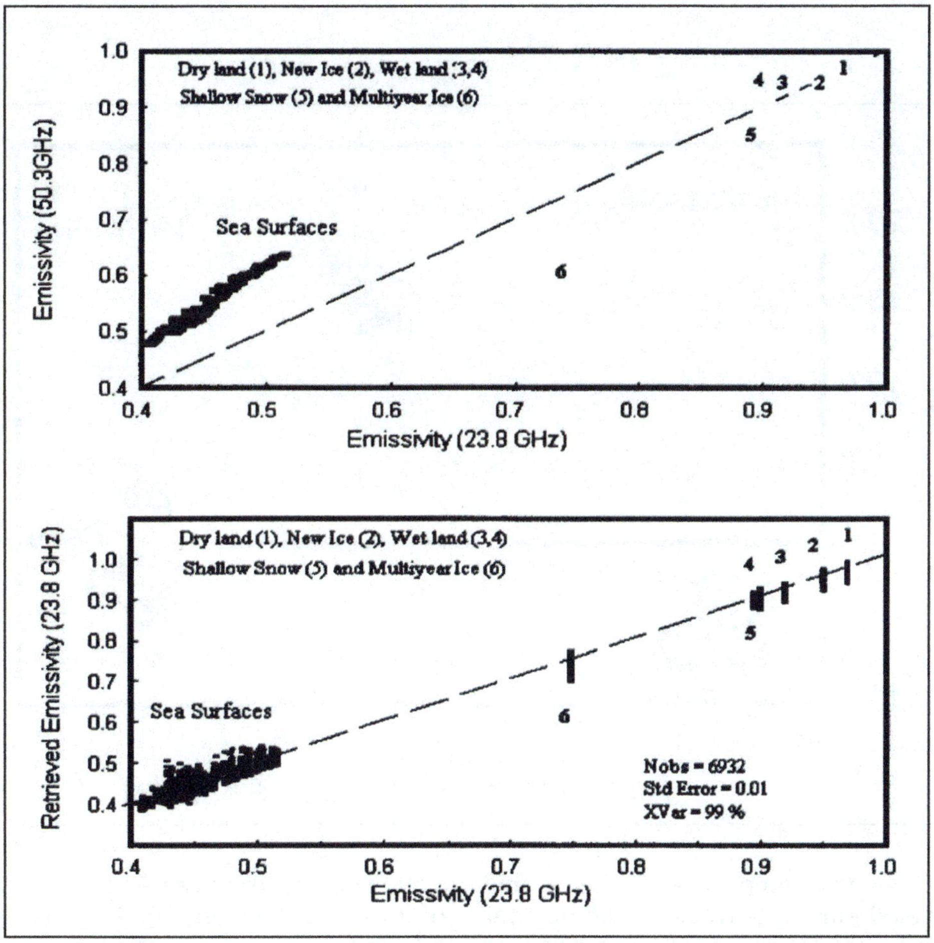

Figure 86 – Shown are the 6 emissivity types (labeled above) in the radiation transfer model. Top shows the emissivities at 50- and 23.8 GHz plotted against each other. Bottom shows the retrieved emissivity at 23.8 GHz using equation (76) plotted against the true emissivity. Note that a standard error of 0.01 is obtained.

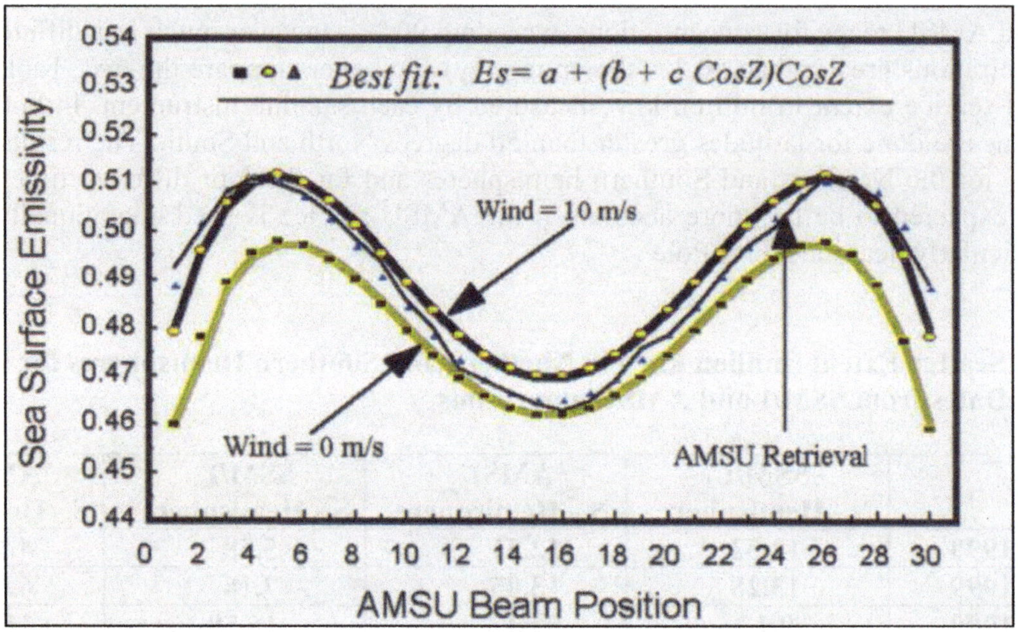

Figure 87 – Comparisons between modeled emissivity for calm seas (Wind =0 m/s), rough seas (Wind = 10 m/s) and that measured by AMSU using equation (76).

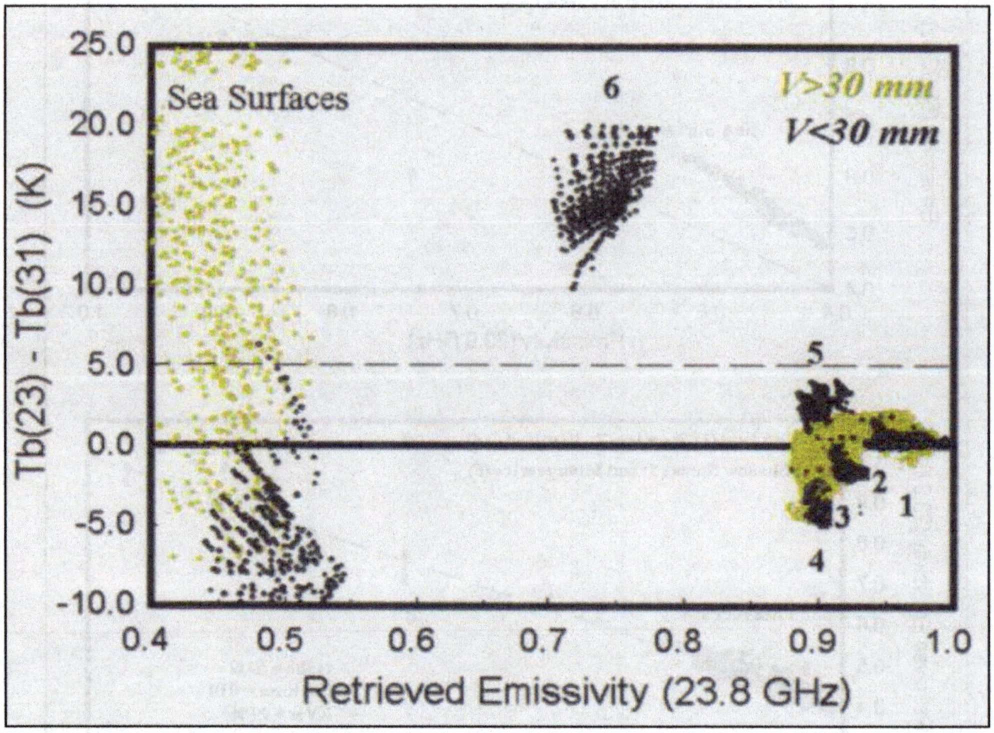

Figure 88. Simulated scattering index Tb(23) – Tb(31) plotted against retrieved 23.8 GHz emissivity. Shown are the 6 - labeled emissivity types whose numbers are indicated in Figure 86. A scattering index greater than 5 K separates multiyear ice (# 6) from sea surfaces as long as the water vapor *V* is below 30 mm.

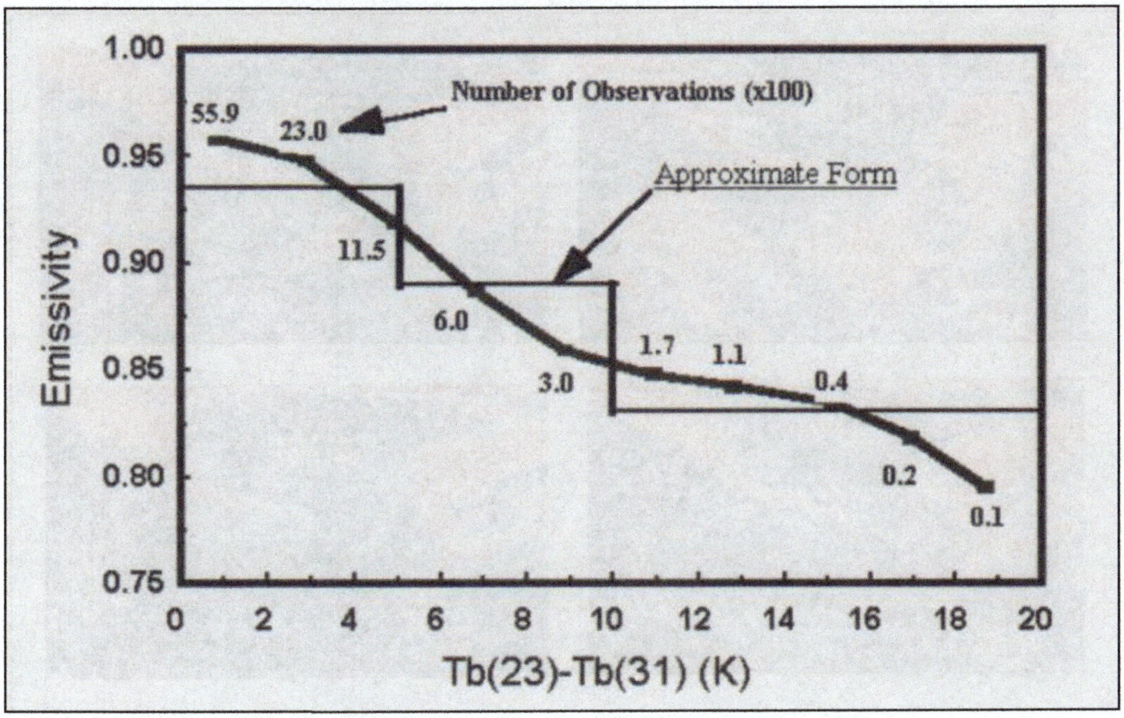

Figure 89. AMSU measured sea ice emissivity as a function of the scattering index. The data was obtained from both hemispheres (60-70 deg. Latitude) and over all beam positions. Also shown is the number of observations and an approximate form (see Text).

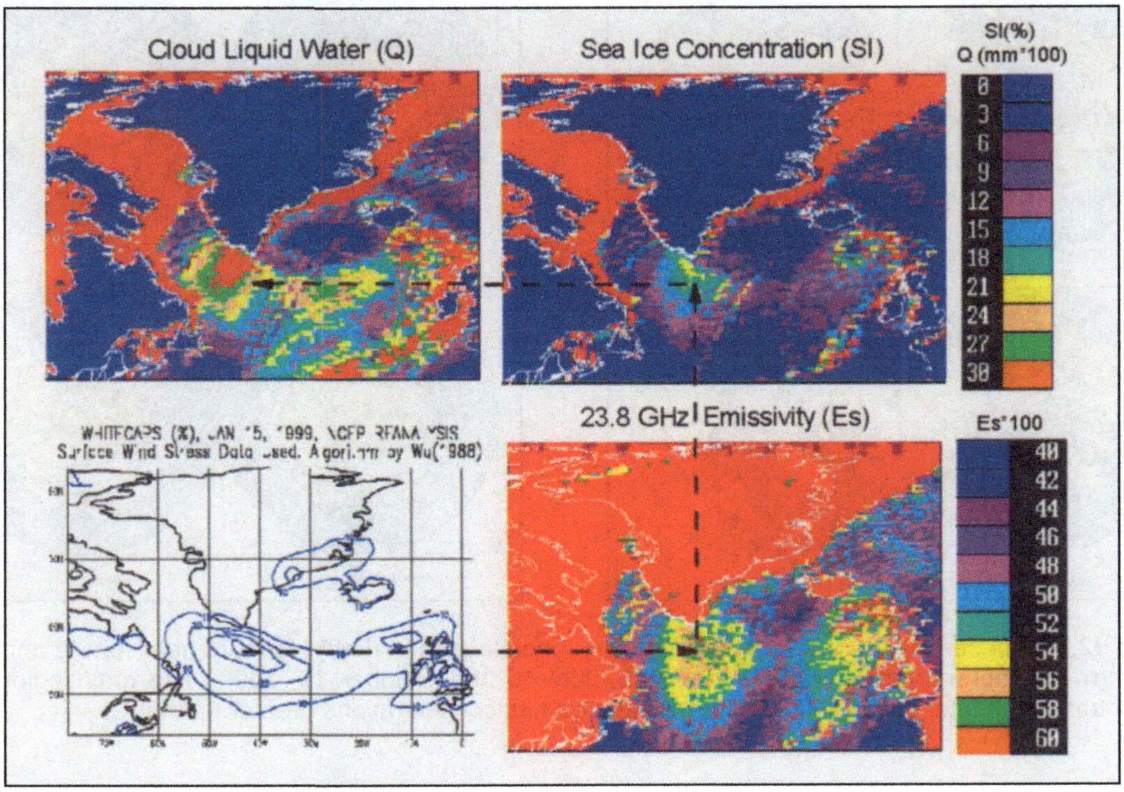

Figure 90. Foam cover (lower–left) from NOAA - NCEP analysis compared with AMSU retrieved emissivity (lower-right), sea ice concentration (top-right) and cloud liquid water (top-left) for January 15, 1999.

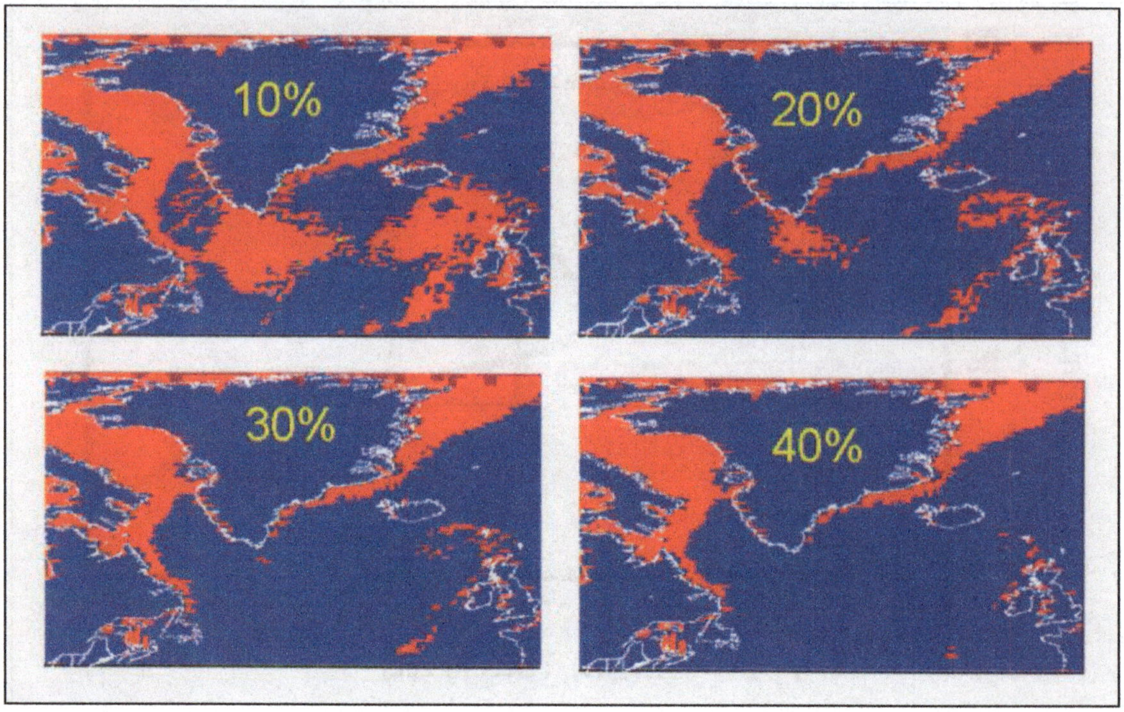

Figure 91.AMSU received sea ice coverage (shown in red) on January 15, 1999 using different ice concentration thresholds (10, 20, 30, 40 %).

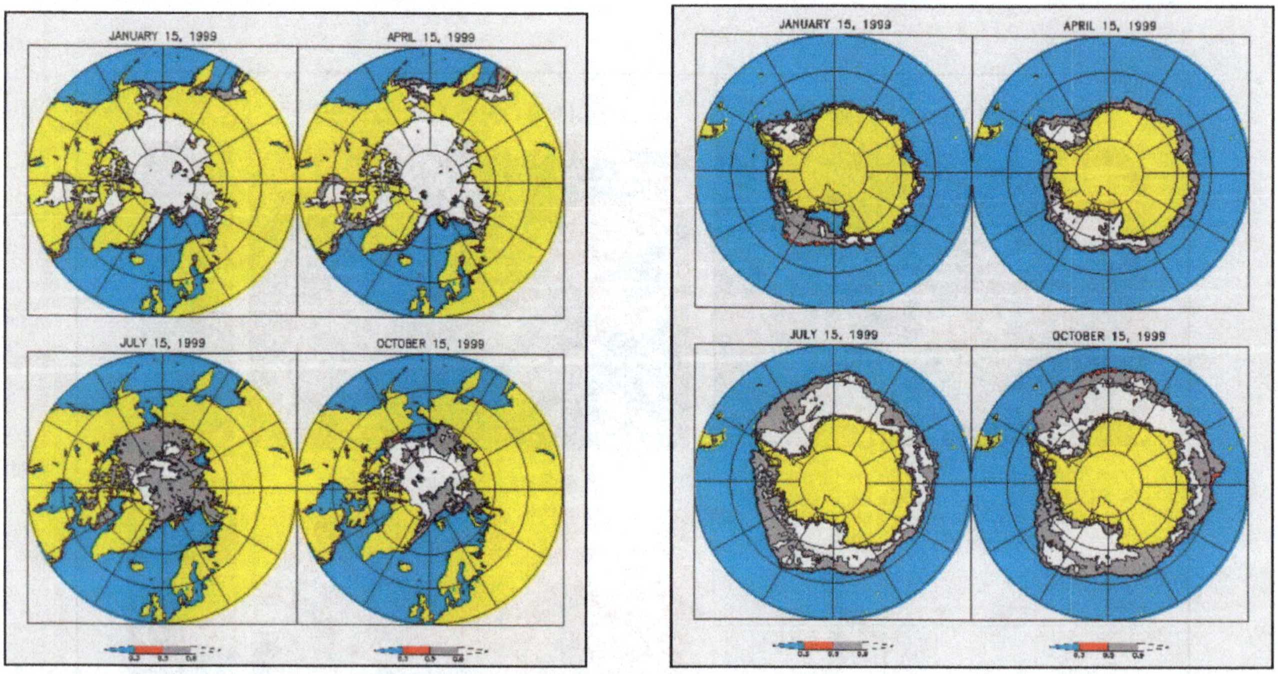

Figure 92. **SSM/I** sea ice analysis. Maps of sea ice concentration in (Left) Northern hemisphere and (Right) Southern Hemisphere for January 15, April 15, July 15 and October 15, 1999. The gray regions have concentrations greater than 50 % while the white area has concentrations greater than 90 %.

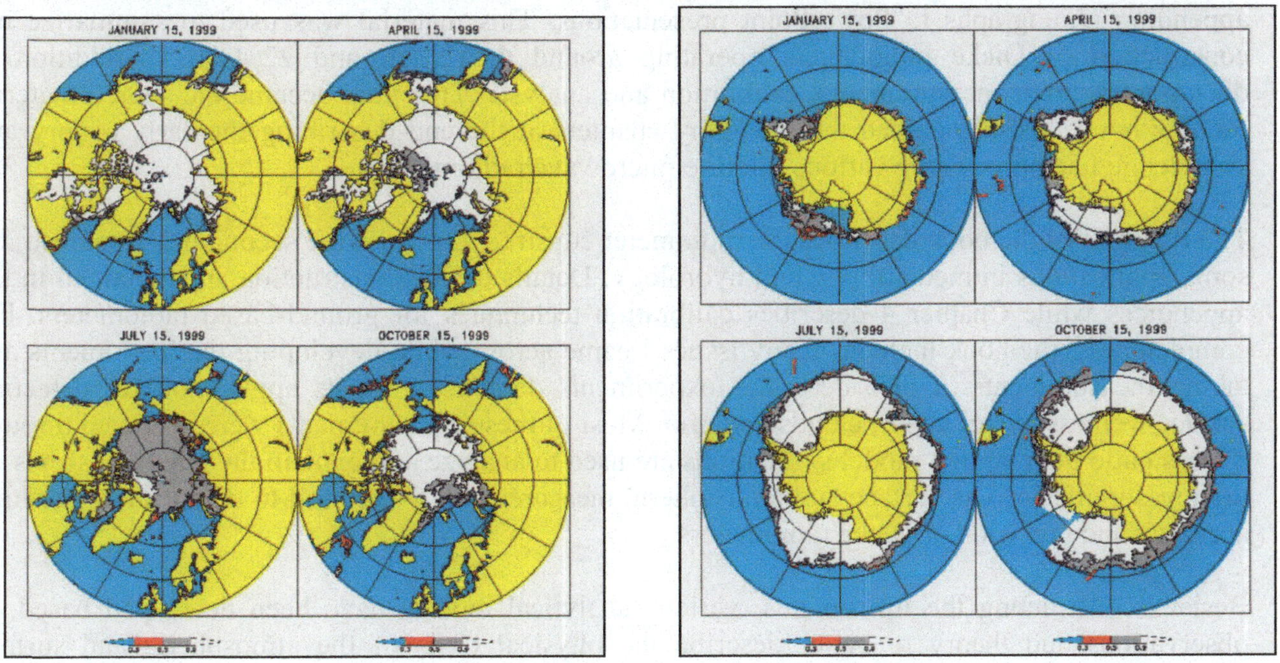

Figure 93. **AMSU** sea ice analysis. Maps of sea ice concentration in (Left) Northern hemisphere and (Right) Southern Hemisphere for January 15, April 15, July 15 and October 15, 1999. The gray regions have concentrations greater than 50 % while the white area has concentrations greater than 90 %.

13. Concluding Remarks

As mentioned in the introduction, although microwave radiometry is less recognized than radar, it is well established in Atmospheric and Earth Sciences due to its unique application in remote sensing. Furthermore, as discovered during this investigation, Dicke radiometers are relatively simple to construct. This is in contrast to radar, which requires digital processing to properly utilize its phase information. Such phase measurements is unique to radar and is particularly important to measure the fall velocity of rainfall and its vertical distribution, which can not be obtained radiometrically. Also obtained from the phase or return time of radar echo's is the identification and distribution of stratified ice layers in snow. In this case, however, Figures 83 and 84 shows how *radiometers can also detect such ice stratification indirectly by its inverted microwave spectrum*. Moreover, as discussed in Chapter 11, radiometers provide the only means of measuring surface and atmospheric temperature remotely. However, as pointed out to me by Dr. Chris Ruff, unlike radar, radiometry does not have the luxury of turning up the transmit power when it needs more signal to noise, or matching the receiver bandwidth to the transmit signal to limit receiver noise contributions.

My first experience with microwave radiometers began in 1971 when I worked at NASA and then at NOAA to help in the design of instruments, evaluate their performance and develop algorithms for deriving atmospheric and surface parameters from their measurements. These parameters are used by various organizations within the United States, England and throughout Europe to monitor, analyze and forecast weather and climate. Except for sea surface winds, I was directly involved in the algorithm development and product evaluation from most of the NASA, NOAA and DMSP microwave radiometers (*e.g.*, SCAMS, MSU, AMSU, SSM/I, SSM/T). Upon retiring in 2005 from NOAA I first considered developing improved algorithms but instead found it more interesting to build microwave radiometers. As the project evolved I documented my progress by appending photographs to PowerPoint presentations. This material was used to summarize my construction of Dicke radiometers operating around 4, 12, 20 and 22 GHz in addition to documenting their measurements, calibration and analysis. This then became the source material for this book. I also included three lengthy chapters at the end describing the early history and knowledge I acquired using various satellite microwave radiometers.

The first half of the book describes the radiometer construction while the second half demonstrates some applications in meteorology and hydrology. Details on the construction are contained in the appendices while Chapter 4 describes calibration techniques for ground-based radiometers. For completeness, the book includes many issues I came across while developing the instruments and taking measurements. It also describes experiments demonstrating its application for detecting surface wetness, water vapor, clouds and rain. Most of these measurements were observed through a glass patio door so that model simulations are used to analyze and explain the effect glass has on the measurements. The surface and atmospheric measurements were used to demonstrate the high accuracy and precision of Dicke radiometers.

Besides constructing the instruments, various analytical models have been developed based on observations and theory to better describe the physical basis of the atmospheric and surface measurements. However, while much is known regarding atmospheric parameters, less is understood about surfaces. The surface emissivity characteristics are briefly discussed in Section 7.1 with examples displayed in Figure 22 based on satellite, aircraft and ground-based radiometer measurements. In contrast to these far-field measurements, Appendix A17 describes the lessons learned when attempting to measure emissivity using near-field measurements. Lastly, Appendix A16 summarizes the different approaches used to model emissivity while Chapter 12 discusses the findings and unresolved satellite measurements discovered over deserts and snow covered surfaces. It is evident that although much has been learned since the first satellite radiometers were launched, the above mentioned chapter and appendices show that *gaps sill remain in our physical understanding of the measurements and modeling of surfaces*. This is in contrast to the design and analysis of microwave radiometers whose design has been well established and operation is well understood.

The rain, clouds, water vapor and surface wetness measurements demonstrated here are a few of the many products derived operationally under nearly all weather conditions using ground-based and earth-viewing satellite microwave radiometers. A reference to these applications in addition to other topics is given in the 1993 book "Atmospheric remote sensing by microwave radiometry", Edited by M. A. Janssen [7]. In fact, Figure 69 in Chapter 12 is a composite picture from the book's cover of sea ice, snow cover, rain rate and cloud liquid water measured using an instrument called the Special Sensor Microwave Imager (SSM/I), which was first launched aboard a polar orbiting satellite in 1987. The SSM/I is a dual polarized multifrequency (19, 22, 37 and 85 GHz) radiometer developed by the Navy to measure these parameters in addition to water vapor, sea surface temperature and wind speed. This instrument as well as others is discussed in the chapters covering satellite radiometry.

Of particular importance for climate monitoring is the more than 40 year record of global atmospheric temperature soundings obtained from a different series of satellite radiometers operating in the 50 to 60 GHz oxygen band whose operational instruments were developed by NOAA. As discussed in Chapter 11, *none of this could be obtained without the high reliability and precision of microwave radiometers to obtain measurements under clear and cloudy conditions.* As shown in Figure 26, even my instruments operate over long periods without any drift or changes in calibration. This is due to the Dicke radiometer design which reduces spurious noise and minimizes the drift due to gain variations. A brief analysis is given in Appendix A15 while a review of radiometer designs is given by M. E. Tiuri in the 1966 historic book "Radio Astronomy" by John. D. Kraus [40].

Historically, beginning in the 1970's, earth viewing satellite radiometers have employed the Dicke design to avoid the need for frequent calibration. However, beginning with the SSM/I and AMSU instruments listed in Table 7, many satellite instruments have opted to use the simpler total power radiometer design. This was based on data analysis, showing high stability as long as the radiometer can step quickly (*e.g.*, every 30 seconds) through a calibration cycle whereby its antenna views the cosmic background and a high emissivity temperature monitored target. Ground based radiometers on the other hand still generally use the Dicke approach since it's difficult to ensure cloud-free calibration measurements in a timely manner. As a secondary standard a high temperature warm target[5] or a precision noise diode input has been used to supplement the cold space calibration measurements for both ground-based and satellite instruments. Additionally, for satellite radiometers, vicarious calibration using high and low emissivity earth targets has also been used in addition to the onboard targets. As discussed in Appendix A18, this has been used to improve the calibration by more accurately estimating the detectors nonlinearity. Lastly, as mentioned in Chapter 5 and in Section 6.1, computer aided techniques can be used to improve the radiometer performance by processing the waveforms illustrated in Figure 3 and 4 digitally. However, due to its complexity only analog radiometers are used here.

It should also be noted that although microwave technology has advanced considerably since the 1940's when Dr. Robert Dicke developed his instrument using vacuum tubes at the MIT Radiation Laboratory, the basic design hasn't changed much until the recent advent of high speed analog to digital converters leading to digital radiometers. As discussed in Chapter 5, digital analysis has recently been used prior to detection to mitigate *RFI* and even replace square law detectors. Also, as mentioned above, the application of microwave radiometers has well surpassed Dicke's original use, which was to measure the atmospheric absorption from water vapor and oxygen. These observations together with radiation measurements of the moon, the sun, and the first estimate (~20 *K*) of the cosmic background microwave radiation were briefly mentioned in his 1946 landmark paper [1]. It was some 16 years later while studying cosmology in the early 1960's that Dicke decided to revisit his early radiometer experiment and improve on his cosmic radiation measurement. This work was however surpassed in 1964 by Drs. Arno Penzias and Robert Wilson at Bell Laboratories, who accidentally found the radiation while studying the noise in a satellite receiver. They both received Nobel Prizes for their very accurate 2.7 *K* cosmic radiation measurement at 4 GHz.

This book concludes by discussing satellite microwave radiometers, which as mentioned above measures the earth's atmosphere and surface under nearly all weather conditions. Consequently, beginning in the 1970's this application of microwave radiometers launched a new field of study called Satellite Microwave Radiometry. Being multidisciplinary, it draws on areas of electrical engineering, oceanography, geophysics, atmospheric and earth sciences. After many years of development, microwave radiometers have matured so they are now considered the primary instruments for measuring the earth's surface and atmosphere from space. Because of their nearly all weather capability, they routinely provide data to numerical models for analysis and prediction of both short-term and long-range weather forecasting. Many of the most important applications of satellite microwave radiometers are highlighted in the last three Chapters and Appendices of this second edition of the book. Although not a standard textbook, the book is written for those wanting a more in-depth description of the operation, analysis and measurements from satellite-launched and ground-based microwave radiometers.

14. Acknowledgments

In conclusion, I am grateful to have had the opportunity to work with many colleagues at NOAA in applying this technology to measure the earth's atmosphere and its underlying surface from space. However, as mentioned in the Preface, I must begin by thanking Michael Fletcher. It was his Radio Astronomy web-site that got me started in constructing the radiometers. I am also grateful for the correspondence I had with Dr. Philip Rosenkranz from MIT who reviewed much of the analysis in the book and corrected some errors. Also acknowledged is Dr. Konstantin Vinnikov from the University of Maryland, who I collaborated with on the climate analysis of MSU data reported on in Chapter 11. He also provided support in developing and validating the AMSU sea ice algorithm described in Section 12.3. Furthermore, upon finally building radiometers after many years of developing algorithms and analyzing data, I can fully appreciate Dicke's remarkable achievement. In fact, Wikipedia states that some believe Robert Dicke deserved a Nobel Prize just for the invention of this powerful measuring device which started a new field of study Satellite Microwave Radiometry. Hopefully, this will encourage others to construct such instruments for sensing the earth as well as astronomical sources as it was first envisioned back in the 1930's by Karl Jansky and Grote Reber.

15. References (*in order of appearance*)

[1] R. Dicke, R. Beringer, R. Kyle and A. Vane, "Atmospheric absorption measurements with a microwave radiometer," *Physical Review*, Vol **70**, No. 5, Sept 1, 1946, pp 340-349.

[2] Swenson & Yang, "An Amateur Radio Telescope-V," *Sky and Telescope*, 56, **3**, pp 201-205, Sept 1978.

[3] Neils Skou "Microwave Radiometer Systems: Design & Analysis," *Artech House, Inc*, 162 pgs., 1989.

[4] D. M. Pozar,"Microwave Engineering," 3rd Edition, John Wiley & Sons, 700 pgs., 2005

[5] R. Nordlund and A.Kassab, "Non-Fourier heat conduction: A boundary element solution of the hyperbolic heat conduction equation," *Trans. on modeling and simulation*, Vol. **10**, pp 279-286, 1995.

[6] A. Stogryn, "The brightness temperature for a vertically structured medium," Radio Science, Vol. **5**, pp 1397-1406, 1970

[7] Michael Janssen (Ed.), "Microwave Remote Sensing by Microwave Radiometry," *John Wiley & Sons*, 592 pgs., 1993

[8] C. Ruf, S. Gross and S. Misra, "RFI detection and mitigation for microwave radiometry with an agile digital detector," *IEEE Trans. Geosci.. Remote Sens.*, Vol **44**, pp 694-706, 2006.

[9] P.G. Tsybulev, M.V. Dugin, A. B. Berlin, N. A. Nizhelskij, D. V. Kratov and R. Yu. Udovitskiy, "1/*f* type noise in a total power radiometer," Astrophysical Bulletin, Vol. 69, No. **2**, 2014.

[10] N. Grody, "Classification of snow cover and precipitation using the SSM/I," *J. Geophys. Res.*, Vol. **96**, pp 7423-7435, 1991.

[11] Donald Kerr (Ed.), "Propagation of Short Radio Waves," MIT, Vol **13**, *McGraw Hill*, 728 pgs., 1951, (see Section 6).

[12] D.C. Hogg, F. Guiraud, F. Snider, M. Decker E. Westwater, "A Steerable Dual-Channel Microwave Radiometer for Measurements of Water Vapor and Liquid in the Troposphere," *Jour of Climate & Applied. Meteorology*, **22**, pp 789-806, 1983.

[13] F.O. Guiraud, J. Howard and D. C. Hogg, "A Dual-Channel Microwave Radiometer for Measurement of Precipitable Water Vapor and Liquid," *IEEE Trans. on Geosci. Elect,* GE-12, 4, pp 129-136, 1979.

[14] D.C.Hogg, F. Guiraud, J. Howard, A. Newell, D. Kremer and A. Repjir, "An antenna for dual-wavelength radiometry at 21 and 32 GHz," *IEEE Trans. on Antenna and Propagation*, Vol. AP-17, No **6**, pp 764-771, November 9, 1979.

[15] Ed. R. Westwater, "The accuracy of water vapor and cloud liquid water determination of dual- frequency ground-based microwave radiometry, " *Radio Sci*, V13, No.**4**, pp 677-685, July 1978**.**

[16] J. C. Blinn, III, J. C. Conel and J.G. Quade, "Microwave emission from geological materials: Observations of interference effects," *Jour. Geophys. Res*., Vol **77**, pp 4366-4378, 1972.

[17] C. Prigent, C. Rossow, E. Mathews and B. Marticorena, "Microwave radiometric signatures of different surface types in deserts," *Jour. Geophys. Res*., Vol. GE-24, pp 220 – 231, 1986.

[18] C. Matzler, "Microwave permittivity of dry sand," *IEEE Trans. Geosci. Remote Sens*., Vol **36**, pp 317-319, 1998.

[19] N. Grody and F. Weng, "Microwave emission and scattering from deserts: Theory compared with satellite measurements," *IEEE Trans. Geosci. Remote Sens*., Vol. **46**, pp 361-375, 2008.

[20] K.S. Shifrin, "Transfer of microwave radiation in the atmosphere," NASA Technical Translation., TTF-590, 185 pgs., 1969.

[21] N. Grody, C. Hayden, W. Shen, P. Rosenkranz and D. Staelin, "Typhoon June Winds Estimate from scanning Microwave Spectrometer measurements at 55.45 GHz," *Jour. Geophys. Res*., **84**, pp 3684-3695, 1979.

[22] N. Grody, and P. Pelligrino, "Synoptic Scale Studies using the Nimbus 6 Scanning Microwave Spectrometer," *Jour. Appl. Meteor*., **16**, pp 816-826, 1977.

[23] K.Y. Vinnikov, N. Grody, A. Robock, R. Stouffer, P. Jones and M. Goldberg, "Temperature trends at the surface and the troposphere", Jour. Geophy. Res. Vol. **11**, D03106, 2006.

[24] S. Kidder, M. Goldberg, R. Zehr, M. DeMaria, J. Purdom, C.Veldon, N. Grody and S. Kusselson, "Satellite analysis of tropical cyclones using the Advanced Microwave Sounding Unit (AMSU)," *Bull. Amer. Meteor. Soc*., **81**, pp 1241 – 1259, 2000.

[25] N. Grody, D. Gray, C. Novak, J. Prasad, M. Piepgrass and C. Dean, "Temperature soundings from the DMSP microwave sounder," Advances in Remote Sensing Retrieval Methods, A. Deepak Publishing, pp 249-267, 1985.

[26] K. F. Kunzi, A.D. Fisher, D.H. Staelin and J. W. Waters, "Snow and ice-surfaces measured by the Nimbus-5 microwave spectrometer," *Jour. Geophys. Res.*, Vol **81**, pp 4965-4980, 1976

[27] R.W. Spencer, W.S. Olson, Wu Rongzhang, D. Martin, J.A. Weinman and D.A. Santek, "Heavy thunderstorms observed land by the Nimbus-7 Scanning Multichannel Microwave Radiometer," Jour. Climate Appl. Meteror. **22**, pp 1041-1046, 1983.

[28] N. Grody, "Relationship between snow parameters and microwave satellite measurements: Theory compared with AMSU observations from 23 to 150 GHz, " *J. Geophys. Res.*, **113,** doi: 10/2009JD009685, 2008.

[29] L. Tsang, J. Kong and R. Shin, "Theory of Microwave Remote Sensing," Wiley Series in Remote Sensing, 613 pgs., 1985 (see page 498).

[30] C. Matzler, "Passive microwave signatures in landscapes in winter," Meteor. Amos. Phys., **54**, pp 241-260, 1994.

[31] N. Grody and A. Basist: "Interpretation of SSM/I measurements over Greenland," *IEEE Trans. Geosci. Remote Sen.*, **35**, pp 360-366, 1997.

[32] C. L. Parkinson, D. J. Cavalieri, P. Gloerson, H. J. Zwally and J. C. Comiso," Artic Sea Ice, 1973 – 1976," NASA SP-489, Available from Government Printing Office, pgs. 295, 1987.

[33] K. Y. Vinnikov, A. Robock, R. Stouffer, J. Walsh, C. Parkinson, D. Cavalieri , J. Mitchell, D. Garrett and V. Zakharov, "Global Warming and Northern Hemisphere sea ice extent," Science, **286**, pp 1934-1937, 1999.

[34] J. C. Comiso, "Sea Ice effective emissivities from satellite passive microwave and infrared observations," *J. Geophys. Res.*,**88**, pp 7686-7704, 1983.

[35] T. Greenfell and J. C. Comiso, "Multifrequency passive microwave observations of first-year sea ice grown in a tank," *IEEE Trans. Geosci. Remote Sens.*, **24**, pp 826-831, 1986.

[36] J. Wu, "Variations in whitecap coverage with wind stress and water temperature," J. Phys. Oceanography, **18**, pp 1488-1453, 1988.

[37] R. W. Grumbine "Automated Passive Microwave Sea Ice Concentration Analysis at NCEP," DOC/ NOAA/ NWS/ EMS/ OMB Technical Note 120, 13 pgs, 1996.

[38] D. J. Cavalieri, "Sea Ice Algorithm in NASA Sea Ice Validation Program for the Defense Meteorological Satellite Program Special Sensor Microwave Imager," Final Report NASA Technical Memorandum 104559, pp 25-32, 1992.

[39] P. Gloerson and D.J. Cavalieri, "Reduction of Weather Effect in the Calibration of Sea Ice Concentration from Microwave Radiances,"JGR, **91**, pp 3913- 3919, 1986.

[40] John D. Kraus, "Radio Astronomy," *McGraw Hill*, 485 pgs., 1966.

[41] K.M. Watson, "Multiple scattering of electromagnetic waves in an underdense plasma," J. Math. Phys., **10**, pp 688-702, 1969.

[42] D. L. Tien, "Thermal radiation in packed and fluidized beds," Jour. Heat Transfer, Vol. **110**, pp 1230 -1242, Nov. 1988

[43] A. Stogryn, "A study of the microwave brightness temperature of snow from the point of view of strong fluctuation theory," *IEEE Trans. Geosci. Remote Sens.*, Vol. **46**, pp 361-375, 1986.

[44] A Stogryn and M. A.. Karam, "The multiple scattering contributions in the Strong Fluctuation Theory to the microwave brightness temperature," Office of Naval Research, Department of the Navy, N00014-91-C0140, Report No. 9617, 47 pgs., Feb.1994.

[45] S. Rice, "Reflection of electromagnetic waves from slightly rough surfaces," Communications on Pure and Applied Mathematics, Vol **3**, No 2/3, pp 351 -378, 1951.

[46] S.M. Anlage, V. V. Talanov and A. R. Schwartz, "Principles of Near-Field Microwave Microscopy, "*Scanning Probe Microscopy, Electrical and Electromechanical Phenomena at the Nanoscale*, Springer, N.Y., pp 215-253, 2007

[47] T. Mo, "A study of the Microwave Sounding Unit on the NOAA-12 Satellite," *IEEE Trans. Geosci. Remote Sen.*, **33**, pp 1141-1152, 1995.

[48] N. Grody, K. Vinnikov, M. Goldberg, J. Sullivan and D. Tarpley, "Calibration of multisatellite observations for climate studies: Microwave Sounding Unit (MSU)," *J. Geophys. Res.*, **109,** doi: 1029/2004JD005079, 2004.

[49] N. Grody, "Severe storm observations using the Microwave Sounding Unit," *J. Climate and Applied Meteorology*, Vol. **22**, No. 4, pp 609-625, 1983.

[50] N. Grody, "Satellite-based microwave retrievals of temperature and thermal winds: Effect of channel selection and A-Priori Mean on retrieval accuracy," *Remote Sensing of Atmospheres and Oceans*, Academic Press, Inc., pp 381- 409, 1980.

[51] P. W. Rosenkranz, "Interference coefficients for overlapping oxygen lines in air," J. *Quant. Spectroscopic. Radiat. Transfer*, **17**, pp 287-297, , 1988.

[52] W. M. Frank, "The structure and energetics of the tropical cyclone, I, Storm structure," *Mon. Weather. Rev.*, **105**, pp 1119-1135, 1977.

[53] N. Grody and W. Shen, "Observations of Hurricane David (1979) using the Microwave Sounding Unit," *NOAA Technical Report*, NESS **88**, 51 pgs., 1982.

[54] G. Williams, "Microwave radiometry of the ocean and the possibility of marine wind velocity determination from satellite observations," *J. Geophys. Res*, **74** (18), p 4591, 1969.

[55] W. Nordberg, Conaway, J., and P. Thaddeous, "Microwave observations of sea state from aircraft," *Quart, J. Roy. Meteorol. Soc.*, **95**, 408, 1969

[56] W. L. Jones, P. G. Black, V. E. Delnore and C. T. Swift, "Airborne microwave remote-sensing measurements in Hurricane Allen," *Science*, **214**, pp 274-280, 1981.

Appendices

The following appendices describe in more detail subjects discussed in the preceding chapters. These appendices also provide additional information on the construction, measurement and analysis of ground-based radiometers. This second edition also adds Appendices A19, A20 and A21, to describe the algorithms for retrieving atmospheric temperature and wind profiles using satellite radiometers. Case studies of different meteorological systems are then used, comparing the retrieval accuracy of microwave sounders against conventional data.

A1. Total Power Radiometer

My first attempt in building a radiometer is briefly described here along with its measurements. As shown in Figure A1-1, a total power radiometer was constructed using a 12 GHz Ka band LNB whose output was connected to a satellite finder. The satellite finder is a commercial unit that contains a preamplifier to amplify the down-converted LNB intermediate frequency (*IF*) output which is between 1 to 2 GHz (*i.e.*, L-Band). In addition to the *IF* amplifier, the satellite finder also contains a detector to provide a DC output. To further increase the output and reduce instrumental noise, the detector is connected to a DC amplifier followed by an RC integrating circuit, with both constructed using standard operational amplifier circuits. The circuit is similar to the AC amplifier described in Appendix A7 but without the input capacitor. Also, an RC integrating circuit is formed by replacing the 1 K resistor between pins 6 and 3 with a 1 Meg resistor, and adding a 0.1 μf capacitor between pin 3 and ground to obtain a 0.1 second integration time.

Figure A1-2 shows the satellite finder, DC amplifier and integrator enclosed in a metal cabinet. The completed radiometer is also shown in the Figure. These components along with the LNB are all powered using a single regulated power supply. The LNB waveguide output is fitted to a crude circular horn that views the atmosphere or ground using a rotated flat reflector. For convenience, the radiometer was placed in my house as shown in Figure A1-3, where the antenna views the scene through a glass patio door. The figure also shows rain being simulated using a garden sprinkler. To save the radiometer data, its output is connected to a PC's USB port using an analog to digital converter. The resulting measurements are shown in Figure A1-4. Note the periodic increase in radiometer voltage as the water from the sprinkler passes in front of the antenna. You also see the increased voltage as I walked in front of the antenna. Both increases in signal voltage are a result of thermal emission, where the inserted diagram in Figure A1-4 gives the equation describing the event. The equation expresses the radiometer brightness temperature (proportional to voltage) in terms of the emissivity or absorption $1-\tau$ and its mean radiating temperature T_M. For water droplets its transmittance τ is a function of the liquid water content, Q, and frequency dependent parameter, $Q(\nu)$. Both quantities are discussed in Chapter 7.3 and given by equations (16b) and (16c). As such, this simple experiment can be expanded upon to study the relationship between radiometer measurements and precipitation parameters such as rainfall, drop size and liquid water content.

In addition to the voltage increase due to emission by water droplets, Figure A1-4 also displays a gradual voltage increase or slow drift due to changes in the radiometer gain from self heating. Such gain changes can result from the LNB, satellite finder, DC amplifier as well as the detector. All of these components can change their characteristics due to temperature variations. Unfortunately, even very small temperature induced change is detectable by the high radiometer gain (120 dB). After trying different components and thermally insulating the radiometer with Styrofoam, I concluded that the only way to reduce such effects is to periodically calibrate a total power radiometer every minute or less. In fact, such an approach is used for satellite instruments as discussed in Chapter 10, where the antenna reflector is rotated to sequentially view cold space, an ambient temperature target and the scene. Alternatively, noise diodes have been used in place of the cold space and the ambient targets, which are viewed by switching the radiometer input as in the front end of a Dicke radiometer. Both of these approaches require digital processing to compensate for drift as well as calibrate the radiometer. However, many have opted to use the Dicke approach which uses synchronous detection to greatly reduce the effect of gain variations.

As an example of this feature, Figures 26 and 35 show the long time performance of the 4, 12 and 20 GHz Dicke radiometers when viewing rain and clouds. Note that for more than 8 hours, these radiometers operate unattended with no noticeable drift while measuring the thermal radiation emanating from rain and clouds. Another example is shown in Figure A1-5 where the 12 GHz Dicke radiometer views clear skies through a glass patio door for 6 hours with no drift and a calculated standard deviation of about 10 mv. Also, the *NEΔT* of the radiometer is 0.3 *K* since the radiometric gain is about 30 *K*/Volt. Besides its excellent stability and historical significance, these Dicke radiometers provide an analog output without the need of digital processing. Furthermore, it provides a unique application of lock-in amplifiers to detect very small signals buried in the noise.

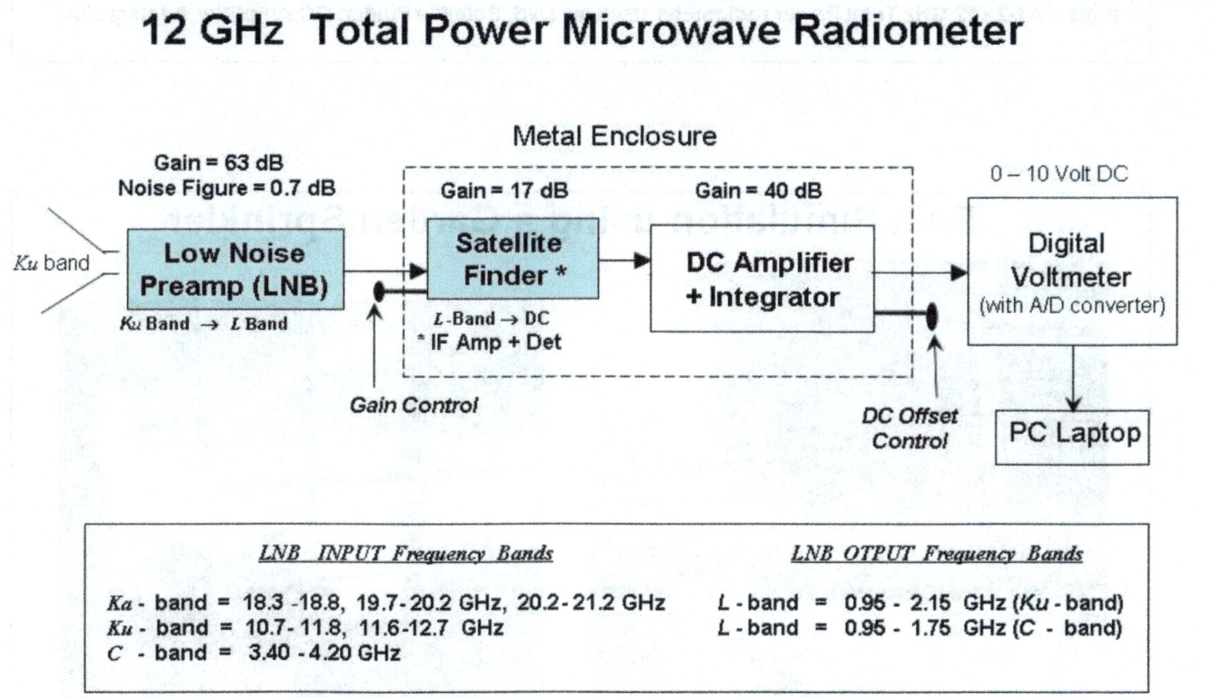

Figure A1-1 - Ku-Band radiometer built using a TV satellite receiver (LNB), a satellite finder (*IF* amp + det) and DC amplifier plus integrator. Not shown is the 12 volt regulated power supply + multiplexer. The radiometer sensitivity is 30 mv/K and also works at C and Ka -Band by changing the LNB.

Figure A1-2 - 12 GHz Total Power radiometer uses an *LNB*, Satellite Finder, DC Amplifier & Integrator

Figure A1-3 - 12 GHz Total Power radiometer measuring water droplets from a garden sprinkler.

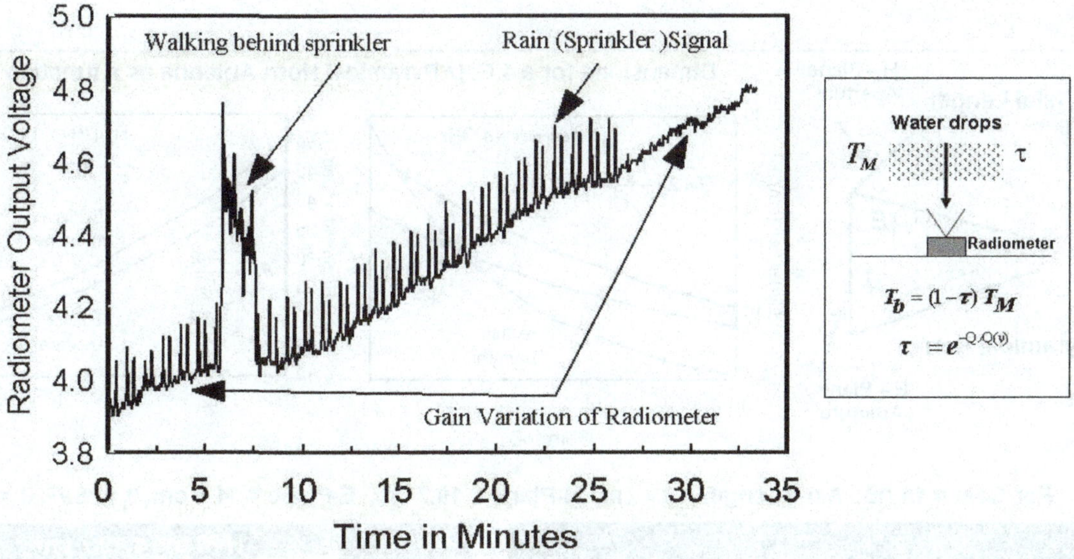

Figure A1-4 Shown is a periodic increase of the 12 GHz radiometer voltage due to thermal emission (1- τ)T_M by water drops from the sprinkler. Also seen is a gradual increase due to gain variations of the amplifiers. To reduce gain variation effect, the total power radiometer must be calibrated every minute or less.

DICKE RADIOMETER

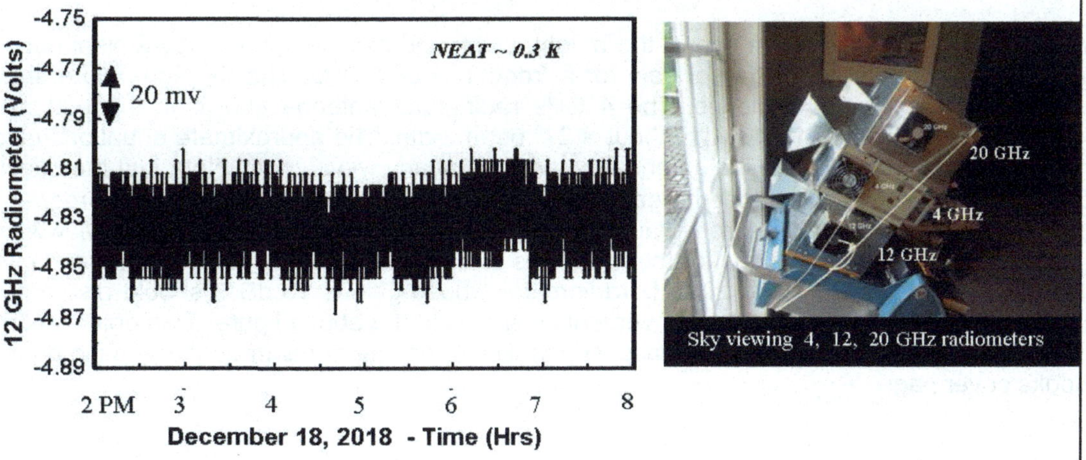

Figure A1-5 - Clear sky 12 GHz Dicke radiometer measurements (Dec. 18, 2018) between 2 and 8 pm. An integration time of 0.1 seconds was used for these measurements.

A2. Pyramidal Horn Antenna

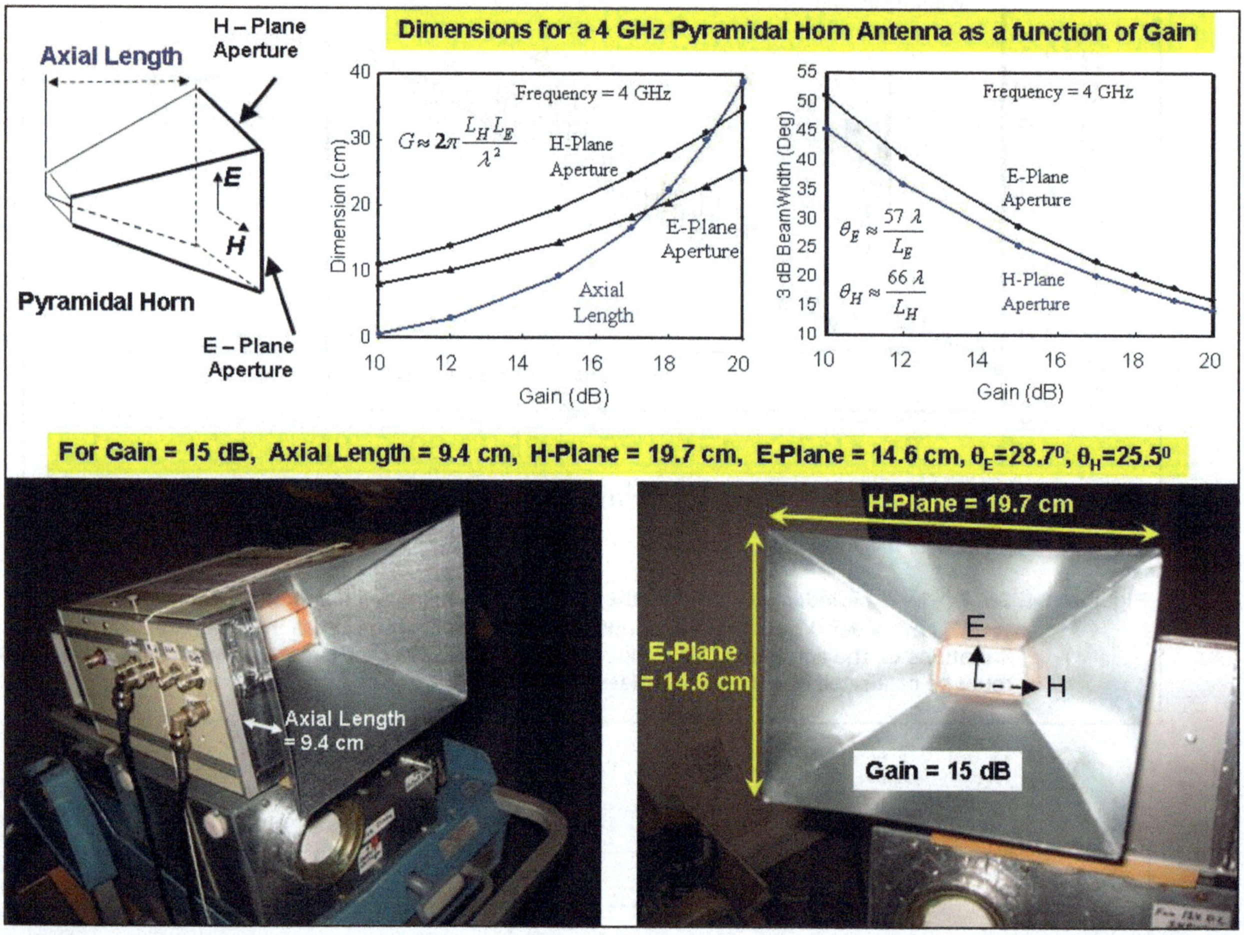

Figure A2. The top-middle figure shows the height, width and axial length of a pyramidal horn antenna. All parameters are plotted as function of gain for a frequency of 4 GHz. The top-right figure shows the 3 dB beamwidths in the E- and H-planes. The 4 GHz radiometer antenna shown in the bottom figures was designed to have a gain of 15 dB with about a 27^0 beamwidth. The approximate equations given in the top figures can be used to calculate the antenna dimensions at any wavelength. Note that the effective aperture is about 50% of its area, A, so that the gain is ½($4\pi A/\lambda^2$). The equations as well as a program to calculate exact dimensions were obtained from Paul Wade's online microwave web site at http://www.w1ghz.org/. It was used to obtain the dimensions for the 4 GHz as well as the 12 GHz radiometer antenna whose gain is 19 dB with a 16^0 beamwidth and 20 GHz radiometer whose gain is 20 dB. I should also mention that the polarization used by both radiometers is vertical as shown in the above figure. This polarization is preferred in order to minimize the reflections by glass when taking measurements through my patio door as shown in the books cover page.

A3. Temperature Controlled Fan

Temperature - Controlled Fan Speed

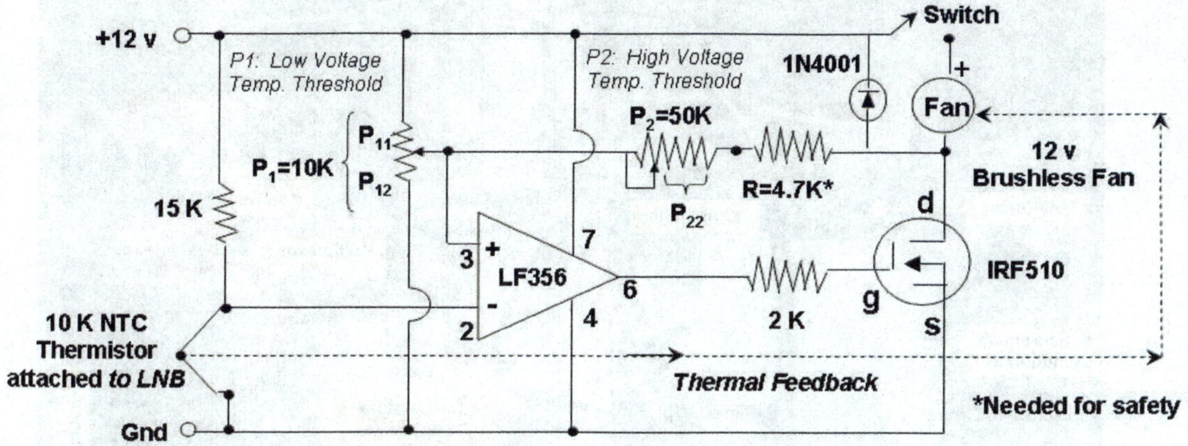

P_1 = Sets the low speed (voltage) of the fan at the cool temperature.
P_2 = Sets gain of circuit; How fast the voltage will rise to full output when temp is higher.

As the Thermistor (LNB) temperature increases the fan speed increases, thereby reducing the LNB temperature, i.e., negative feedback between LNB temperature and Fan Speed. Because the ambient temperature is lower than the LNB temperature, the fan speed can run slower and still keep the LNB at a cooler temperature.

CALIBRATION - PROCEDURE:

1 - Have Thermistor at room temperature. Adjust P_1 for the low speed you want fan to run.
2 - Heat Thermistor to the temperature you want fan to run at full speed. Adjust P_2 until fan is at full speed. Then adjust P_2 until the fans speed just begins to drop off.
3 - For testing purposes replace the Thermistor with a 100 K resistor and a 1 K resistor.

Figure A3. A temperature controlled fan speed circuit is used to cool the LNB by increasing the ambient temperature flow. The circuit uses a difference amplifier (LF356) whose negative voltage depends on the thermistor (10 K NTC) resistance, while the positive (reference) voltage is set using a 10 K resistive trimmer (P_1). Its differential output goes to the MOSFET (IRF510) which injects more or less current into the brushless fan depending on the thermistor resistance. The thermistor is attached to the LNB so that thermal feedback between the fan speed and LNB temperature is used to keep the differential input voltage constant. The difference amplifier gain set by resistor P_2 determines how fast the circuit responds to temperature change.

A4. 12 GHz Radiometer

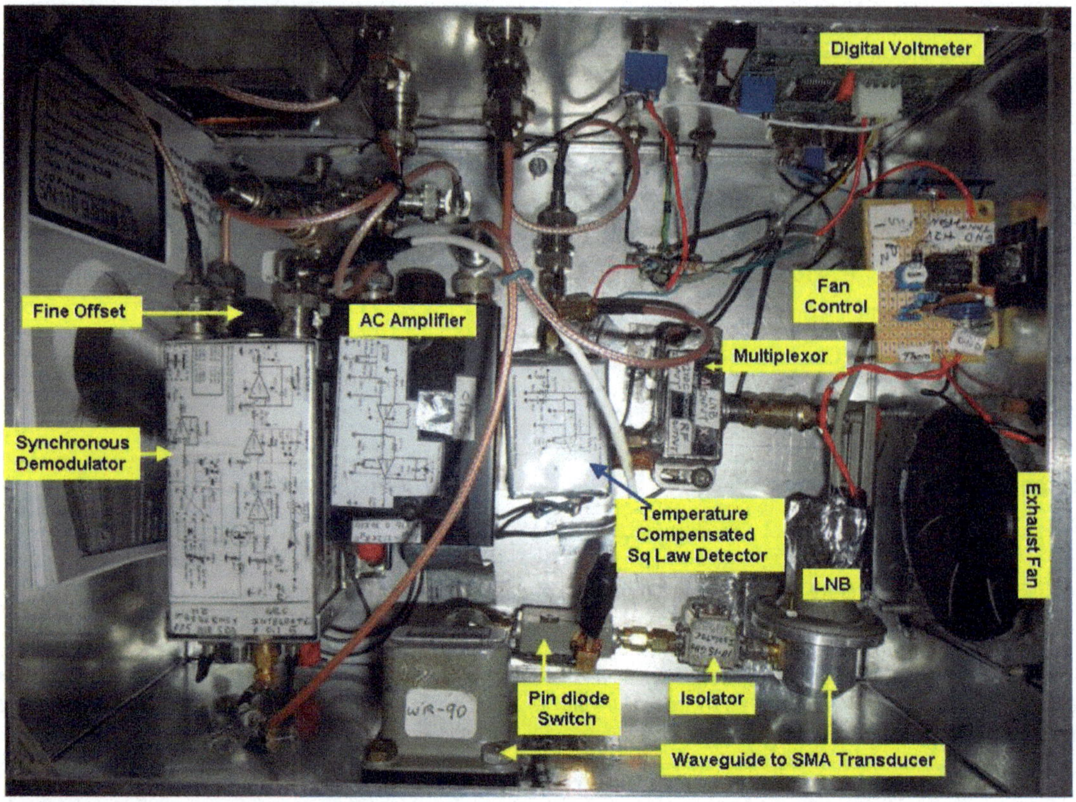

Figure A4. The top lid of the 12 GHz radiometer is opened (see Figure 2) to show the components. These are the same components referred to in the block diagram of Figure 3.

A5. Synchronous Demodulator

Figure A5-1 is the schematic diagram of the synchronous demodulator used in the 12 GHz radiometer. It also shows the waveforms at the different input and output stages of the circuit. A picture of the demodulator with its top opened and components labeled is shown in Figure A5-2. It is similar to the ones used in the 4 GHz and 20 GHz radiometers. Beginning from the right side of Figure A5-1, the input signal, V_{IN}, is the output from the AC amplifier. Note that V_{IN} is a square wave of voltage Ref when the pin diode switch in Figure 3 connects to its internal resistive load, followed in time with voltage Ant when the switch connects to the antenna input. The 10 μf input capacitor in Figure A5-1 removes any DC level from the AC amplifier output so that the unbiased waveform varies from (Ref-Ant)/2 to –(Ref-Ant)/2. This signal located at tap TP_1 goes to the demodulator, which is a unit gain difference amplifier constructed using half of a dual operational amplifier (OP2111).

The bimodal input to the operational amplifier at Pin 3 is switched from open to ground using a J177 p-channel depletion mode MOSFET connected without any bias circuit. The MOSFET switching action is obtained by operating it between its saturation and cut-off regions by driving its gate using the NE555 clock generator. This is the same clock that energizes the pin diode switch. Note that the clock frequency can be set at 172 Hz, 500 Hz or 1000 Hz although the lower frequency clock is found to be sufficient. *Also, at such low frequencies the pin diode switch is excited without the need for any specialized drivers and clock generators.*

Due to synchronization, the difference amplifier reverses the polarity of the negative input signal at TP_1 so that its output at TP_2 is (Ref–Ant)/2. For reference, Figure A5-3 provides the circuit analysis of the synchronous demodulator input stage as well as the output amplifier stage in Figure A5-1. As an example of measurements, Figure 5 shows the waveform at TP_2 when the 4 GHz radiometer views cold space. The demodulator output, TP_2, is next connected to an integrator (OP2111) that smoothes the signal according to equation (7b). Note in Figure 14 that T_b fluctuations are reduced by increasing the integration time from 0.1, 1.0 or 5.0 seconds. The integrator also reverses the input polarity at its output, and provides an adjustable offset to the last stage which is a DC amplifier of adjustable gain (1 to 6). As such, the DC amplifier output becomes G_2(Ant - Ref)/2 where G_2 is the amplifier gain. The DC amplifier uses an AD711 operational amplifier to assure very low noise and very small effects due to temperature drift. However, I must mention that an AD630 integrated circuit (IC) synchronous demodulator is available. This IC contains the difference amplifier, switch and integrator shown in Figure A5-1. It should then be easy to add a clock generator, AC amplifier and DC offset on a PCB to construct the radiometer back end. Only the front end components of antenna, switch, LNB and detector is then needed to construct a small radiometer.

Figure A5-2 shows a picture of the synchronous demodulator, which fits inside a small aluminum box. Small coax connectors are used to access the TP1 and TP2 diagnostic outputs while four BNC connectors are used for the input and output signals as well as the two clock signals, one having positive and the other negative voltage. The negative clock is obtained using the inverter circuit shown in Figure A5-1. It is needed to drive the Hewlett Packard pin diode switch (33102A) in Figure A4. However, the 4 GHz and 20 GHz radiometers use a General Microwave pin diode switch (M862B) in Figure 27 which only requires a positive switching voltage. As such, the inverter stage is omitted in the 4 and 20 GHz synchronous demodulators. I also found that a 500 Ω external resistor can be added to reduce the current that turns the M862B switch off from its specification of 37 ma to 16 ma at 12 volts without affecting its performance.

In Figure 5 of Chapter 3, the square wave clock signal was used to synchronize the sweep of an oscilloscope to observe the radiometer output signals from the synchronous demodulator, AC amplifier and detector. Also, to test and even improve the demodulator performance, the clock signal can be used to simulate the radiometer measurement for very cold or very hot sources. In this application the clock signal is connected to the demodulator input while its output is displayed on the scope. The negative clock input signal represents the maximum amplitude of about -10 volts when the radiometer views cold space. The demodulator output waveforms from TP_1 and TP_2 then appear similar to that shown in Figure 5. Conversely, the positive clock input signal represents the demodulator input when the radiometer views an extremely hot target. Compared to the negative clock input, the positive clock input reverses the phase of the TP_1 and TP_2 waveforms. Also, the demodulator output following its integrator stage becomes a smooth positive or negative voltage level depending on the polarity of the input clock signal.

Lastly, it was later discovered that the small jump in the TP_2 output shown in Figure 5 results from an un-balance of the difference amplifier stage gain. Although these jumps get smoothed out by the integrator stage, they can also be reduced using better gain equalization, *i.e.*, reducing the 100 K resistance between pin 2 and TP_1 in Figure A5-1 (R_1 in Figure A5-3) to about 90 K. Consequently, for an improved demodulator, one would replace resistor R_1 with an 80 K resistor in series with a small variable resistor of 20 K that is set for optimal performance.

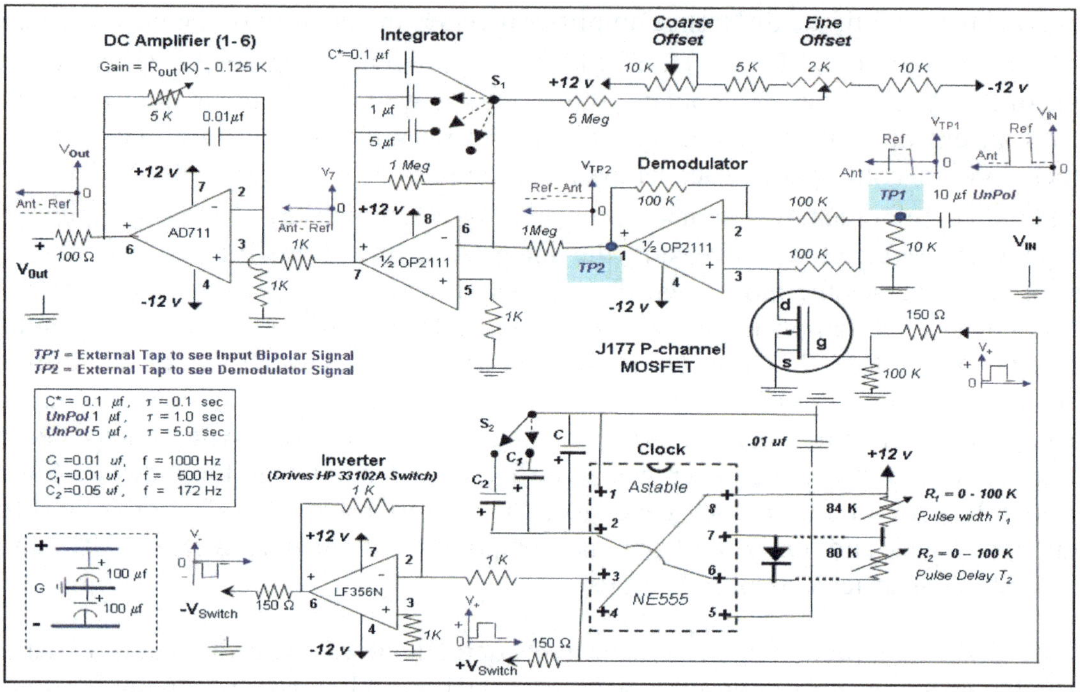

Figure A5 -1. Synchronous demodulator for 12 GHz radiometer. Shown is the circuit and waveforms at each stage. Starting from the input on the top right, it uses op amps connected as a difference amplifier, integrator and DC amplifier. It also contains a NE555 clock generator and J177 MOSFET whose gate (g) is driven by the clock to switch the difference amplifier gain to +/-1. This operation is derived in Figure A5-3. Finally, the bottom left op amp is connected as an inverter to provide a negative clock signal to drive the HP 33102A pin diode switch shown in Figure A4. The 4 GHz and 20 GHz radiometer circuits do not have the inverter since it uses a General Microwave M862B pin diode switch.

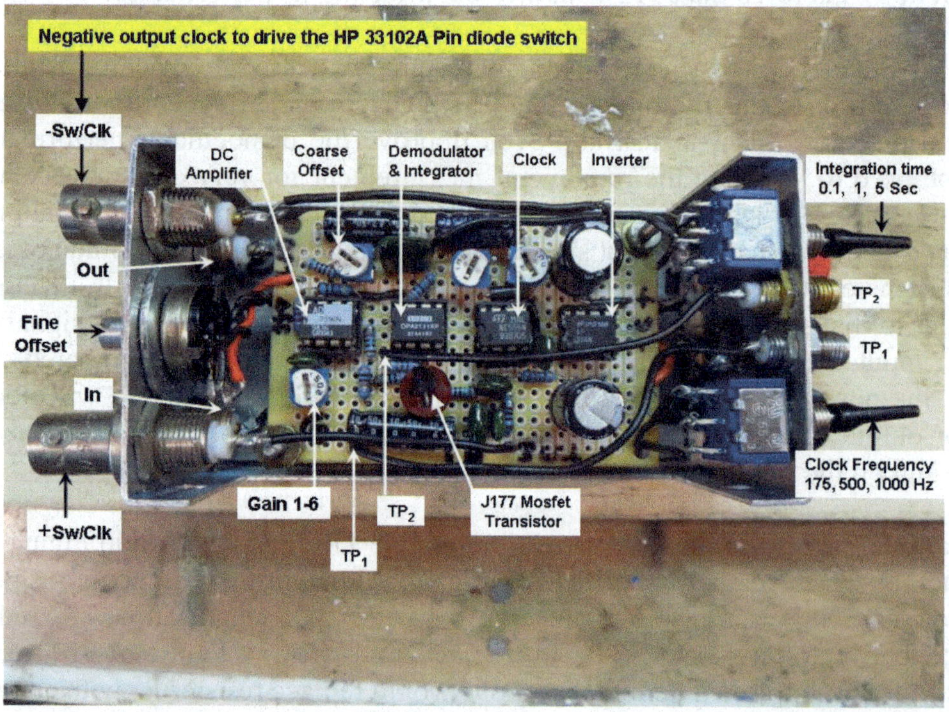

Figure A5-2. Synchronous demodulator used in 12 GHz radiometer. Figure A5-1 shows the schematic diagram. The case is opened to view the components along with input and output connectors. The circuit board contains the variable resistors used to set the DC amplifier gain and coarse offset while the fine offset is outside the case. The outside switches on the right set the integration time and clock.

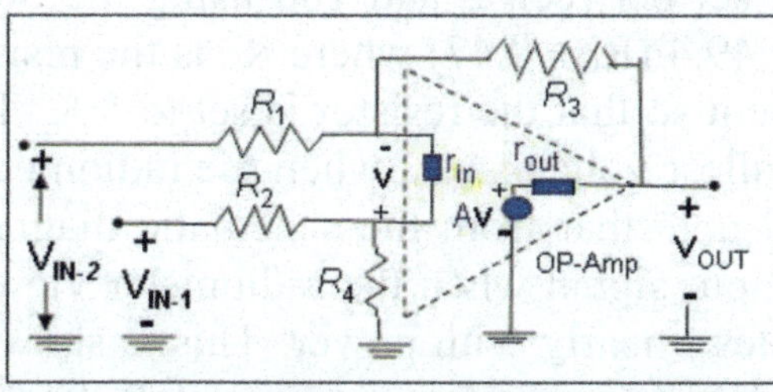

General configuration for a Difference Amplifier.

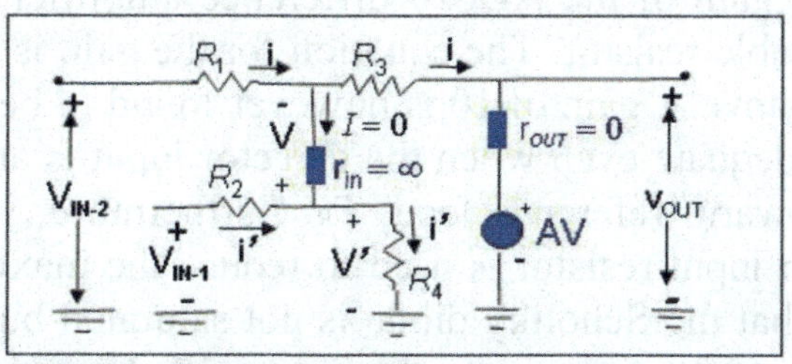

Ideal Op-Amp representation ($r_{in} = \infty$, $r_{OUT} = 0$).

For an ideal Op - Amp, $r_{in} = \infty$ **so** $I = 0$, **and** $r_{out} = 0$ **so** $V_{OUT} = AV$.

Therefore, $V = \frac{V_{OUT}}{A} = (iR_1 - i'R_2) + (V_{IN-1} - V_{IN-2})$ **with** $i = \frac{V_{IN-2} - V_{OUT}}{R_1 + R_3}$ **and** $i' = \frac{V_{IN-1}}{R_2 + R_4}$

$$V_{OUT} = \frac{V_{IN-1}\frac{R_4}{R_2+R_4} - V_{IN-2}\frac{R_3}{R_1+R_3}}{\frac{1}{A} + \frac{R_1}{R_1+R_3}} \quad \textbf{so for } A >> \frac{R_3}{R_1}, \quad V_{OUT} \cong V_{IN-1}\frac{R_4}{R_2+R_4}\left[1+\frac{R_3}{R_1}\right] - V_{IN-2}\frac{R_3}{R_1}$$

Synchronous Demodulator $V_{IN-1} = V_{IN-2} = V_{IN}$ **and** $R_1 = R_3$ **so that** $\frac{V_{OUT}}{V_{IN}} = \frac{R_4 - R_2}{R_4 + R_2}$ **where** $\frac{V_{OUT}}{V_{IN}} = \begin{cases} +1 \text{ for } R_4 >> R_2 \\ -1 \text{ for } R_4 << R_2 \end{cases}$

Non - Inverted Amplifier $V_{IN-2} = 0$ **and** $R_4 >> R_2$ **so that** $\frac{V_{OUT}}{V_{IN-1}} = \left[1 + \frac{R_3}{R_1}\right]$

Inverted Amplifier $V_{IN-1} = 0$ **so that** $\frac{V_{OUT}}{V_{IN-2}} = -\frac{R_3}{R_1}$

Figure A5-3. Analysis of the synchronous demodulator front end stage which transforms the bimodal input ($V_{IN-1}=V_{IN-2}=V_{IN}$) at TP_1 to the unimodal output (V_{OUT}) at TP_2 in Figure A5-1. Note that R_4 in the above circuit denotes the resistance of the J177 MOSFET switching transistor. Due to the clock signal, the MOSFET resistance becomes small when the gate voltage is near zero and large when the voltage is more than 2 volts. Also derived above is the transfer function of the non-inverted and inverted amplifiers used in the synchronous demodulator of Figure A5-1 as well as in the AC amplifier shown in Figure A7.

A6. Temperature Compensated Detector

Figure A6 shows the temperature compensated detector used in the 4 GHz radiometer. The circuit uses a matched pair of Schottky diodes (HSMS - 282P) connected to a well balanced difference amplifier in the form of an instrumentation amplifier (AD620) as shown in Figure A6. One of the diode sections is used as reference while the other is connected to the input, which is the LNB *IF* output signal. The difference output then approximately cancels the temperature effect since both diodes operate at nearly the same temperature. To assure that the detector's output current varies approximately as the input voltage squared, or power, a very small current bias of 25 μa is applied. This stems from the fact that the diode response is greater than square law at low signal levels and is closer to a linear voltage detector and rectifies at high levels. Only between the noise level and around -20 dBm does a Schottky diode respond linearly to power, as shown in Figure 15. This is the square law region where the detector and radiometer responds linear to power or brightness temperature.

The gain of the AD620 difference amplifier can be set between 2 and 100 using the 50 K ohm variable resistor. The equation for the gain is $G_d = 1 + 49.4/(R_G + 0.47)$ where R_G is the resistance in kilohms. A gain of 10 is however found to be sufficient so that the resistor is set to 5 K. This gain is adequate even when the detector input is at its smallest voltage, *i.e.*, when the radiometer views the warm reference load, T_R. Furthermore, you will note that from the schematic diagram, a 47 ohm input resistor is used to reduce the maximum input signal when the radiometer views space, so that the Schottky diode is not saturated but operates linearly with power. This is shown by the detector measurements in Figure 19. Note that the detector operates approximately linearly with input power, or voltage squared, with errors less than +/- 0.2 dBm for power input between -30 dBm to -12 dBm. The 12 GHz detector is similar to that shown in Figure A5 except that it uses an AD711 operational amplifier with a fixed gain of 10 rather than the instrumentation amplifier. It also uses an HSMS 2825 matched pair Schottky diode rather than the HSMS 282P. The connections between the Schottky diode and difference amplifier inputs should be kept as short as possible (< 1 cm) to minimize radiation loss and coupling effects. However, even when minimizing the connections, Figure 29 shows a narrower frequency response of the 20 GHz radiometer detector beyond that of the individual element when the diode is connected using bulk circuit components.

As an alternative, many chip manufacturers offer RF detectors that operate over an 80 dB dynamic range of input power and 100 C range of temperature (*e.g.*, ADL5513 by Analog Devices). To obtain this very large dynamic range the units are configured as a logarithmic detector by cascading a number of detector elements. Therefore, for use in radiometers the detector output can only be used over a small range of brightness temperature or alternatively, the logarithmic output must be transformed to a linear power response. While not as common, I did find a square law detector by Linear Technology, the LTC5509 chip which is specified to operate between 0.3 to 3 GHz with input power between -30 and 6 dBm. It uses a dual Schottky diode that is temperature compensated using a circuit similar to that of Figure A6 but only has a buffer amplifier with a gain of two. I measured the detectors sensitivity to be 0.56 mv/μw at 1.4 GHz. This is nearly a factor of 10 less than the sensitivity given by equation (9b) which uses the circuit described above. Therefore, an additional amplifier with a gain of 10 is needed after the detector to increase the output of my 4 GHz radiometer when viewing space from about -1 volt to -10 volts. As such, the LTC5509 detector plus amplifier performs similar to that in Figure A6, which also uses an amplifier whose voltage gain is set to 10.

I should also mention that commercially produced wideband (0.1 to 3.2 GHz) square law detectors are sold on eBay for under $10 which uses the voltage doubler circuit shown in Figure 15. The detector is shown in Figure 16 and is wideband by using the optimally designed layout of components on the printed circuit board rather than the homebuilt unit shown in Figure 15. Upon testing the detector its sensitivity was measured to be about 3 mv/μw. The sensitivity is found to vary with temperature with a variation of about 0.1 mv/^{0}F. This variation is increased by the radiometer amplifiers so it is greater than a temperature compensated detector whose radiometer output was shown to be 10 mv/^{0}F in Figure 20. Therefore, for stable temperatures this eBay detector can serve as a viable alternative when requiring a wideband detector. It was used in constructing the dual frequency (21.2, 22.2 GHz) radiometer described in Appendix A14 whose detectors measure the *IF* output frequencies of the Norsat 9000D LNB at 0.95 GHz and 1.95 GHz.

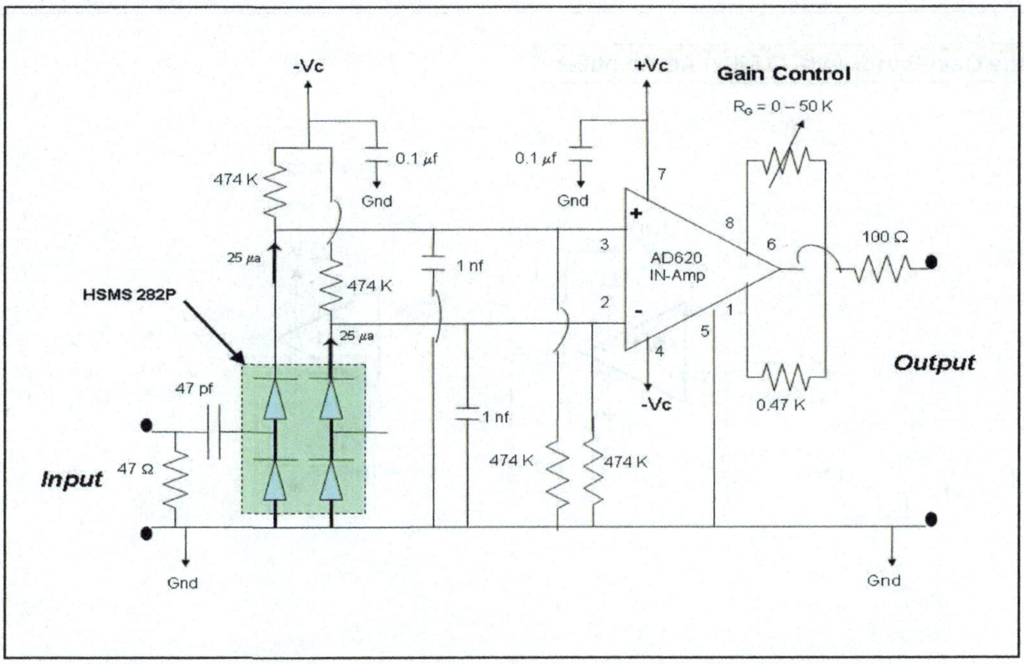

Figure A6. Temperature compensated square law detector for the 4 GHz radiometer uses an HSMS 282P balanced Schottky diode and an AD620 instrumentation amplifier powered using a supply voltage, V_C, of +/-12 volts. The amplifier gain can be varied from 2 to 100 by setting the resistor R_G according to the equation $G_d = 1 + 49.4/(R_G + 0.47)$. However, a gain of 10 is found to be sufficient. Also, to optimize the range, the diodes are forward biased (25μa).

A7. AC Amplifier

The AC amplifier in Figures 3 and 4 was designed using two AD711 operational amplifiers or op-amps connected as shown in the Figure A7. These AD711 op-amps were chosen because of their very small effect due to temperature drift, low noise and large open loop gain bandwidth product of 3 MHz. Each amplifier stage in Figure A7 is configured as non-inverted amplifiers whose closed loop gain is derived in Figure A5-3. As such, the gain of each stage is $1 + R_f / R_l$ where R_f is the feedback resistor between pins 2 and 6 and R_l is the input resistor between pin 2 and ground. Also, the AD711 has an internal circuit to remove any internal DC offset. This offset adjustment is implemented for the second stage op-amp using the 10 K trim resistor shown in Figure A7. The first stage acts as a buffer with a gain of 2 since $R_f = R_l = 100$ K, while the second stage has an adjustable gain from 1 to 1000 using a 1 megohm variable feedback resistor, R_f, with $R_l = 1$ K. Therefore, the total amplifier gain can be set from 2 to 2000. This gain can be determined by measuring the DC output resistance of the amplifier and using the equation Gain = $2R_{OUT}$ with R_{OUT} in kilohms. The only change I would make in constructing the amplifier is to replace the 1 megohm variable resistor with a precision stepped attenuator of fixed resistors, since I found that the variable resistor can sometimes change its value slightly over long time periods.

Large Gain-Bandwidth (3 MHz) AC Amplifier

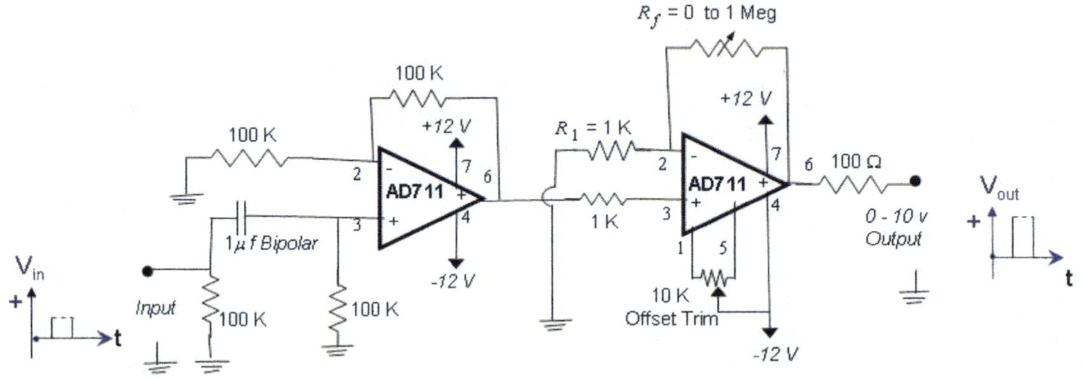

Figure A7. The AD711 operational amplifier (Op Amp) has a large gain bandwidth (GBW) of 3 MHz enabling high Dicke switching frequencies. As an example, with the second stage set to its maximum gain of 1000 the Op Amp bandwidth becomes 3 KHz. The AD711 amplifier also has an internal circuit to remove the Op Amp DC offset using an external 10 K variable resistor. The input stage gain is G1 = 1 + (100K /100K) = 2 while the output stage gain is 1 + (R_f / R_1) where R_f = 1 Meg variable resistor and R_1 = 1 K. As such, the amplifier gain can be varied from 2 to 2000 by setting R_1. Also since R_1 = 1 K, the amplifier gain can be written as 2 R_{Out} where R_{out} is the DC output resistance in kilohms. Therefore, the gain can be determined by measuring the DC output resistance in kilohms and multiplying it by 2.

A8. 4 GHz Radiometer Isolator

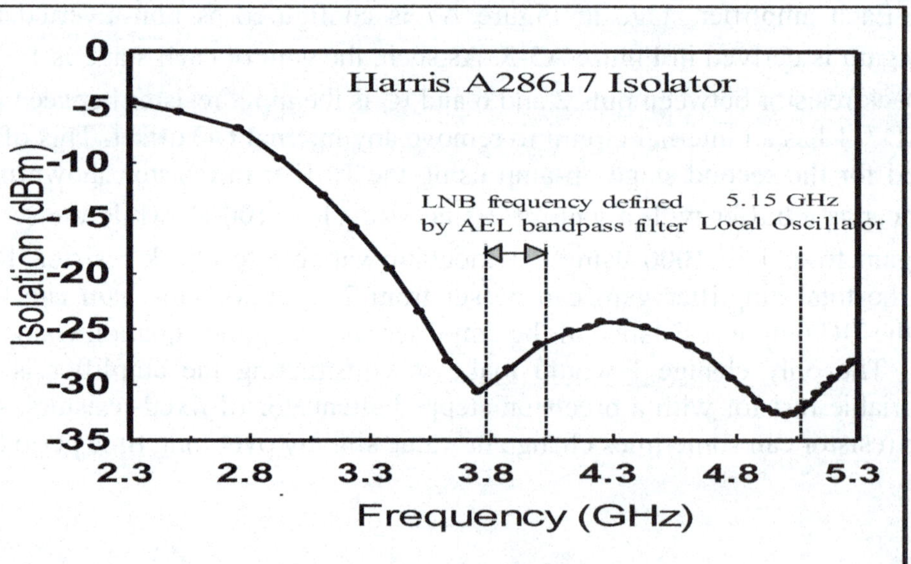

Figure A8. The isolator produces 30 dB of isolation at the C-Band LNB 5.15 GHz *LO*. It also provides similar isolation at the LNB frequencies of 3.76 to 4.01 GHz (defined by AEL filter).

A9. 4 GHz Radiometer *RFI* Filter

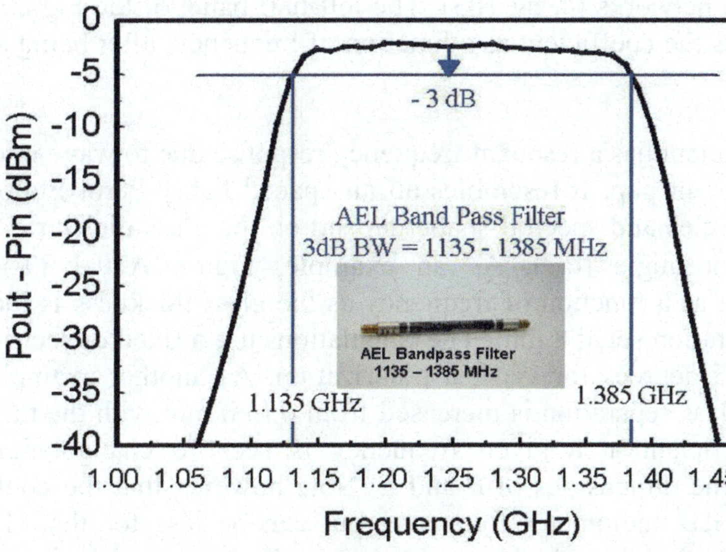

Figure A9. The bandpass filter is used in the 4 GHz radiometer to eliminate interference from WiFi, radar, aircraft altimeters. The filter has a 3 dB bandpass between 1.135 to 1.385 GHz (*i.e.* BW = 250 MHz) with a center frequency of 1.260 GHz and a 2 dB insertion loss in the passband. The filter bandpass corresponds to LNB input frequencies between 3.75 and 4.01 GHz.

A10. Glass Door Reflection and Transmission

Chapters 7 and 8 describe experiments whereby the radiometers view the sky and ground through my basement glass patio door. As discussed throughout these chapters, the glass reflectivity greatly affects the radiometer measurements. Of particular interest here is the slope of 0.52 shown in Figure 34 between the 22.2 and 20.5 GHz measurements on December 29, 2019. The brightness temperature seen by the skyward viewing radiometers at these frequencies is given by equation (15a), *i.e.*,

$$T_b(\nu) = \Im_g(\nu)[1-\tau^{\mathrm{Sec}\,\theta}]T_M + R_g(\nu)\,T\ ,$$

where R_g is the glass door reflection coefficient and $\Im_g = 1 - R_g$ is the transmission coefficient. Since the atmospheric transmittance is similar at the above mentioned frequencies, the slope is mainly due to the transmission coefficient. As shown in this Appendix, the transmission coefficient is a highly variable function of frequency due to wave interference by the glass interfaces.

Model simulations are used to analyze the effect glass has on the radiometer measurements. Unlike the simple model used in Chapter 9 to represent sand, a more extensive model is needed which includes multiple reflections within the glass and its glass-air interface[7]. Such a model is given in Figure A10-1 (Top-Left). As illustrated in the Figure, the door contains two glass panes having dielectric constant ε with thickness d and separation distance L. At microwave frequencies the molecules comprising glass can not move freely so the modes of excitation mainly begin in the

[7] Such coherent interference effects occur for smooth stratified surfaces, which are rarely seen in nature.

infrared region. Glass is therefore considered lossless with a dielectric constant having no dissipation that only reflects the incident microwave radiation with no thermal emission. The reflection and transmission coefficient is derived using the ABCD transmission matrix formulation of cascaded microwave networks [4, pg 183]. The left-half panel of the Figure outlines the model while the right half plots the coefficient as a function of frequency, after being averaged over a 200 MHz bandwidth.

The transmission coefficient has a resonant frequency response due to wave interference within the two glass sheets and its air gap. It resembles an air-spaced Fabry-Perot Etalon or interferometer whose resonant frequencies and spectral shape depend on the glass dielectric constant, thickness, separation and incident angle (θ_{Inc}). As an example, Figure A10-1 (Top-Right) plots the transmission coefficient as a function of frequency as the glass thickness is increased from 4 to 5 mm with the glass separation set at 8 mm. The calculations use a fixed dielectric constant of 6 with an incident angle of 25 degrees for vertical polarization. As another example, the bottom right shows the plot as the glass separation is increased from 6 to 8 mm with the thickness set at 4 mm. The transmission coefficient at a given frequency is seen to change significantly even for millimeter changes in the dimensions of *d* and *L*. Note however that the coefficient is generally very high at the 3.7 GHz radiometer frequency but can be less for the 11.9 GHz radiometer frequency depending on the glass thickness. However, the largest variation occurs at the higher radiometer frequencies of 20.5 and 22.2 GHz. For example, the top right shows the 22.2 GHz transmission smaller than at 20.5 GHz for a glass separation of 8 mm, with a thickness of 4.0 and 4.5 mm. This is consistent with the measurements in Figure 34. However, the bottom right shows the opposite frequency response for a glass separation less than 8 mm, which is inconsistent with the measurements. Unfortunately, the exact glass door dimensions were unavailable to compare the calculations with the slope of 0.52 obtained from the radiometer measurements.

To further examine these features, the model is simplified by considering a single glass plate with no air gap, *i.e.*, $L = 0$. The transmission coefficient then becomes

$$\Im = \frac{1+\mathrm{Tan}^2 k'd'}{1+\rho^2\,\mathrm{Tan}^2 k'd'} \quad \text{where} \quad 2\rho = \frac{Z_1}{Z_L} + \frac{Z_L}{Z_1}$$

$$\text{with} \quad \left.\frac{Z_L}{Z_1}\right]_{V-POL} = \sqrt{\frac{\varepsilon}{\varepsilon_0}\frac{\mathrm{Cos}\,\theta_{Inc}}{\mathrm{Cos}\,\theta'}} \quad \text{and} \quad \left.\frac{Z_L}{Z_1}\right]_{H-POL} = \sqrt{\frac{\varepsilon}{\varepsilon_0}\frac{\mathrm{Cos}\,\theta'}{\mathrm{Cos}\,\theta_{Inc}}} \,.$$

The transmission coefficient contains parameters ρ and $k'd' = (2\pi d/\lambda)\sqrt{\varepsilon/\varepsilon_0}\ \mathrm{Cos}\ \theta'$ where λ is the free space wavelength and *d* is the glass thickness. The ρ parameter depends on the polarization (V, H), dielectric constant ε and cosine of transmitted angle in the glass θ' which is related to the incident angle θ_{Inc} by Snell's law $\mathrm{Sin}\ \theta' = \sqrt{\varepsilon_0/\varepsilon}\ \mathrm{Sin}\ \theta_{Inc}$. Figure A10-2 plots $\Im$ for $\theta_{Inc} = 0^0$ as a function of $k'd'$ for $\varepsilon = 3$ and $\varepsilon = 6$, which represents sand and glass, respectively.

Note that $\Im$ becomes unity when $k'd' = \eta\pi$ or $d = \frac{\eta}{2}\lambda\left(\sqrt{\varepsilon_0/\varepsilon}\ \mathrm{Sec}\ \theta'\right)$ where η = 0, 1, 2, *etc.* This high transmission and low reflection is particular useful in designing radar domes (radomes). Conversely, $\Im$ is minimum when $k'd' = (2\eta+1)\pi/2$ with a value that decreases for increasing dielectric constant and incident angle. However, Figure A10-1 shows that wave interference within the air gap significantly alters the spectrum from that of a single slab. The pronounced effect of the air gap is further shown in Figure A10-3, which plots the transmission coefficient as the glass separation is increased from 0 to 3 mm which is more than 10 times less than the free space microwave wavelength. As an experiment, it would be interesting to observe these features at the different frequencies using a setup similar to that described in Chapter 9 but using glass sheets of different separation and thickness rather than the desert sand.

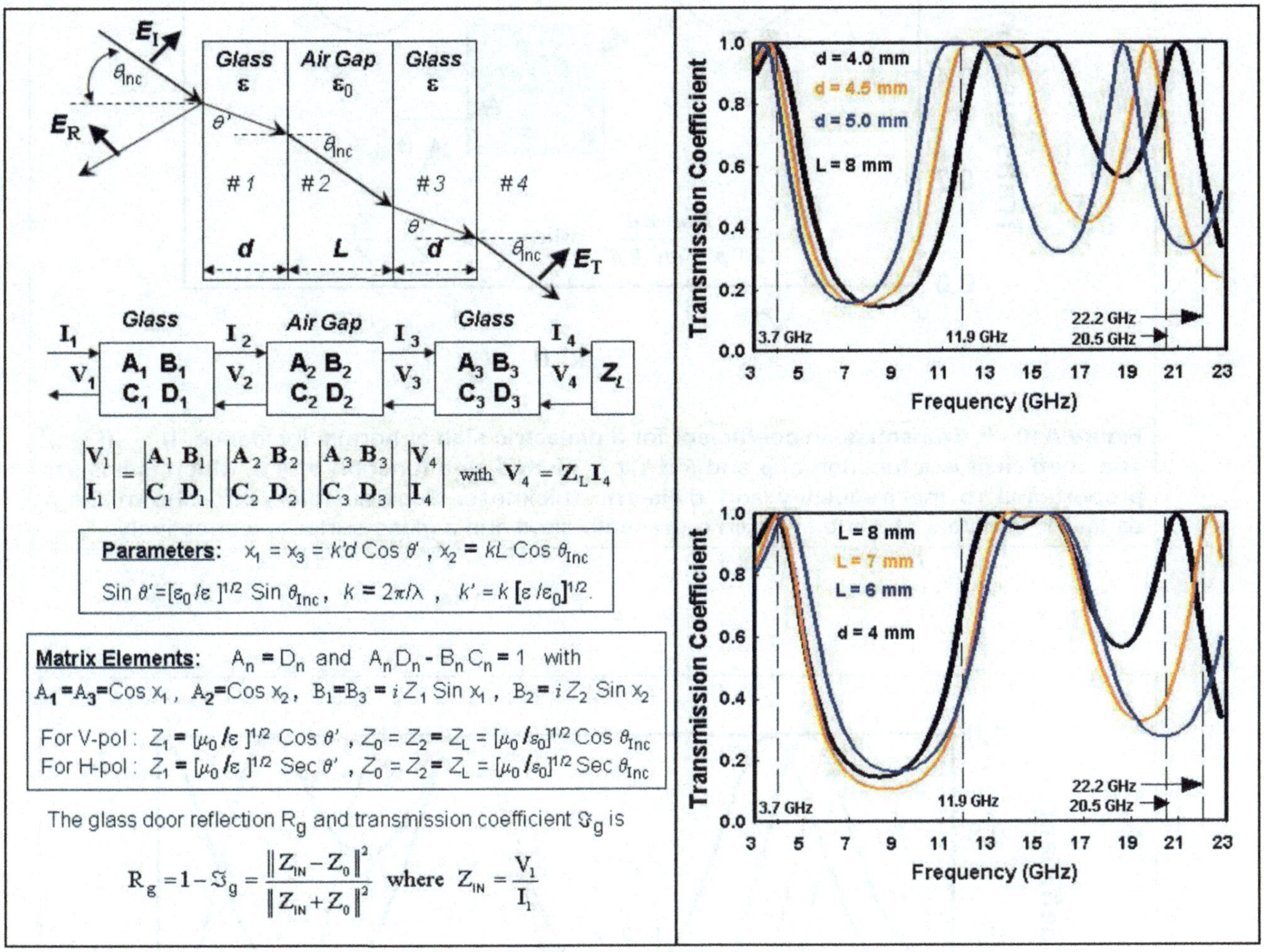

Figure A10-1. Dual polarized transmission coefficient $\Im_g$ of a double pane glass door derived using the ABCD transmission matrix formulation of cascaded microwave networks. The analysis is summarized on the Left panel and plotted on the Right panel as a function of frequency for vertical polarization with $\varepsilon = 6$, $\theta_{inc} = 25^0$ and BW = 200 MHz. Radiometer frequencies are indicated by the vertical dashed lines. The top-right plot shows the calculations for different glass thickness *d* with a fixed separation *L* of 8 mm. Similarly, the bottom right shows the plot for different separation distance *L* with a fixed glass thickness *d* of 4 mm.

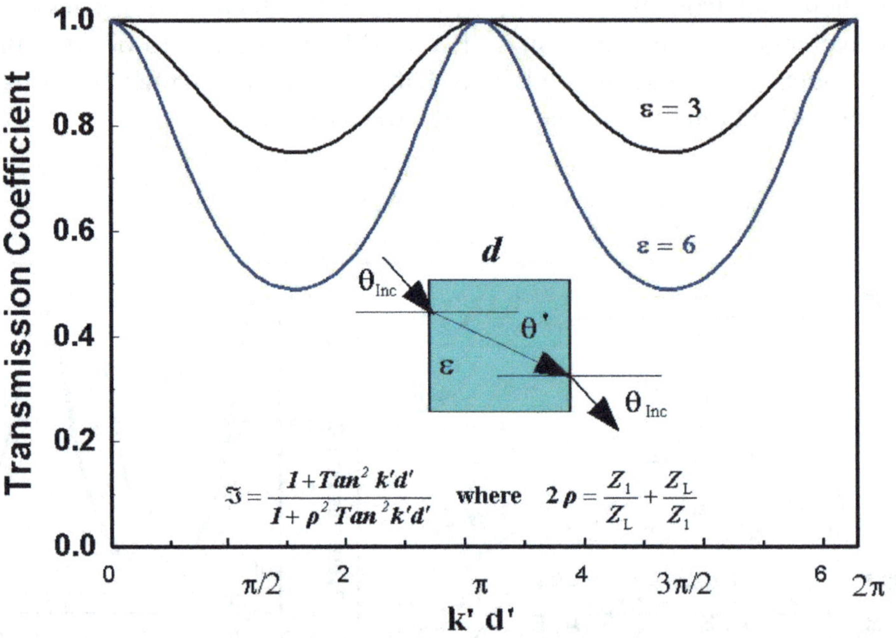

Figure A10 - 2. Transmission coefficient for a dielectric slab at normal Incidence, $\theta_{Inc} = \theta' = 0^0$. The coefficient is a function of ρ and *k'd '*. It is plotted as a function of *k*'d' which is linearly proportional to the frequency and dielectric thickness. Separate plots are shown for a dielectric constant of 3 and 6, which represents sand and a glass surface, respectively.

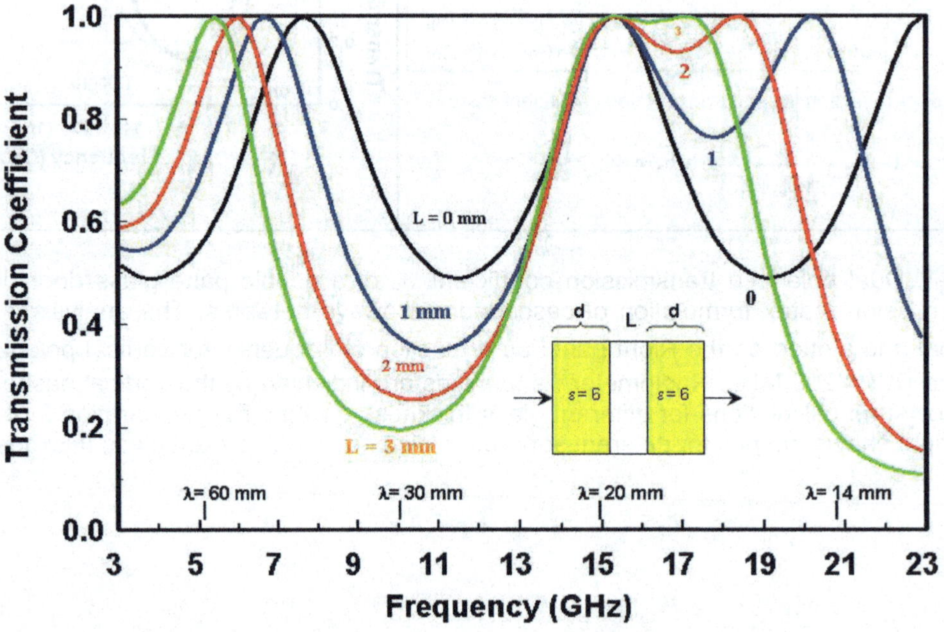

Figure A10 - 3. As in Figure A10 - 1, the transmission coefficient is calculated for a glass thickness, *d*, of 4 mm with a dielectric constant of 6. However, in this case the separation distance, *L*, is varied from 0 to 3 mm with the incident angle set to 0^0

A11. Glass Door Insertion Loss Measurements

For convenience the surface and atmospheric features measured by the radiometers are viewed through a glass patio door. Chapter 7 and 8 describes the measurements and the effect of glass reflection. Appendix A10 models the glass reflection coefficient while this Appendix uses insertion loss measurements to measure it at 4 and 12 GHz. To obtain the coefficient, the radiometers measure the scene brightness temperature with the glass patio door opened and closed. As indicated below in Figure A11, the brightness temperatures are then used to determine the reflection coefficient.

Figure A11 shows the equations used to obtain the reflection measurements. Since the glass door absorption is very small, the brightness temperature is accurately expressed as $T_b = (1 - R_g)\,T_0 + R_g T$. A similar result is also given by equation (15a) in Section 7.3. The reflection coefficient R_g then becomes

$$R_g = \frac{T_b - T_0}{T - T_0}$$

where T_0 is the brightness temperature measured with the glass door opened (*i.e.*, $R_g = 0$) while T_b is the measurement with the door closed. Also contained in the above equation is the in-house radiation at temperature *T*, which is also reflected by the glass door. Using this equation and radiometer measurements, Figure A11 calculates the reflection coefficients to be 0.20 at 4 GHz and 0.64 at 12 GHz.

Section 7.1 discusses these measurements while Appendix A10 simulates the reflection and transmission coefficients as a function of frequency and glass door parameters. In addition to the reflection coefficient, Figure A11 shows the open door brightness temperature, T_0, to be 77 K at 4 GHz and 105 K at 12 GHz. These measurements are much larger than the 2.7 *K* cosmic background due to the nearly horizontal viewing angle used to make the insertion loss measurements. Note that this is different then when performing calibration measurements in Sections 4.3 and 4.4, where the radiometer antenna is directed skyward to avoid viewing any surrounding earth radiation. However, the effect of this terrestrial radiation is not a problem since it results in the same scene radiation, T_0, when opening and closing the door.

Insertion Loss Measurements

Calibration Equations: $T_b]_{4GHz} = 289.5 + 32.1\ V$, $T_b]_{12GHz} = 291.8 + 31.5\ V$

Voltages and calibrated brightness temperatures

	4 GHz		12 GHz		12 GHz minus 4 GHz	
	V	T_b	V	T_b	V	T_b
Door Closed	- 5.25 V	120 K	- 2.13 V	224 K	3.12 V	104 K
Door Opened	- 6.62 V	77 K	- 5.93 V	105 K	0.69 V	28 K

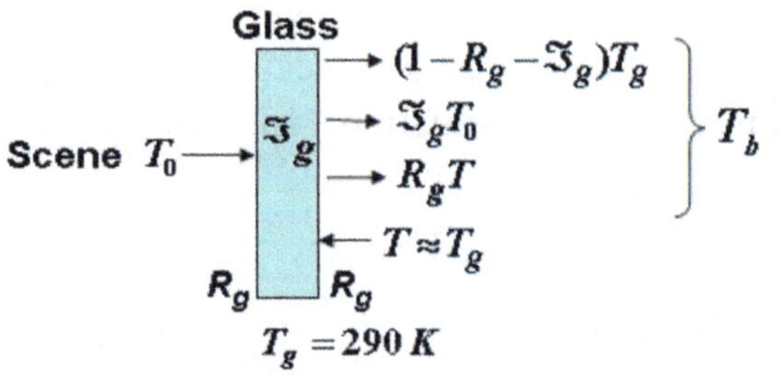

$$T_b = \Im_g T_0 + (1 - R_g - \Im_g)T_g + R_g T$$

With no absorption, $\Im_g = 1 - R_g$

so that, $T_b = (1 - R_g)T_0 + R_g T$

Therefore, $R_g = \dfrac{T_b - T_0}{T - T_0}$

Parameters; $T_0 = T_b]_{R_g=0}$ = Scene Radiation, T = House Temperature, R_g = Glass Reflection Coefficient

4 GHz: $T - T_0 = 213$ K and $T_b - T_0 = 43$ K so $R_g = 43/213 = 0.20$
12 GHz: $T - T_0 = 185$ K and $T_b - T_0 = 118$ K so $R_g = 118/185 = 0.64$

At 4 GHz the glass reflection coefficient $R_g = 0.20$
At 12 GHz the glass reflection coefficient $R_g = 0.64$

Figure A11. Insertion loss of a glass door is obtained from radiometer measurements with the door opened and closed. Neglecting absorption, the glass reflection coefficient is calculated to be 0.20 at 4 GHz and 0.64 at 12 GHz.

A12. Severe Storm Measurements

As with the rain event on June 12, 2014, discussed in Section 7.2, measurements were taken on February 24, 2016 for another storm using the smallest integration time of 0.1 seconds. These measurements were also obtained with the radiometer viewing the event through my basement glass patio door. For reference, the local radar and enhanced satellite images were observed on my laptop computer and pictures were taken. The radiometer data was also displayed on my computer in real time using software provided with the analog to digital converter mentioned in Section 4.1. A composite picture of the radiometer measurement together with the weather radar and satellite images is shown in Figure A12. Of particular significance is the large increase in the 12 GHz radiometer voltage during the most intense rain period. This is accompanied by a relatively large increase in the 4 GHz radiometer voltage as well. Note that the scale used to display the 4 GHz radiometer measurements is at its minimum dynamic range of 0.45 volts while the scale of the 12 GHz measurements is expanded to a 4.8 volt range to display the full extent of its observations.

Figure A12 shows the largest radiometer increase at 12 GHz is 4.8 volts, while the corresponding voltage increase at 4 GHz is 0.375 volts. This voltage ratio of 12.8 for the two frequency measurements is nearly the same value found for the rain event discussed in Chapter 7. Other rain events showed the voltage ratio to vary between 10 and 15. Also, when the 12 GHz output was increased by 4.8 volts its output reached 0.3 volts. For even heavier rain events the 12 GHz measurement has been observed to saturate at nearly zero volts while the 4 GHz showed no sign of saturation. Also note the abrupt voltage increases in the 4 GHz measurements. The spikes are seen sporadically throughout the measurement period although they are most pronounced during the most intense rain period when I heard and observed lightning. However, the 12 GHz radiometer measurements show no change associated with lightning discharge.

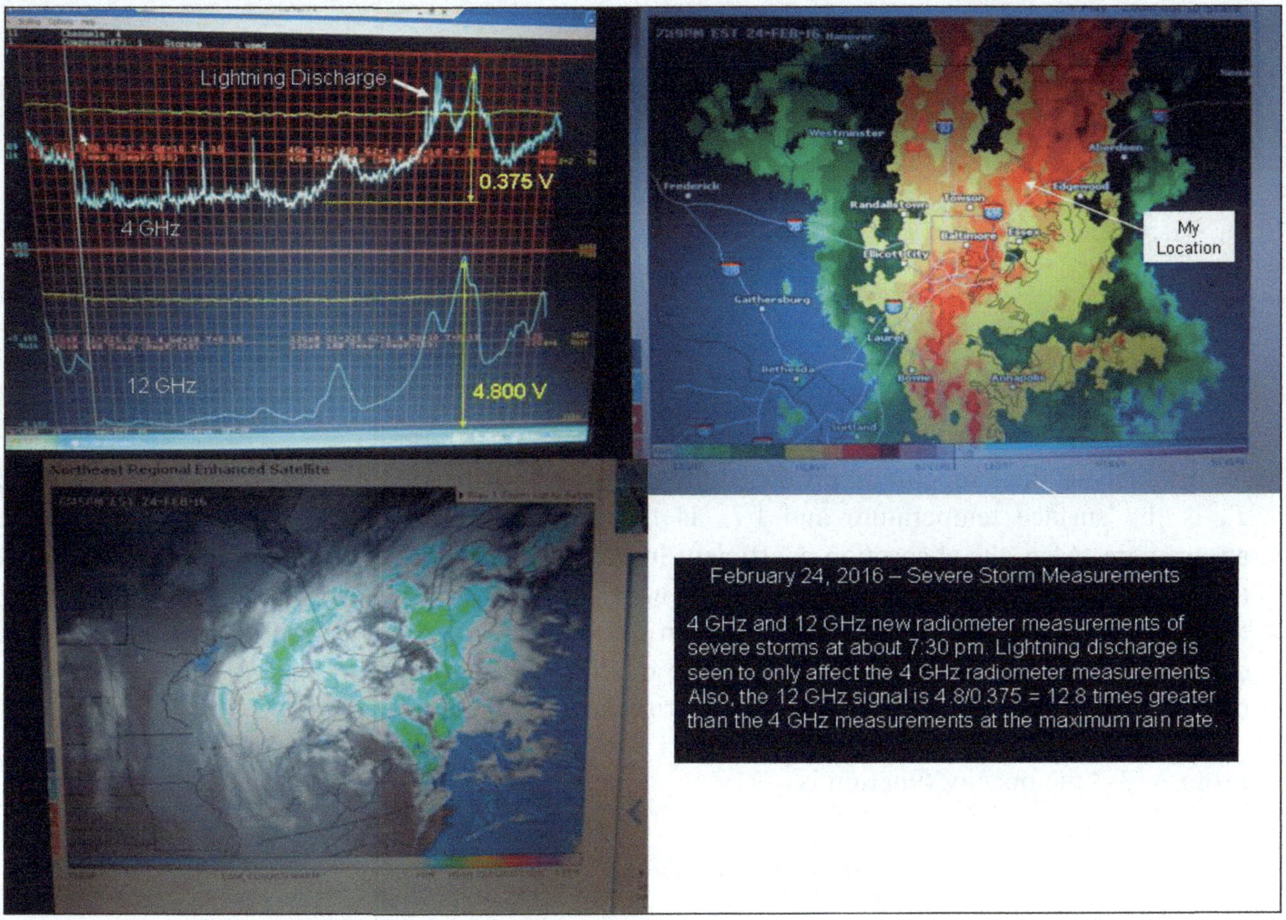

Figure A12. Radiometer measurements of a severe storm on Feb. 24, 2016. Shown are the 12 and 4 GHz measurements taken through a glass door as the storm passed through my area. Also shown is the online weather radar and satellite images at nearly the same time of my measurements, which was at 7:30 pm. For the maximum rain event, the ratio of the 12 to 4 GHz measurements is 12.8, which is nearly the same found in Figure 26 for a different rain system.

A13. Tipping Curve Analysis

This Appendix analyzes the tipping curve procedure used in Section 8.4 to measure the atmospheric opacity and calibrate the 20 GHz radiometer. The procedure is based on the radiation transfer equation for the downwelling radiation which was first introduced in Section 4.3 and given by equation (5), *i.e.*,

$$T_{SKY}(\theta) = \tau^{Sec\theta}\, T_{CB} + \left(1 - \tau^{Sec\theta}\right) T_M \quad . \qquad \text{(A13-1)}$$

The equation assumes the radiometer antenna has an unobstructed view of space. It also assumes a horizontally stratified atmosphere where τ is the atmospheric transmittance whose exponent Sec θ

accounts for the larger path length at zenith angle θ. The equation also contains the mean radiating temperature T_M and cosmic background radiation T_{CB} of 2.7 K. We begin the analysis by first studying the mean temperature and its angular variation. This term was chosen since it generally results in the largest tipping curve errors at the largest scan angles.

Mean Radiating Temperature

The mean radiating temperature, T_M, in A13-1 is given by equation (6a) in Section 4.3 where T(z) is the vertical temperature profile and τ(z) is the transmittance function, *i.e.*,

$$T_M = \frac{\int_0^\infty T(z)\frac{d\tau(z)^{Sec\theta}}{dz}dz}{\int_0^\infty \frac{d\tau(z)^{Sec\theta}}{dz}dz} \qquad \text{(A13-2)}$$

where

$$\tau(z) = e^{-\alpha(z)} = e^{-\int_0^z \gamma(z')\,dz'} . \qquad \text{(A13-3)}$$

In this analysis the temperature profile in the troposphere is represented as $T(z) = T_S(1 - \Gamma z)$ where T_S is the surface temperature and ΓT_S is the lapse rate which is 6.5 K/km for a standard atmosphere. Also, the absorption coefficient in the transmittance function τ(z) is approximated by the empirical equation $\gamma(z) = \gamma_0\, Exp(-z/H)$ where γ_0 is the surface absorption coefficient and H is the scale height. Both parameters depend on frequency, where H is about 2 km for frequencies dominated by water vapor absorption. These model functions for T(z) and γ(z) are used below to calculate the mean temperature T_M and sky brightness temperature T_{SKY}.

From A13-3 the opacity function is

$$\alpha(z) = \int_0^z \gamma(z)\,dz = \gamma_0 H(1-e^{-z/H}) \qquad \text{(A13-4)}$$

where $\alpha(\infty) = \gamma_0 H$ is the total absorption in a vertical column. Also, the weighting function in A13-2 becomes

$$-\frac{d\,\tau(z)^{Sec\theta}}{dz} = \left[\gamma_0\, Sec\theta\; e^{-z/H}\right] e^{-(1-e^{-z/H})\gamma_0 H\, Sec\theta} , \qquad \text{(A13-5)}$$

which contains the absorption coefficient in brackets multiplied by the transmittance function. The weighting function peaks at the surface and decreases exponentially with altitude due mainly to the absorption coefficient term.

When normalized to the peak value occurring at the surface the weighting function becomes

$$\frac{d\,\tau(z)^{Sec\theta}}{dz} = e^{-z/H}\, e^{-(1-e^{-z/H})\gamma_0 H\, Sec\theta} . \qquad \text{(A13-6)}$$

Figure A13-1 (Top-Right) shows the normalized weighting function plotted as a function of z/H for zenith angles of 0^0 and 70^0. The plot is obtained for $\gamma_0 H = 0.22$ or $\tau(\infty) = e^{-\gamma_0 H} = 0.8$. It shows the weighting function decreasing exponentially with height. This feature results from the leading term $e^{-z/H}$ in A13-6, which is the normalized absorption coefficient, and is the main weighting function contribution since $\gamma_0 H$ Secθ < 1. This simplification was first mentioned by Dr. Robert Dicke [1] *but never explained.* It is for this reason I performed the above analysis where Figure A13-1 compares the approximation with the exact weighting function for zenith angles of $0^0$and 70^0. Note that the leading term $e^{-z/H}$ in A13-6 is only slightly broader than the exact weighting function for θ =0^0. However, a larger difference is seen for θ = 70^0, where the weighting function becomes even narrower due to the second exponential term in A13-6 which is the transmittance function and contains $\gamma_0 H$ Secθ. Therefore, the exact weighting function is used next to study the mean temperature and sky brightness temperature variation with zenith angle.

Substituting A13-5 into A13-2 with $T(z) = T_S(1 - \Gamma z)$, the ratio of mean temperature to surface temperature, Γ_M, becomes

$$\frac{T_M}{T_S} = 1 - \Gamma H \frac{\tau^{\mathrm{Sec}\,\theta}}{1-\tau^{\mathrm{Sec}\,\theta}} \gamma_0 H \,\mathrm{Sec}\theta \int_0^\infty u\, e^{-u}\, e^{\gamma_0 H\, \mathrm{Sec}\,\theta\, e^{-u}}\, du \,. \qquad \text{(A13-7)}$$

where $u = z/H$ and $\tau = \tau(\infty)$.

The integral is obtained using a Taylor series expansion of the 2nd exponential term, *i.e.*,

$$e^{\gamma_0 H\, \mathrm{Sec}\,\theta\, e^{-u}} = \sum_{n=0}^{\infty} \frac{(\gamma_0 H\, \mathrm{Sec}\theta)^n}{n!} e^{-n u} \qquad \text{(A13-8)}$$

so that

$$\int_0^\infty u\, e^{-u}\, e^{\gamma_0 H\, \mathrm{Sec}\,\theta\, e^{-u}}\, du = \sum_{n=0}^{\infty} \frac{(\gamma_0 H\, \mathrm{Sec}\theta)^n}{(n+1)^2\, n!} \,. \qquad \text{(A13-9)}$$

Substituting (A13-9) into A13-7 and using $\gamma_0 H\, \mathrm{Sec}\theta = -\ln \tau^{\mathrm{Sec}\theta}$ we obtain

$$\boxed{\frac{T_M}{T_S} = 1 - \Gamma H \frac{\tau^{\mathrm{Sec}\,\theta}}{1-\tau^{\mathrm{Sec}\,\theta}} \sum_{n=1}^{\infty} \frac{(-\ln \tau^{\mathrm{Sec}\theta})^n}{n^2\, n!}} \,. \qquad \text{(A13-10\}}$$

Equation A13-10 is a rapidly convergent series that generally requires no more than 4 terms and depends on the atmospheric transmittance τ along a vertical path between the radiometer and upper atmosphere. It also depends on the temperature lapse rate parameter Γ, absorption scale height *H*, and zenith angle θ. In general, the proportionality factor $\Gamma_M = T_M/T_S$ and sky brightness temperature increase as either ΓH or $\tau^{\mathrm{Sec}\theta}$ decrease. Using equation A13-10, the mean temperature is plotted in Figure A13-1 (Top-Left) as a function of zenith angle for T_S=300 K and ΓH = (6.5x2/300) = 0.04 with τ = 0.8 and τ = 0.9. The largest angular increase of T_M occurs for the smallest transmittance of 0.8. For this transmittance, T_M increases by 1.2 K as θ increases from 0^0 to 70^0. In contrast to the small increase in T_M, the bottom-left plot shows an 80 K increase in T_{SKY} between these angles due to the large change in emissivity $(1-\tau^{\mathrm{Sec}\theta})$ in equation A13-1. *As such, the following analysis considers T_M constant for zenith angles between* 0 and 70^0.

Measuring Atmospheric Transmittance

This section derives the equation for determining the atmospheric transmittance using angular scan measurements. The equation is based on the radiation transfer equation A13-1. As mentioned above, the Secθ exponent of τ in the equation accounts for the longer path length when viewing at zenith angle θ for a horizontally stratified atmosphere. All of the quantities in the radiation transfer equation (T_M, T_{CB}, τ, θ) are shown symbolically in Figure 12 as well as in Figure A13-2. The Figure also shows the geometry of a radiometer whose reflector redirects the downwelling radiation to its antenna. Figure 37 is a picture of the actual reflector used in the measurements. However, as discussed in Section 8.4, to reduce radiation leakage by the reflector at low elevation, the radiometer was scanned in elevation and azimuth by mounting it on a tripod.

We begin the analysis by considering the radiometer initially calibrated, for example, using the near-field variable target temperature procedure described in Section 4.1. Using equation A13-1, the atmospheric transmittance due to oxygen and water vapor can then be obtained from the sky brightness temperature at any zenith angle, viz.,.

$$\tau = \left[\frac{T_M - T_{SKY}(\theta)}{T_M - T_{CB}}\right]^{\text{Cos}\theta} \qquad \text{(A13-11a)}$$

so that

$$\alpha = -\ln\tau = -\text{Cos}\theta \ln\left[\frac{T_M - T_{SKY}(\theta)}{T_M - T_{CB}}\right]. \qquad \text{(A13-11b)}$$

Equation A13-11b can be used to obtain the opacity, α, and transmittance from a single sky measurement given the radiation temperatures T_M and T_{CB}. Furthermore, if we measure the brightness temperature at two viewing angles θ_1 and θ_2 we can eliminate the quantity T_{CB} -T_M by neglecting the angle dependence of T_M as described above. The opacity then becomes,

$$\alpha = \frac{1}{Sec\theta_2 - Sec\theta_1}\ln\left[\frac{T_M - T_{SKY}(\theta_1)}{T_M - T_{SKY}(\theta_2)}\right] \quad \text{where} \quad \tau = e^{-\alpha}. \qquad \text{(A13-12)}$$

Although any two zenith angles can be used, for simplicity we use $\theta_1 = 0^0$ and $\theta_2 = 60^0$ so that the opacity is

$$\alpha = \ln\left[\frac{T_M - T_{SKY}(0^0)}{T_M - T_{SKY}(60^0)}\right]. \qquad \text{(A13-13)}$$

Lastly, if we measure the sky brightness temperature at a third angle θ_3, then T_M becomes

$$\left[\frac{T_M - T_{SKY}(\theta_1)}{T_M - T_{SKY}(\theta_2)}\right]^{\eta} = \frac{T_M - T_{SKY}(\theta_1)}{T_M - T_{SKY}(\theta_3)} \quad \text{where} \quad \eta = \frac{Sec\theta_3 - Sec\theta_1}{Sec\theta_2 - Sec\theta_1}. \qquad \text{(A13-14)}$$

While A13-14 is a nonlinear equation for T_M, it can be linearized for specific angles. For example, if $\eta \equiv 2$, $Sec\theta_2 - Sec\theta_1 = Sec\theta_3 - Sec\theta_2$ so that the solution for T_M reduces to,

$$T_M = \frac{T_{SKY}(\theta_1)T_{SKY}(\theta_3) - T_{SKY}(\theta_2)^2}{T_{SKY}(\theta_1) + T_{SKY}(\theta_3) - 2T_{SKY}(\theta_2)} \qquad \text{(A13-15)}$$

whose three angles can be $\theta_{1,2,3} = 0^0, 30.0^0, 40.2^0$ or $\theta_{1,2,3} = 0^0, 45.0^0, 56.9^0$, etc.,

Furthermore, upon substituting A13-15 into A13-12 we obtain the simplified equation,

$$\alpha = \frac{1}{Sec\theta_2 - Sec\theta_1} \ln\left[\frac{T_{SKY}(\theta_1) - T_{SKY}(\theta_2)}{T_{SKY}(\theta_2) - T_{SKY}(\theta_3)}\right] \qquad \text{(A13-16)}$$

whose opacity is independent of T_M and T_{CB}. It only depends on the sky measurements at three viewing angles defined by $Sec\theta_2 - Sec\theta_1 = Sec\theta_3 - Sec\theta_2$, *i.e., equal spacing between* $Sec\theta$.

Having determined the opacity or transmittance, the cosmic background radiation temperature is obtained from upward viewing measurements using A13-1, *i.e.*,

$$T_{CB} = \tau^{-1}[T_{SKY}(0^0) - (1-\tau)T_M]. \qquad \text{(A13-17)}$$

Since T_{CB} = 2.7 *K*, then A13-17 can serve as a consistency check. Any increase in the calculated value of T_{CB} beyond 2.7 *K* is likely due to obstructions or unaccounted radiation affecting the sky viewing measurements at the different viewing angles. An example of such unaccounted radiation was described in Chapter 4 as due to the surrounding natural thermal emitted radiation scattered within the antennas FOV.

In summary, the above relationships utilize anywhere between one to three angular measurements to determine the atmospheric opacity. It is inherently assumed in deriving equations A13-12 to A13-16 that the opacity is independent of the viewing angle. This assumption is only possible for cloud free atmospheres where the opacity results from the nearly uniform absorption by oxygen and water vapor. However, once the radiometers are calibrated using cloud-free multiangle measurements, the opacity of clouds and rain can be obtained from single angle measurement using equation A13-11b.

Equations A13-11b through A13-16 provides analytic relationships to derive the opacity from angular measurements. An alternate graphical means of determining the opacity is obtained by differentiating A13-11b, so that

$$\boxed{\alpha = -\frac{d\ln[T_M - T_{SKY}(\theta)]}{d\,\mathrm{Sec}\theta}\,.} \qquad \text{(A13-18)}$$

Equation (A13-18) can be used to obtain the opacity by plotting ln [T_M -T_{SKY}] against $Sec\theta$ (referred to as *air mass*) and determine the best fit straight line having slope α. This graphical procedure has been the more traditional way of measuring opacity from radiometer measurements than the analytical solution given by A13-12. In addition to measuring the opacity, the cosmic background T_{CB} is determined by extrapolating the brightness temperature plot to zero air mass, *i.e.*, $Sec\theta$ = 0 in A13-1. This approach was originally developed and applied by Dicke to measure the opacity of oxygen and water vapor and is currently used as the primary technique for measuring opacity and calibrating ground-based radiometers. The same procedure is applied in Section 8.4 to measure the clear atmospheric transmittance and calibrate the 20 GHz radiometer.

While A13-18 uses the slope or first derivative of ln [T_M -T_{SKY}] to determine the opacity, equation A13-16 also requires the curvature or second derivative. This is seen by differentiating A13-18 with respect to Secθ so that

$$\alpha = -\frac{\dfrac{d^2\,T_{SKY}(\theta)}{d\,Sec^2\theta}}{\dfrac{dT_{SKY}(\theta)}{d\,Sec\theta}} \quad . \qquad \text{(A13-19)}$$

Using A13-16 to determine the opacity requires accurate measurements of the slope and curvature of the sky brightness temperature. However, Figure 44 (Left) does not display any noticeable curvature so that a quadratic fit of the data was considered unwarranted to determine α. As such, the opacity was obtained using A13-18 with the mean atmospheric temperature approximated as 285 K. A more accurate estimate of T_M would involve auxiliary data such as surface temperature and humidity. In summary, the tipping curve procedure uses calibrated T_{SKY} measurements to determine opacity, in addition to calibration given the cosmic background radiation T_{CB}. This point is referenced in the 5th footnote on page 60 and described in Section 8.4 using the 20 GHz radiometer.

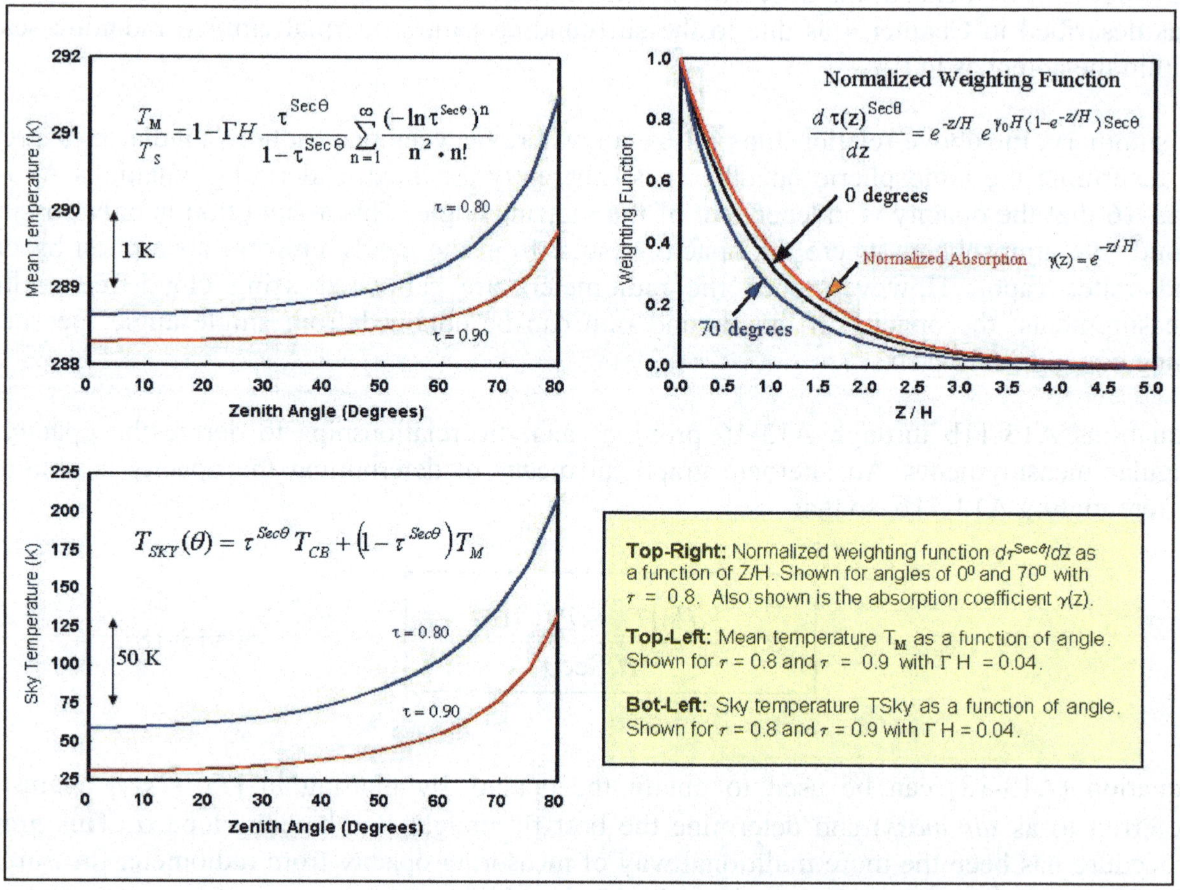

Figure A13-1. Top Right - Weighting function and absorption coefficient plotted against Z/H ratio.
Top Left - Mean temperature T_M plotted against zenith angle with T_S=300 K, ΓH=0.04.
Bot Left - Sky temperature T_{SKY} plotted against zenith angle with T_{CB}=2.7 K, ΓH=0.04.

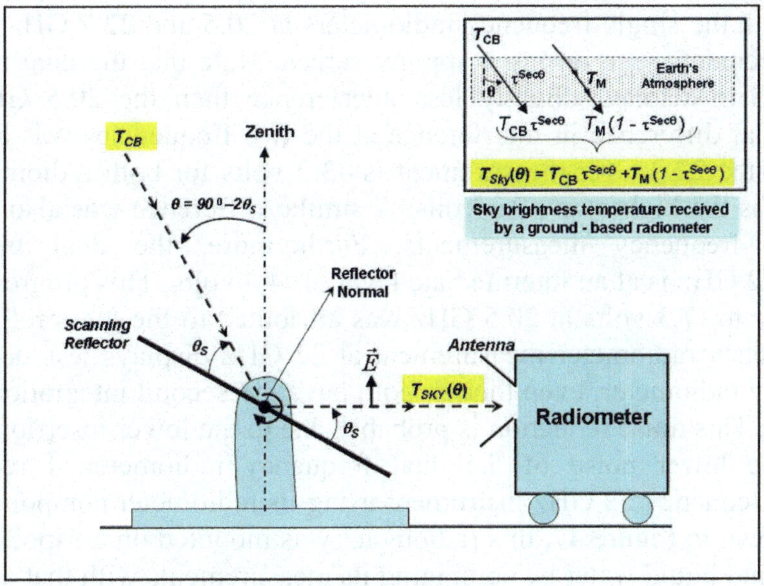

Figure A13-2. Schematic of the tipping curve setup for a radiometer having a linearly polarized antenna where ***E*** is the electric field direction. The antenna views the sky through a reflecting surface that is rotated at scan angle *θ*s. The zenith angle *θ* is related to the scan angle by *θ* = *90*⁰ - 2*θ*s. The inserted figure shows the radiation components whose summation is $T_{SKY}(\theta) = \tau^{Sec\theta}\, T_{CB} + (1-\tau^{Sec\theta})\, T_M$. The actual setup is shown in Figure 38 where Section 8.4 demonstrates its use in calibrating the 20 GHz 1radiometer and measuring the water vapor absorption.

A14. Dual Frequency 21 - 22 GHz Radiometer

This Appendix describes the construction of a dual frequency Dicke radiometer using the Norsat 9000D LNB. As indicated in the radiometer block diagram of Figure A14-1, this LNB amplifies frequencies between 21.2 to 22.2 GHz with 60 dB gain. It accomplishes this using a 20.25 GHz local oscillator to down convert the input radiation to frequencies between 0.95 to 1.95 GHz. The radiometer then uses a power splitter followed by narrowband *IF* filters to obtain separate outputs near 0.95 and 1.95 GHz. The actual filter response is shown in Figure A14-2, which corresponds to input frequencies near 21.2 and 22.2 GHz. These frequencies were chosen to examine their different response to *RFI*, wave interference by the glass door and the effect of the water vapor distribution or vertical profile on the measurements and retrieval accuracy.

In addition to the filters frequency response, Figure A14-2 shows the response of the power splitter and pin diode switch. The power splitter [4, pg 318] is shown to have about a 3.5 dB insertion loss over its frequency range. Compared to the 20.5 GHz radiometer switch in Figure 30, this M862A pin diode switch is shown to have a much lower insertion loss beyond 21 GHz. Also, the modulation due to the switch is detected at each filter output using the wideband Schottky diode detectors mentioned in Appendix A6 and shown in Figure 16, which only has a sensitivity of about 3 mv/μw. This is different then the detectors used in the other radiometers which use a difference amplifier with 20 dB gain to provide temperature compensation. Therefore, to obtain a sufficient output signal, the block diagram of Figure A14-1 has AC amplifiers with 82 dB gain. To obtain this higher gain, the first stage gain in Figure A7 is increased by a factor of 10 by replacing the 100 K feedback resistor with a 1 Megohm resistor. The complete radiometer with its lid opened is shown in Figure A14-3.

As an experiment, Figure A14-4 compares the sky viewing radiometer measurements through the glass patio door with the single frequency radiometers at 20.5 and 22.2 GHz. The Figure shows a snapshot of measurements seen on my computer screen. Note that the dual frequency radiometer response at 21.2 GHz displays slightly less interference than the 20.5 GHz single frequency radiometer. A similar difference in interference at the two frequencies was also noted in Section 8.1. Also note that the 22.2 GHz measurement is –3.2 volts for both radiometers, while the 20.5 GHz measurement is the highest at –7.3 volts. A similar difference was also found in Section 8.2 between the two frequency measurements. Furthermore, the dual frequency radiometer measurement at 21.2 GHz is at an intermediate level of -4.4 volts. This progressive increase from –3.2 volts at 22 GHz to -7.3 volts at 20.5 GHz was attributed to the glass reflectance. Lastly, note that the dual frequency radiometer measurement at 22 GHz displays less noise fluctuations than the single frequency radiometer, even though both have a 1 second integration time and 200 MHz *IF* filter bandwidth. This noise reduction is probably due to the lower insertion loss of its pin diode switch. Due to the lower noise of the dual frequency radiometer, I reconstructed a small lightweight single frequency 22 GHz instrument using its radiometer components. As described in Section 8.6 and shown in Figure 47, this radiometer was mounted on a tripod and used to measure water vapor and cloud liquid water by combining its measurements with that at 12 GHz.

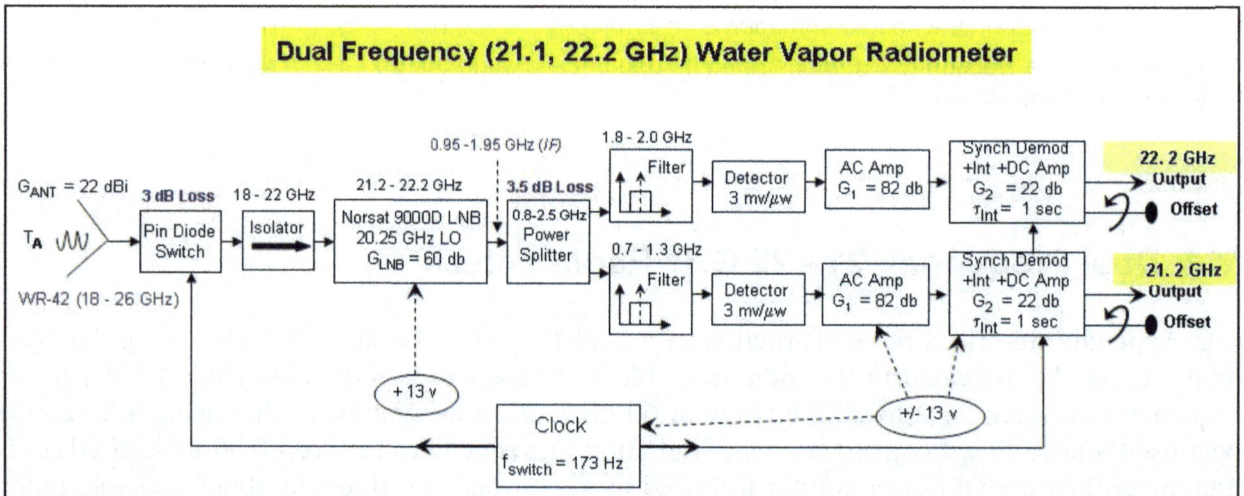

Figure A14-1. The Dicke radiometer uses a Norsat 9000D LNB. It has 60 dB power gain with a 20.25 GHz Local Oscillator (*LO*) and Intermediate Frequencies (*IF*) between 0.95 to 1.95 GHz. Radiometer frequencies (*RF*) are centered at 21.2 and 22.2 GHz using a power splitter, followed by bandpass filters centered at the highest and lowest *IF*, *i.e.*, *RF* = *LO* + *IF*. The next stage uses wideband Schottky diode detectors with a sensitivity of about 3 mv/μw. This is followed by high gain (G_1=82 dB) AC amplifiers. To assure an output between -9 V (Space) to 0 V (Ambient) the DC amplifier gain (G_2) in the synchronous demodulator is set to 22 dB. As such, the total radiometer gain is 60 + 82 + 22 = 164 dB. This high gain is needed to account for the small detector sensitivity and 3.5 dB loss in the power splitter.

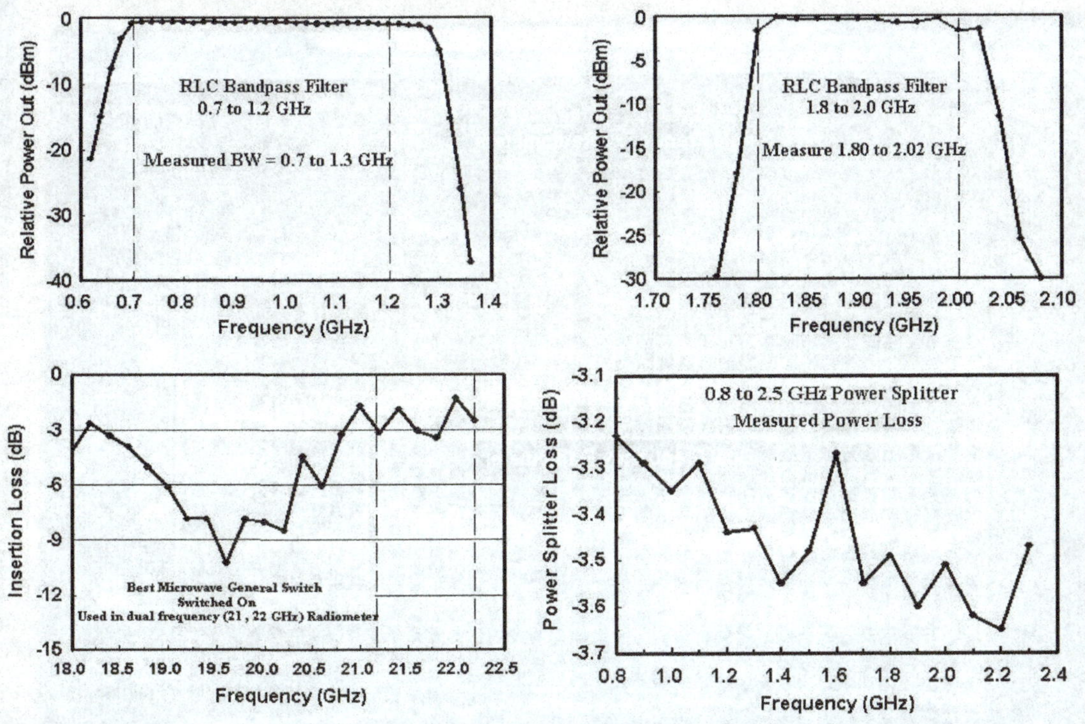

Figure A14-2. Shown is the frequency response of the two IF filters (Top), and the insertion loss of the the pin diode switch (bottom-left) and power splitter (bottom-right).

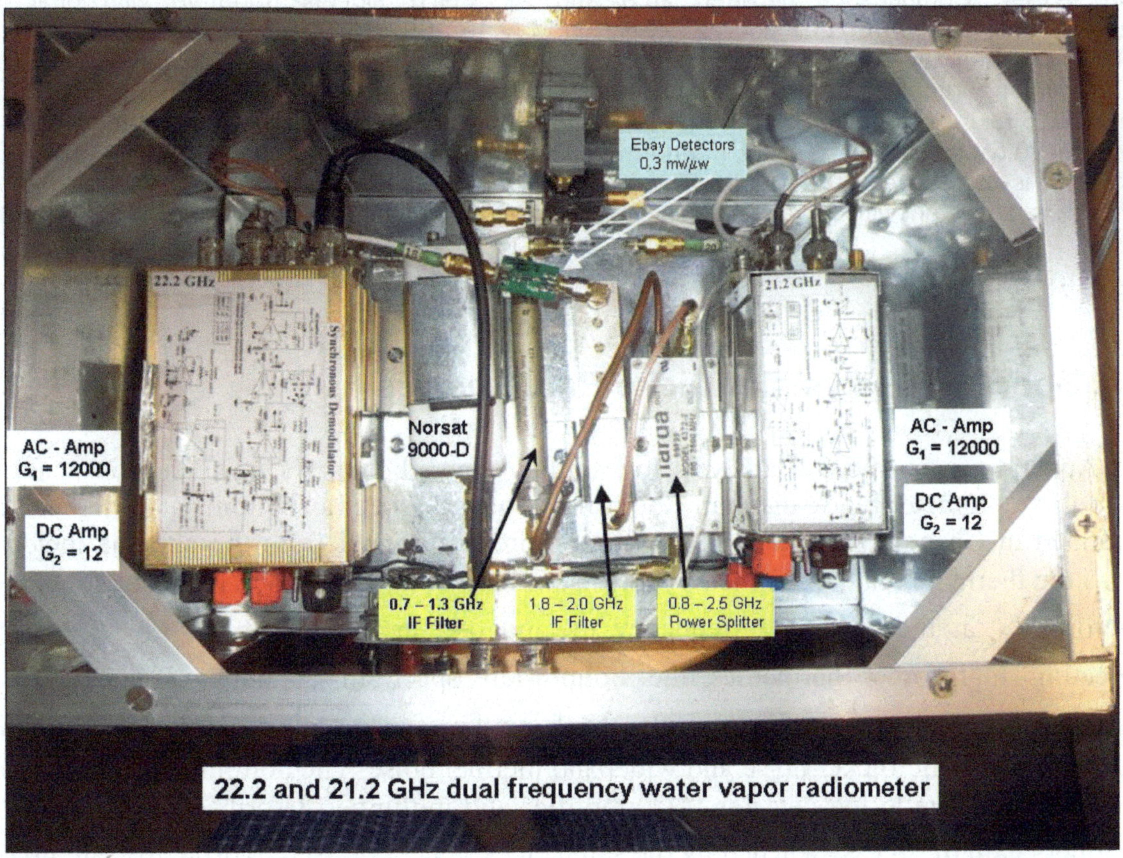

Figure A14-3. Dual 21.2 and 22.2 GHz radiometer with block diagram in Figure A14-1.

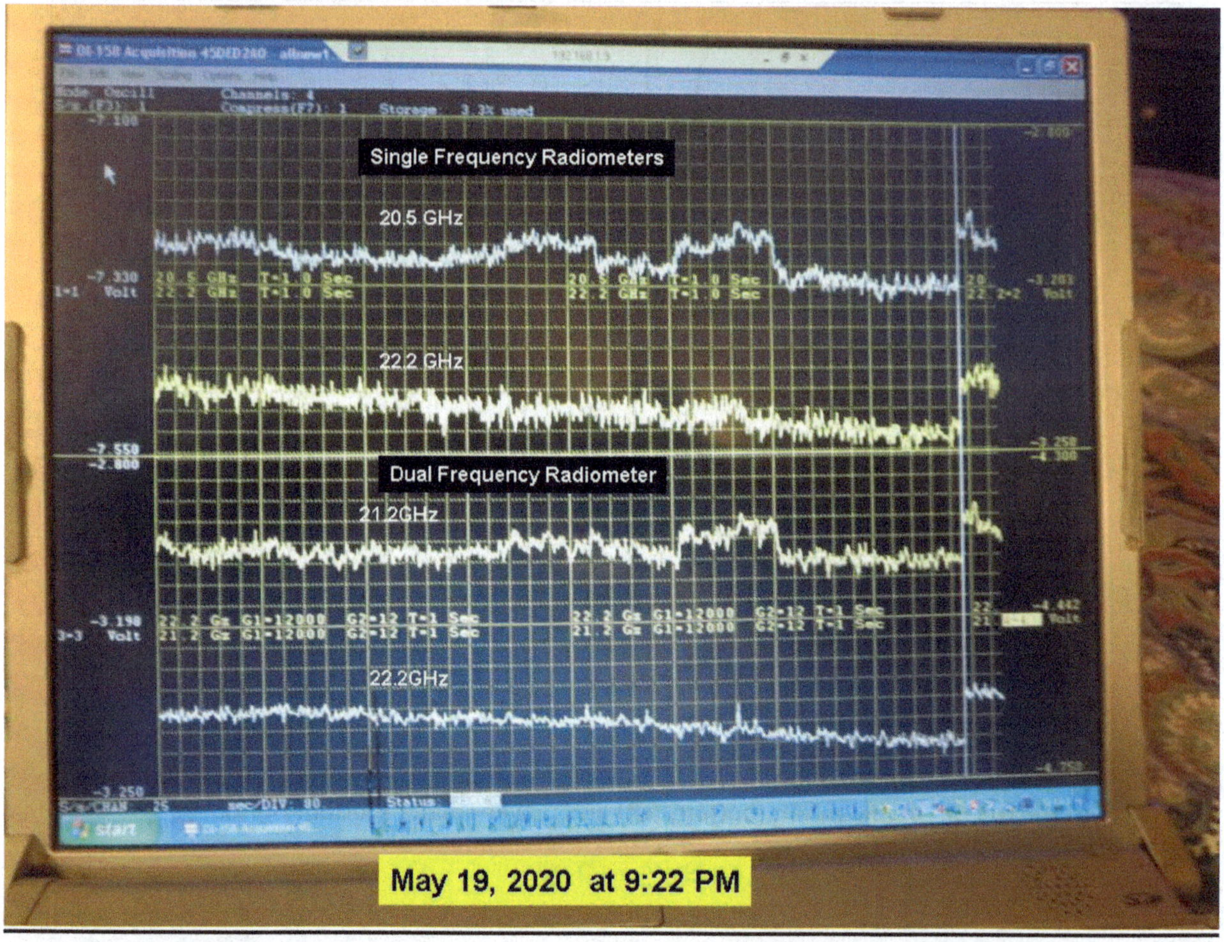

Figure A14-4. Comparisons between 20.5 and 22.2 GHz signals (top-half) with dual frequency measurements (bottom) at 21.2 and 22.2 GHz on May 19 at 9:22 PM. All measurements were obtained using 1 sec integration and displayed using a 0.45 voltage range. Both the 20.5 and 21.2 GHz signals display similar abrupt jumps due to *RFI*, although at other times the 21.2 GHz variation is significantly less than at 20.5 GHz. Also, the 22.2 GHz measurements from both radiometers display the same -3.2 volts while the 20.5 signal is -7.3 volts. Furthermore, the 21.2 GHz measurement from the dual frequency radiometer is in between at -4.4 volts. This progressive decrease in measurements from 20.5, 21.2 and 22.2 GHz is consistent with wave interference due to the glass patio door described in Appendix A10.

A15. Spectral Analysis of Dicke Radiometers

This Appendix uses spectral analysis to examine the radiometer's noise response. Figure A15 shows the radiometer block diagram (top) and frequency spectra (bottom) at different stages. As mentioned in Chapter 3, a pin diode switch re-directs its input from the antenna incident signal V_{IN} to the thermal radiation emitted by a resistor, both at frequency ν. This results in a modulated waveform V_{S1} at the switching frequency ω_S and odd harmonics of the clocks square wave. Neglecting high order harmonics the bottom-left spectrum contains sidebands of V_{IN} at frequencies ν +/- ω_S. This spectrum also contains the effective noise input V_n due to very low frequency electronic flicker noise and wideband thermal noise. The bottom-center spectrum shows the output V_{S2} due to the detector, AC amplifier and multiplier described in Appendix A5. The multipliers gain of +/-1 is switched by the same clock generator used to drive the pin diode switch. This multiplication modulates V_{S1} to produce a spectrum with a frequency shift of the noise $\hat{V}_n$ at

+/- ωs while the input signal $\hat{V}_{IN}$ is shifted at +/- 2ωs, together with a DC component $2\hat{V}_{IN}$. Lastly, a low-pass output filter with cutoff frequency ωc below ωs (bottom-right) suppresses the resulting noise $\langle \hat{V}_n \rangle$ while extracting the DC input component $\langle 2\hat{V}_{IN} \rangle$. The low-pass filter also removes higher order square wave harmonics.

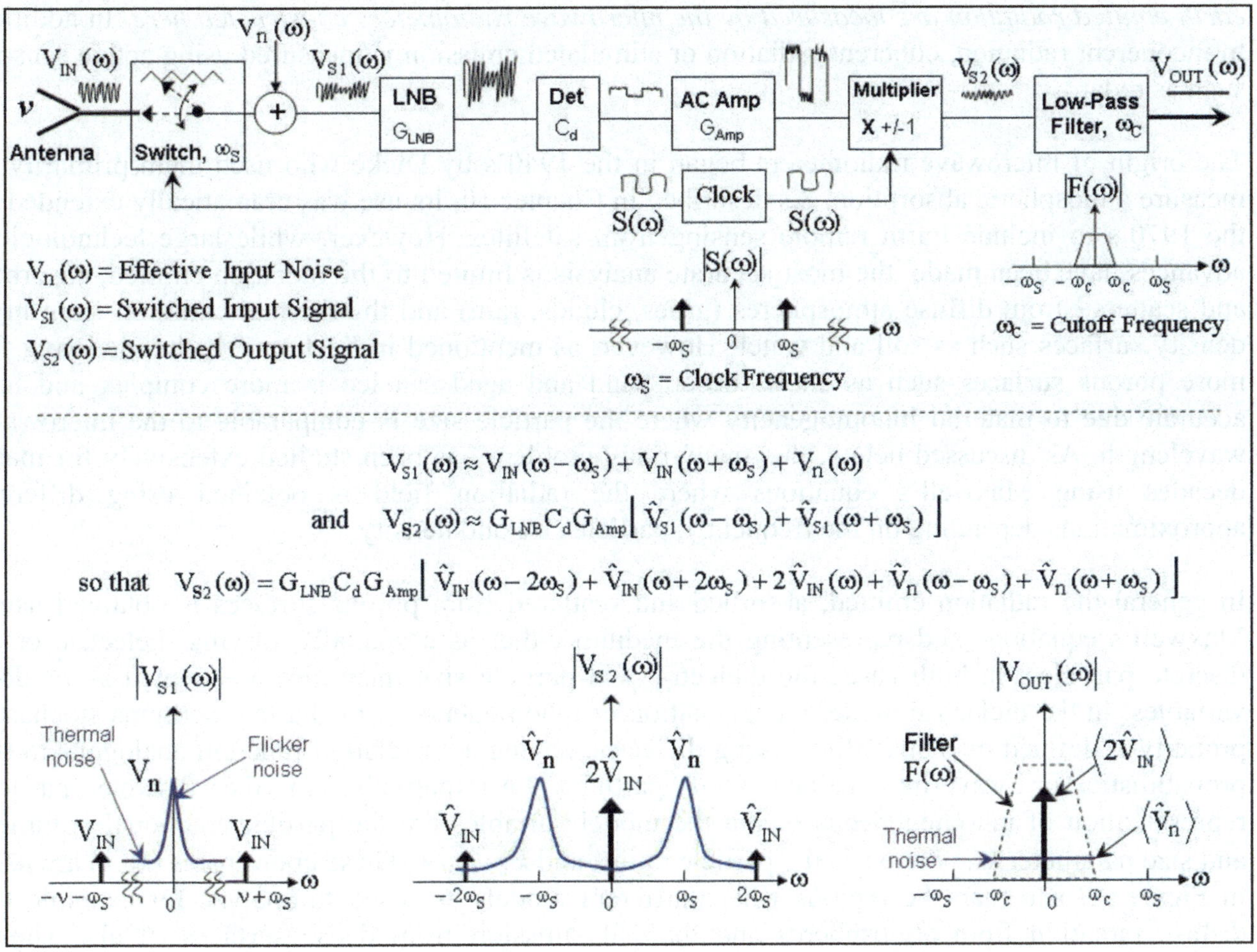

$$V_{S1}(\omega) \approx V_{IN}(\omega-\omega_S) + V_{IN}(\omega+\omega_S) + V_n(\omega)$$

$$\text{and} \quad V_{S2}(\omega) \approx G_{LNB} C_d G_{Amp} \left[\hat{V}_{S1}(\omega-\omega_S) + \hat{V}_{S1}(\omega+\omega_S) \right]$$

$$\text{so that} \quad V_{S2}(\omega) = G_{LNB} C_d G_{Amp} \left[\hat{V}_{IN}(\omega-2\omega_S) + \hat{V}_{IN}(\omega+2\omega_S) + 2\hat{V}_{IN}(\omega) + \hat{V}_n(\omega-\omega_S) + \hat{V}_n(\omega+\omega_S) \right]$$

Figure A15. Waveforms (Top) and frequency spectra (Bottom) of a Dicke radiometer. The effective noise input V_n is low frequency flicker noise superimposed on wideband thermal noise. The microwave signal at frequency ν is switched between a resistive load and antenna input. It results in an amplitude modulated signal V_{S1} shifted by the clock frequency ω_S. The LNB amplified signal is detected, AC amplified and modulated in synchronism to produce the signal V_{S2}. Lastly, a low-pass filter with cutoff ωc below ω_S extracts the time-averaged detected signal $\langle 2\hat{V}_{IN} \rangle$ and suppresses the noise $\langle \hat{V}_n \rangle$.

A16. Modeling Limitations of Random Media

Microwave radiometers have been used to measure the thermally emitted electromagnetic radiation by the earth's atmosphere and its surface up to about 183 GHz. In the atmosphere it is referred to as spontaneous emission due to transitions by the outermost shell of electrons in atoms as described by Einstein, who also theorized stimulated or induced emission. For surfaces the thermally induced radiation has been referred to as Johnson noise after his discovery in electrical components such as resistors. It is also denoted as Nyquist noise, named for his development of the thermal noise equation based on the random motion of free electrons. *These very low levels of earth emitted radiation are measured by the microwave radiometers constructed here.* In addition to incoherent radiation, coherent radiation or stimulated emission is measured using active sensors such as radar.

The origin of microwave radiometers began in the 1940's by Dicke who used them primarily to measure atmospheric absorption. As described in Chapter 10, its use was dramatically extended in the 1970's to include Earth remote sensing from satellites. However, while large technological advances have been made, the most accurate analysis is limited to the radiation emitted, absorbed and scattered from diffuse atmospheres (gases, clouds, rain) and the thermal emission from high density surfaces such as soil and water. However, as mentioned in Section 7.1, the modeling for more porous surfaces such as snow, desert sand and aged sea ice is more complex and less accurate due to material inhomogeneity where the particle size is comparable to the microwave wavelength. As discussed below, this many-body problem has been studied extensively for many decades using Maxwell's equations where the radiation field is obtained using different approximations depending on the frequency, particle size and density.

In general the radiation emitted, absorbed and scattered from porous surfaces is obtained using Maxwell's equations and representing the medium either as a spatially varying dielectric or as discrete particles. In both cases the dielectric and particle size quantities are treated as random variables. In the dielectric model representation of inhomogeneous media the mediums stochastic property is defined by a spatially varying dielectric having a correlation function analogous to the probabilistic or wave-like behavior of particles. Alternatively, in the discrete particle representation of inhomogeneous media the model variables are the particle fractional volume f and size parameter kr, where r is the particle radius and $k = 2\pi/\lambda$. These coordinates (kr, f) are used in Figure A16 to map the regions where different models are used to analyze for example the diffuse radiation from atmospheres and thermal emission from dense surfaces. It also shows regions where different analytical formulations are used, with some requiring complex numerical computations. Also discussed below are ad-hoc methods as well as the model limitations obtained for some of the surfaces at high frequencies.

For isolated spheres where $f = 0$ the classical solution of Maxwell's equations was derived by Gustav Mie in 1908. As indicated in Figure A16 (left), this solution reduces to the Rayleigh limit for small size parameters ($kr < ½$) and can be expressed as (16b) for the water droplets in clouds and rain. Similarly, for large size parameters ($kr > 2$) the solution reduces to the Geometric Optics approximation. However, in general where the fractional volume covers the full range between 0 and 1, various dense media models have been developed that are mostly applicable to small size parameters, $k < 1$. An approximate solution of Maxwell's equations for larger values of kr requires a lengthy perturbation type expansion of terms. To aid in the analysis and physical interpretation, the terms are represented by a series of connected diagrams (*e.g.*, Feynman diagrams) that represent different orders of radiation scattering between particles. For tenuous media where the particles and fractional volume is small, certain connections can be neglected so that the truncated series, referred to as the ladder approximation, can be expressed as a first order integro-differential equation for radiation intensity called the Radiation Transfer Equation. Another interesting derivation of the radiation transfer equation from Maxwell's equations was obtained for underdense plasma by K.M. Watson [41]. Unlike the larger size neutral particles comprising non-ionized media, radiation transfer theory can be extended to higher frequencies in gaseous plasmas since the particles are microscopic in size so $kr << 1$. Also, electrostatic shielding among the electrons and ions in plasmas reduces the particle interaction compared to the neutral particles in non-ionized media.

Figure A16 (right) also indicates the region where particles act as isolated independent scatterers, or as dependent scatterers. These bifurcated regimes are separated by the clearance to wavelength ratio $C/\lambda = ½$ which corresponds to a 5 % deviation of the Mie solution for isolated particles [42]. This analysis by Tien [42] also shows the need for using a *form factor* to account for dependent scattering when $C/\lambda < ½$. For the independent scattering regime when $C/\lambda > ½$ one approach commonly used is the 1st order scattering approximation. This method applies to a sparse random group of particles and is also referred to as the Born approximation, named after Max Born who proposed it in the early days of Quantum Mechanics. It approximates the incident field to the scatterers by the externally applied field. It is accurate when the applied field is much larger than the scattered field and was used to obtain the absorption due to clouds and rain as given by equation (16b), and for diffuse gaseous atmospheres using line shape functions such as (52b). This approximation can also be applied in the case of light Styrofoam containers, where the applied external field can be assumed to be much larger than the scattered field from the plastic particles. A different method is the Radiation Transfer Equation, which as mentioned above is used for sparse discrete random media whose particles interact independently where $C/\lambda > ½$ as denoted in Figure A16 (Left). This formulation was developed and applied in astrophysics at optical wavelengths in the early 1900's, and later used at microwave frequencies in the mid 1940's following the advent of radio astronomy in the 1930's. Equation (5) in Section 4.3 describes an important example of its use in determining the downwelling radiation by the Earths atmosphere as viewed by ground-based radiometers. Radiation transfer theory was also used in Chapter 10 to obtain a similar equation (62) for the upwelling radiation seen by satellite microwave radiometers.

We now discuss higher density dependent scattering media such as snow, multiyear sea ice and desert sand whose emissivity shown in Figure 22 decreases as frequency increases. Modeling these surfaces is difficult since Figure A16 (right) shows that unlike rain, the large fractional volume of scatterers increase the dependent scattering effects. Modeling such effects often require three dimensional numerical solutions of Maxwell's equations, which is an ongoing effort. Another approach that provides a physical interpretation of dependent scattering effects is Strong Fluctuation Theory [43]. In this theory, the Maxwell fields are represented by spatially averaged and random fluctuating components. The coupling between components results in an effective propagation constant for the average field whose absorption coefficient increases by transferring energy to the random component such as indicated in equation (70a). As a result, the negative emissivity slope derived from the average field is much smaller than that from independent scattering theories such as the Born approximation or Radiation Transfer Theory. However, to extend Strong Fluctuation Theory beyond 37 GHz, additional terms are needed to include multiple scattering. This is all described in a lengthy report by Stogryn and Karam [44] whose analytical model covers frequencies up to 90 GHz. Most importantly, the model reduces the emissivity slope of snow, making it compatible with actual measurements. *To obtain such realistic emissivity spectra using Radiation Transfer Theory, dependent scattering effects must be accounted for by using the less rigorous approach of introducing an effective dielectric constant. Similarly, higher order Born approximations must be used to account for changes in the incident field by dense media.*

Lastly, as shown in Figure A16, in the limit of very dense homogeneous surfaces such as water or soils the electrical properties are defined by their bulk dielectric constant. As discussed in Section 7.1, the emissivity for smooth dielectric surfaces is obtained using the Fresnel reflection coefficient shown in the 4th footnote on page 37. However, most natural surfaces do not appear perfectly smooth at microwave frequencies and result in cross-polarization as well as an increase in surface emissivity. These roughness effects increase with frequency and occur for many surfaces, where the most prominent is ocean waves. Analysis of surface roughness is obtained using Maxwell's equations for a dielectric interface whose vertical height and horizontal extent varies randomly about its mean value. One often cited analysis on the subject is the early paper by S.O. Rice [45], which who used small perturbation theory to analyze surface roughness effects.

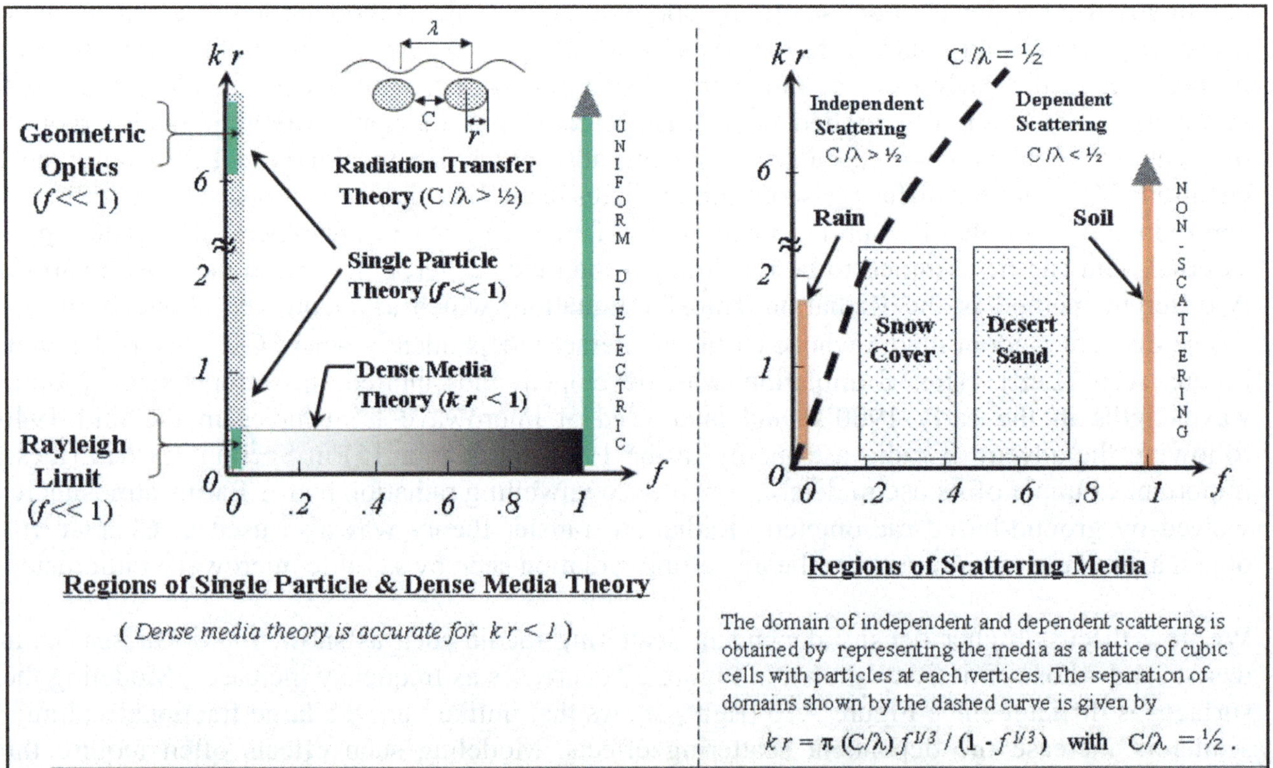

Modeling Parameters	Examples	Size Parameter $= kr = 2\pi r/\lambda$
λ = Wavelength	Rain: $f \ll 0.1$, $C/\lambda > ½$	Clouds: $kr < 0.1$
r = Particle radius	Snow: $f > 0.1$, $C/\lambda < ½$	Rain, Snow, Deserts: $kr > 0.1$
C = Particle separation	Desert: $f > 0.5$, $C/\lambda < ½$	
f = Fractional volume		

Figure A16. The left diagram shows the region of applicability of different theories depending on the size parameter *kr* and fractional volume *f*. Since we are dealing with random media, the particle size and fractional volume should be interpreted as average values. The right-most diagram shows the bifurcation of different scattering regimes (*independent vs. dependent*) also based on the size parameter and fractional volume. Separation between independent and dependent scattering is delineated by the ratio of particle separation, C, and wavelength, λ, *i.e.*, C/λ . Note that the particles in diffuse media such as the molecules in gaseous atmospheres and the water droplets in rain are considered to interact *independently*, while the closely spaced ice particles in snow interact collectively and are therefore considered *dependent*.

A17. Near - Field Emissivity Measurements

This Appendix describes the use of near-field laboratory measurements to determine surface emissivity rather than using far-field radiometer observations from satellites, aircraft or ground-based platforms. While near-field microwave radiometer measurements have been used in medical and plasma diagnostics, and most recently in microwave microscopy [46], they have not been applied to earth surface measurements. Besides the ease of working in a laboratory, another advantage is that the radiometers temperature is stable and only need to be calibrated at ambient temperatures. As such the cold sky and cryogenic target measurements used for calibration are not required. Instead, it uses the near-field procedure discussed in Section 4.1. Unfortunately, however, near field observations can alter the field distribution so it is generally much more difficult to apply than the more common far field approaches.

A brief description of a table top near-field measurement was discussed in Section 9.1, where this Appendix is more detailed. While no final measurements have yet been performed, I felt it instructive to describe some lessons learned from preliminary observations. In this regard, I must acknowledge Dr. Phil Rosenkranz and Dr. Al Gasiewski for forwarding some challenging issues pertaining to near field measurements which I will discuss here. To begin, Figure A17-1 shows the time series of measurements performed for four targets using the 20 GHz radiometer. As shown in Figure 56, each object is measured by first mounting it on an Aluminum plate. The object is then heated to about 337 K and placed over the horn aperture. Figure A17-1 shows the measured radiometer voltage (Top-Left) as each target is sequentially heated, after which it cools to ambient temperature. A more complete description of the process is described in Section 4.1. Two sets of measurements were taken to check the repeatability. Also, to follow the quick initial temperature change the shortest integration time of 0.1 second is used. The top-right shows the brightness temperature obtained using equation (42) as well as the thermocouple-monitored target temperature. For analysis, the bottom-left shows these two temperatures plotted against each other. Note that the four targets are measured over a 4.5 hour period. The first target is a 1 inch square piece of Eccosorb attached to the Aluminum plate. This is followed by the Aluminum plate without the target attached. Next shown are the measurements of a quartz sample attached to the Aluminum. The last measurement is the Eccosorb calibration target which fully encloses the horn aperture. For each target, Figure A17-2 also plots the best fit line between brightness temperature T_B and its physical temperature T. As shown next, the slope of each line is $\Delta T_B/\Delta T = f\,\varepsilon$ where ε is the targets emissivity and f is the effective fractional area seen by the radiometer.

1. Modeling Aspects

Rather than use Maxwell's field equations this analysis uses radiation transfer theory to determine the brightness temperature measured by a radiometer. As an example, Figure A17-3 shows the three radiation components for a quartz sample attached to the Aluminum plate. The first component is the thermally emitted radiation by the radiometer T_R which is reflected by the quartz free portion of the plate. As mentioned in Section 4.2, this radiation is primarily due to the radiometers LNB which is the major heat source. The next component is the radiation emitted by quartz at its temperature T_Q with emissivity ε_Q. In addition to emission, the quartz also reflects and transmits the thermal radiation T_R to the back plate. This reflectivity R_Q is a composite quantity that includes the effect of quartz in addition to metal backing. The radiation received by the horn contains the three components weighted by their view factor, f, *i.e.*

$$T_B = (\varepsilon_Q T_Q + R_Q T_R)\, f \; + \; T_R\,(1 - f)\cdot \qquad \text{(A17 − 1)}$$

Furthermore, based on energy conservation, upon setting $T_Q = T_R$ then $T_B = T_Q = T_R$. This results in the relationship $\varepsilon_Q = 1 - R_Q$ so that the above equation becomes

$$T_B = f\,\varepsilon_Q T_Q + (1-f)T_R \,. \qquad \text{(A17 – 2)}$$

Lastly, since the radiometer's thermal emission T_R is constant in time, the slope of brightness temperature when plotted against the target temperatures in Figure A17-2 is given by

$$\frac{\Delta T_B}{\Delta T_Q} = f\,\varepsilon_Q \,. \qquad \text{(A17 – 3)}$$

Equation (A17-3) is a general relationship that in principal can be used to derive emissivity from the brightness temperature slope given the parameter *f*. This parameter would be unity if all the radiation is captured by the horn. However, as discussed below, when a target partially fills the horns aperture *f* is measured to be less than unity. This reduction in *f* is not only due to its fractional area, but as discussed next, results from perturbations due to near-field effects.

The radiometers WR-42 rectangular waveguide adapter is transitioned into a flared pyramidal shaped horn to provide a gradual transition from the waveguide impedance of about 500 Ω to the free space impedance[8] of 377 Ω. Within the horn's aperture the *E* and *H* fields then vary approximately as the TE_{10} rectangular waveguide mode [4, pg 112]. The field amplitudes therefore vary as Cos(πx/a) where "a" is the horn's aperture width of 7.5 cm and "x" is the distance along the width beginning at the center where x = 0. However, even when including this non-uniform power distribution of $\text{Cos}^2(\pi x/a)$, the calculated *f* parameter (~ 0.16) is not large enough to explain the large slopes of 0.58 and 0.66 for the partially filled Quartz and Eccosorb targets in Figure A17-2. *These slopes are in fact about a factor of 4 times larger than that calculated from the targets small fractional area of about 0.16.* To resolve this ambiguity, the perturbed field due to multimodes resulting from the partially filled target must be included. To further complicate the analysis, the spatial variation or modes derived from Maxwell's equations for the horn geometry are spherical waves having their origin at the apex of the horn; a point referred to as the phase center. Therefore, a large phase difference occurs at the center and outer edge of the horns aperture for wide flare angles. As such, mixed modes are also generated when using non-spherical targets. This is different than for rectangular waveguides whose modes are plane waves having constant phase in the aperture. However, even for waveguides, except for a uniform object filling the aperture, the presence of small or irregular shaped targets generates higher order modes so that the *f* parameter must be adjusted to account for the altered field distribution. To minimize any field distortion, microwave microscopy [46] uses near-field probes much smaller than the wavelength. At the other extreme, the field distribution is also unperturbed by viewing objects at far field distances from the antenna. As discussed next, a large simplification occurs for objects seen at large distances from the horn.

---

[8] For TE_{mn} modes $Z_G = (\lambda_G/\lambda)\sqrt{\mu/\varepsilon}$ where $\lambda_G/\lambda = 1/\sqrt{1-(f_C/f)^2}$ and $f_C = (1/\sqrt{\mu\,\varepsilon})\sqrt{(m/2a)^2+(n/2b)^2}$ is the cutoff frequency where a is the waveguide width and b its height. Waveguides operate as high-pass filters which for air filled WR-42 guides, the lowest mode TE_{10} has $f_C = 1/2a\sqrt{\mu_0\varepsilon_0}$ =14 GHz. The next mode TE_{20} has f_C = 28 GHz. Also, $\sqrt{\mu_0/\varepsilon_0}$ =377 Ω is the far field impedance, which is the ratio of E to H field which become in phase.

2. Near Field Effects

For observations beyond the far field distance $2D^2/\lambda$, where D is the aperture size and λ is the wavelength, the angular spread over a target is small. As such, the phase variation seen by the 20 GHz horn having a 7.5 cm aperture width and 5 cm height is small for distances greater than 110 cm since $D = \sqrt{7.5^2 + 5^2}$. Therefore, beyond this distance the horn radiation received from any object is effectively a plane wave rather than spherical. Furthermore, the perturbation of the horn's field distribution resulting from the target is negligibly small so the targets radiation received by the horn is simply weighted by its area *A* relative to the antennas footprint or *FOV* so $f = A/FOV$. *This view factor then becomes the fractional area.* It was used when analyzing radiometer measurements of surfaces in Section 7.1 and sky viewing measurements in Section 7.2. However, as mentioned above this equation for *f* is not applicable when objects reside in the near field or horn aperture. Mode mixing then occurs between the plane waves emitted and reflected by a flat target and the spherical wave modes resulting from the horn geometry. This is in addition to the multimodes generated by the target. Only when the target fills the aperture and has high emissivity or low reflectivity can such effects be minimized. The interface then appears as free space radiating at the targets temperature rather than cold space with no reflections from other sources. For this reason, the Eccosorb target placed over the horn's aperture is used for calibration in Section 4.1 since then $T_B = T_{Eccosorb}$ and represents a *unique application of near-field antenna measurements in radiometry*. Furthermore, $f = 1$ is expected when a surface filling the aperture has the same curvature as the modes, which is spherical for horns. As discussed next, some of the adverse near-field effects are exhibited by the measurements in Figure A17-2.

Eccosorb has high emissivity and small reflectivity so its fractional area can be considered unity when covering the full horn aperture. As such, according to (A17-3) the 1.035 slope shown in the bottom-right of Figure A17-2 should correspond to its emissivity. The slightly larger than unity slope is due to slight changes in calibration. As mentioned in Sections 3.1 and 4.2, this can result from changes in LNB gain and detector sensitivity with temperature so that the calibration equation (42), $T_b = 299.1 + 39.57$ V now becomes $T_b = 296.4 + 38.18$ V using the latest Eccosorb measurements in Figure A17-2. Furthermore, when using the small Eccosorb sample, the reflection from the surrounding Aluminum plate results in a slope of 0.658. As mentioned above, this slope is about a factor of 4 times large than based on its fractional area of 0.16. This effect is presumably due to mixed modes, which is further enhanced when the Aluminum fully covers the aperture. The measured slope shown in the top-right of the Figure is then reduced from 0.658 to 0.174 due to the metals low emissivity. However, the slope obtained for metal surfaces should be even smaller so the excess radiation resulting in its larger slope is also evidence of mixed mode effects. Similarly, in the case of the quartz sample its slope of 0.585 shown on the bottom-left is similar to that of the small Eccosorb sample. Again, its slope can not be equated directly to its emissivity due to the complicated effect of mode mixing by the Aluminum plate as well as additional modes resulting from the quartz sample which distort the field distribution in the horn's aperture.

As discussed above, in order to use such near field horn measurements to determine the emissivity it is best to use spherically curved surfaces to match the modes excited by the horn. This limitation was not originally anticipated when developing the procedure. However, another approach mentioned above is to replace the horn antenna with a rectangular waveguide. This simplifies the emissivity determination since the fields reflected by the aluminum plate have the same plane wave structure as the normal waveguide modes. In fact, the equivalent network then becomes a short circuited transmission line. More importantly, any flat object completely covering the Aluminum plate would not excite higher order modes. Therefore, a dielectric with metal backing becomes a short circuited transmission line covered by a dielectric slab connected to a waveguide. Moreover, when the waveguide and dielectric slab are operated at frequencies corresponding to the dominant TE_{10} mode[8] they can be depicted as a single transmission line so the equivalent network is shown in Figure A17-4.

3. Transmission Model

Based on transmission line theory [4, pg 60], the input impedance of the metal backed dielectric slab is

$$Z_{IN} = Z_T \text{Tanh}\, \gamma_T d, \quad Z_T = \sqrt{\frac{\mu_0}{\varepsilon_T}} \quad \text{and} \quad \gamma_T = \frac{2\pi}{\lambda}\sqrt{\frac{\varepsilon_T}{\varepsilon_0}} \qquad (A17-4)$$

where d is the slabs thickness, γ_T is its propagation constant, Z_T its characteristic impedance and λ is the free space wavelength. Both γ_T and Z_T are functions of the targets dielectric constant ε_T which is generally a complex variable expressed as $\varepsilon_T = \varepsilon' + i\,\varepsilon''$ where ε' and ε'' are the real and imaginary components. As such, γ_T in (A17-4) is also a complex variable written as $\gamma_T = \alpha + i\beta$ where $\alpha = \text{Re}\{\gamma_T\}$ and $\beta = \text{Im}\{\gamma_T\}$

Equation (A17-2) was derived using the fact that the target's emissivity is one minus the reflection coefficient, *i.e.*, $\varepsilon_Q = 1 - R_Q$. Furthermore, its reflection coefficient can be written in terms of impedance so

$$E_T = 1 - \left|\frac{Z_{IN} - Z_T}{Z_{IN} + Z_T}\right|^2 . \qquad (A17-5)$$

where the emissivity is denoted as E_T so it is not confused with the dielectric constant ε_T.

Therefore, after substituting (A17-4) into (A17-5) the emissivity becomes

$$E_T = \frac{4\, \text{Tanh}\, \alpha d}{(1 + \text{Tanh}\, \alpha d)^2} = 1 - \text{Exp}\{-4\alpha d\}, \qquad (A17\text{–}6a)$$

$$\text{where } \alpha = \text{Re}\{\gamma_T\} = \frac{2\pi}{\lambda}\sqrt{\frac{\varepsilon'}{2\varepsilon_0} + \sqrt{\left(\frac{\varepsilon'}{2\varepsilon_0}\right)^2 + \frac{\varepsilon''}{2\varepsilon_0}}} \qquad (A17\text{–}6b)$$

Note that E_T only depends on the real part of the propagation constant α. However, even for lossless media such as glass or quartz where $\varepsilon'' \approx 0$, $E_T > 0$ due to insertion loss resulting from an impedance mismatch. Only when d=0 in (A17-6a) is $E_T = 0$ due to the metal backed surface.

As shown in Figure A17-4, to best transfer the targets emitted radiation to the radiometer, isolators are used to minimize reflections at the two waveguide interfaces. This effect of impedance mismatch is analyzed next. Also, the dielectric sample covering the metal plate must fully enclose the waveguide aperture so the viewing factor f is unity in (A17-3). This is not a problem when using the Eccosorb or metal plate targets which are flat, but is difficult to achieve for irregular shaped objects such as quartz which must fit the small WR-42 waveguide opening. Provisions are also needed to monitor their temperature using a thermistor after heating the sample. Except for using low frequency radiometers, some of this can be alleviated using a smooth transition to a larger waveguide or a tapered fixture to hold the object.

The issue of impedance mismatch at the waveguide interface is minimized using isolators at both ends. Without such isolators the target emissivity would be different than that seen by the radiometer. To show this, transmission line theory is used to obtain the impedance at the waveguide input Z'_{IN}, which is also the radiometer input, *i.e.,*

$$Z'_{\mathrm{IN}} = Z_G \frac{Z_{\mathrm{IN}} + i Z_G \mathrm{Tan}\,(2\pi L/\lambda_G)}{Z_G + i Z_{\mathrm{IN}} \mathrm{Tan}\,(2\pi L/\lambda_G)} \qquad \text{(A17-7)}$$

This equation is for a lossless waveguide, and transfers the impedance at the target location Z_{IN} to that seen by the radiometer Z'_{IN}. It contains the waveguide length L, its wavelength λ_G and characteristic impedance[26], Z_G. The corresponding emissivity measured at the radiometer input E'_{IN} is therefore

$$E'_{\mathrm{IN}} = 1 - \left| \frac{Z'_{\mathrm{IN}} - Z_G}{Z'_{\mathrm{IN}} + Z_G} \right|^2 \qquad \text{(A17-8)}$$

so after substituting (A17-7) in (A17-8) we obtain

$$E'_{\mathrm{IN}} = 1 - \left| \frac{Z_{\mathrm{IN}} - Z_G}{Z_{\mathrm{IN}} + Z_G} \right|^2 \qquad \text{(A17-9)}$$

This input emissivity seen by the radiometer is similar to the target (A17-5), with the only difference being that the characteristic impedance is now the waveguide impedance Z_G. In retrospect, this impedance mismatch is possibly the largest error affecting the emissivity measurements using the horn antenna setup in Figure A17-3. In fact, when substituting (A17-4) into (A17-9) the input emissivity becomes

$$E'_{\mathrm{IN}} = \frac{4\eta \mathrm{Tanh}\,\alpha d\,(1 + \mathrm{Tan}^2 \beta d)}{(\mathrm{Tanh}\,\alpha d + \eta)^2 + (1 + \eta \mathrm{Tanh}\,\alpha d)^2 \mathrm{Tan}^2 \beta d} \quad \text{with} \quad \eta = \frac{Z_G}{Z_T} \quad . \qquad \text{(A17-10)}$$

Also, upon substituting (A17-6a) for $\mathrm{Tanh}\,\alpha d$, (A17-10) can be written as

$$E'_{\mathrm{IN}} = \frac{4\eta\, E_T (1 + \mathrm{Tan}^2 \beta d)}{\left[(\eta+1) + (\eta-1)\sqrt{1 - E_T}\right]^2 + \left[(\eta+1) - (\eta-1)\sqrt{1 - E_T}\right]^2 \mathrm{Tan}^2 \beta d} \quad . \qquad \text{(A17-11)}$$

In general, E'_{IN} is less than E_T since $\eta \approx \sqrt{\varepsilon_T / \varepsilon_0}$ is larger than unity. As an example, for high loss materials $E_T \approx 1$ so

$$E'_{\mathrm{IN}} = \frac{4\eta}{(\eta+1)^2} < 1 \quad . \qquad \text{(A17-12)}$$

Also, for low loss materials where $\mathrm{Tan}\,\beta d < 1$. The input emissivity becomes

$$E'_{\mathrm{IN}} = \frac{4\eta E_T}{\left[(\eta+1) + (\eta-1)\sqrt{1 - E_T}\right]^2} < E_T \quad . \qquad \text{(A17-13)}$$

Impedance mismatch errors can be reduced using the isolators shown in Figure A17-4. The non-reciprocal property of isolators reduces interface reflections so that the emissivity seen by the radiometer is nearly the same as the target's emissivity (A17-6a). However, instead of an impedance mismatch, the difference between E'_{IN} and E_T is then due to the isolator's insertion loss. Fortunately, this effect can be accounted for by multiplying the emissivity in (A17-3) by the isolator transmittance. The isolator's transmittance can then be obtained using insertion loss measurements as done for the pin diode switches in Sections 8.1 and Appendix A14.

4. Summary

In summary, the waveguide approach appears to be best suited to reduce the near-field distortion due to multimode excitation and reflections due to impedance mismatch. A metal backed flange having the same opening as the waveguide aperture can be used to contain the sample and small thermocouple to measure its temperature. While this sample holder may be easy to use for targets such as Eccosorb, it is more difficult for irregular shaped objects such as quartz which must be sized to fit the flange opening. As such, the idea of measuring emissivity in a laboratory setting is more involved than originally anticipated. In particular, near field measurements requires knowledge of the aperture field when using the horn setup. However, the approach was shown to be a good technique for calibration when using the high emissivity Eccosorb target that fully covers the aperture so that $f = 1$ in A17-3. This results from its high absorption which reduces the effects of impedance mismatch and multimodes.

Lastly, on September 22nd, preliminary measurements were obtained using the waveguide setup illustrated in Figure A17-4. As an important test of its effectiveness, a slope of only 0.02 was obtained when placing the aluminum target over the waveguide aperture. The resulting emissivity of 0.02 is very reasonable and almost an order of magnitude less than the 0.173 value shown in Figure A17-2 using the horn setup. Lastly, as an exercise for the student of waveguide network theory, I have yet to find any analysis of the impedance for a metal backed, partially covered dielectric slab using Maxwell's equations. In particular, it would be very valuable to know how the target impedance depends on the dielectric dimension or fractional area as well as the dielectric constant of the slab. In this way one could use the setup illustrated in Figure A17-4 to determine the emissivity for dielectric material of any size from impedance measurements. I believe that this type of analysis would help answer the question whether this table-top laboratory approach is a viable means of measuring emissivity as opposed to the far-field measurements currently being used. As far as I can tell, only in the case of microwave microscopy [46] has extensive analysis been done in the use of near-field probes to measure surface properties at high microwave frequencies.

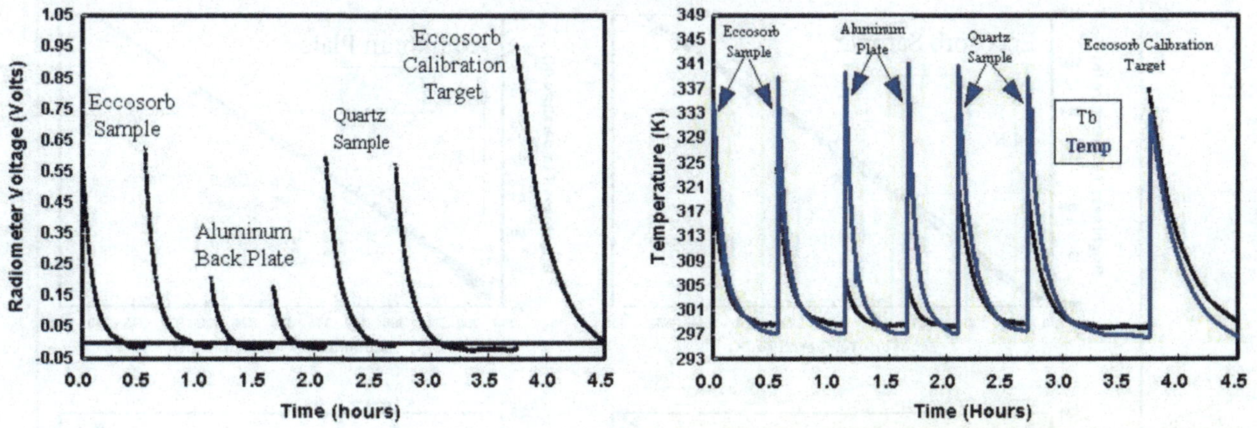

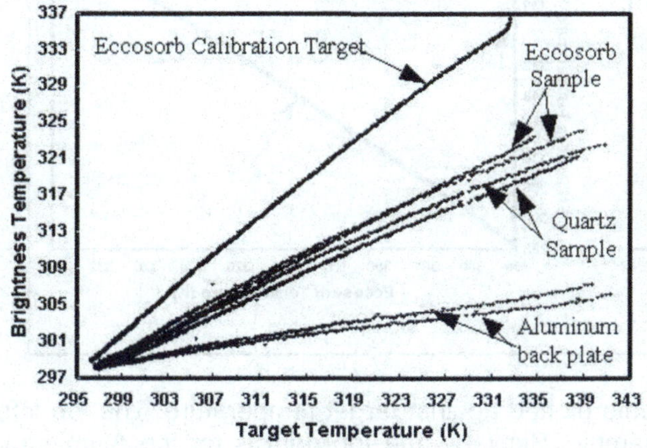

August 30, 2021

Top-left shows the radiometer voltage V when the antenna views the Eccosorb sample, metal plate, quartz sample and calibration target. Integration time is set at 0.1 seconds. The top-right shows the corresponding brightness temperature Tb and sample temperature using a thermocouple attached to each sample. The brightness temperature is obtained using the calibration equation Tb=299 + 39.57 V. The bottom-left shows Tb plotted against the target or sample temperature. All plots show a linear relationship between brightness temperature and target temperature. The next Figure A17-2 shows the best fit linear equation for each material.

Figure A17-1. 20 GHz radiometer viewing four targets (Eccosorb, Aluminum, Quartz and Calibration Target). The top-left shows the radiometer voltage after initially heating the targets to about 337 K. The top-right shows the calibrated brightness temperatures and target temperature while the bottom-left plots the temperatures against each other.

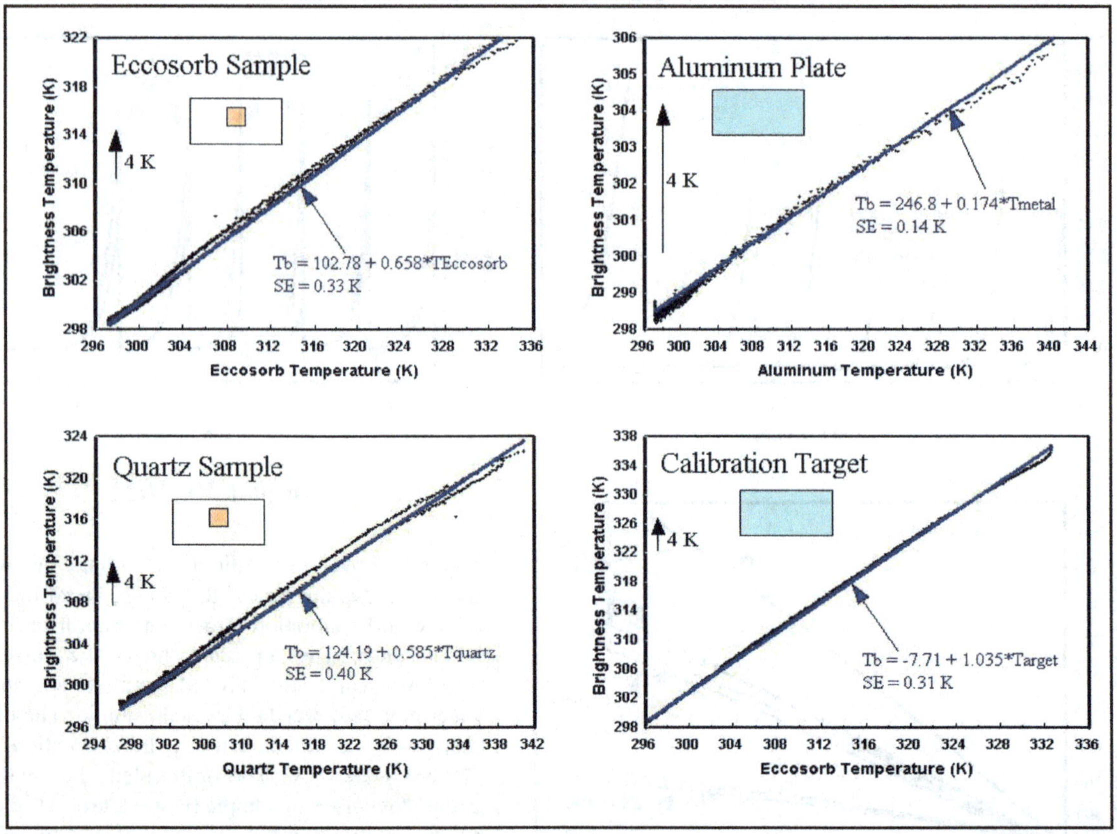

Figure A17-2. Brightness temperature measurements plotted against target temperature. The top-left is for Eccosorb while the bottom-left is for the quartz sample. Similarly, the top right is for the Aluminum plate while the bottom-right is for the Eccosorb Calibration Target. Each plot also shows the best straight line fit and its regression equation.

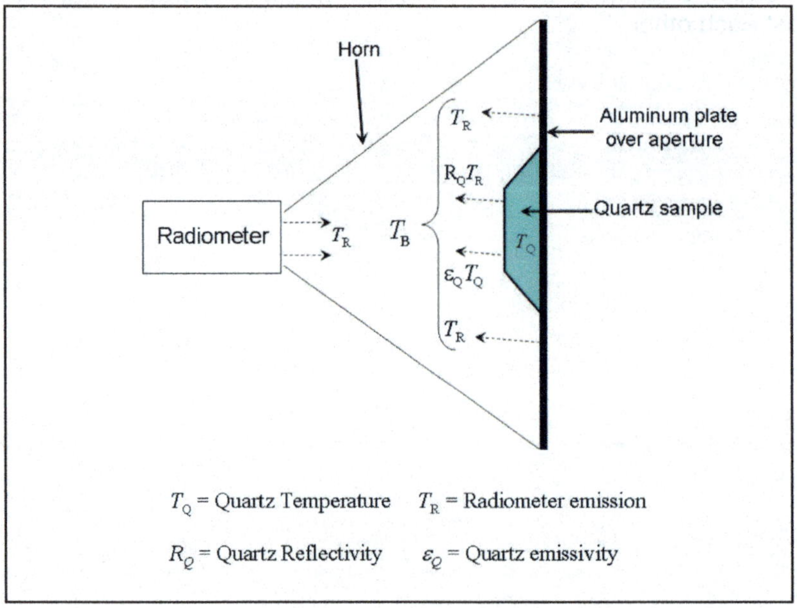

Figure A17-3. Brightness temperature components seen by a horn antenna. Shown is the thermally emitted radiation by the radiometer T_R and reflected radiation by the Aluminum plate. Also shown is the thermally emitted radiation from quartz with emissivity ε_Q and temperature T_Q.

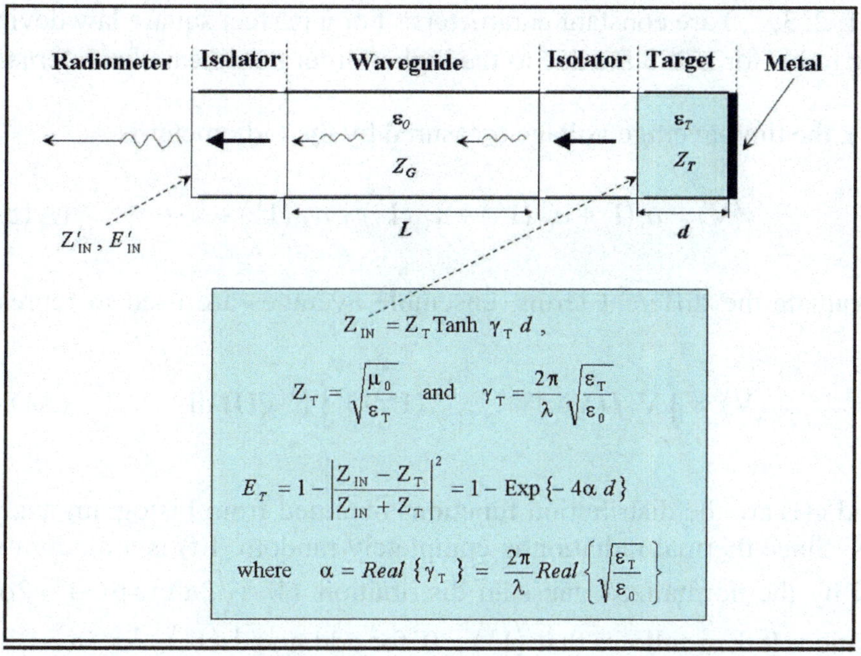

Figure A17-4. Compared to Figure A17-3, this improved setup to measure emissivity uses a waveguide rather than the horn antenna. Furthermore, the aperture is fully covered by the target so that its emissivity E_T is obtained using an equation similar to (A17-3) with $f = 1$. It also uses waveguide isolators at both ends of the waveguide to minimize reflections due to impedance mismatch by the target and radiometer. The insert shows the resulting target emissivity E_T, which depends on its thickness d and dielectric constant ε_T (see text).

A18. Nonlinear Calibration Equation

.

This Appendix analyzes the detector response and its effect on the radiometer calibration. As mentioned in Chapter 6, the randomly emitted thermal radiation collected by an antenna is amplified over a prescribed spectral region and detected using solid-state detector elements. Microwave radiometers use square law detectors to transform the random input signal (having a zero time average value) to a non-zero power level output or brightness temperature. This Appendix develops the calibration equation, containing the effect of the detector's nonlinearity due to an imperfect square law response. The derivation is based on an unpublished document previously acquired from the late Dr. Alex Stogryn, whose theoretical analysis have greatly extended our understanding pertaining to microwave radiometry (*e.g.*, see Appendix A16). This Appendix also discusses some important findings when using the nonlinear calibration equation for microwave temperature sounders.

As discussed in Chapter 6, the output current, I, from a solid-state detector varies exponentially with applied voltage, V. The current through the detector element $I = I_S\left[\exp(V/\eta V_{th}) - 1\right]$ can be expressed in a Taylor expansion of voltage with few terms since $V/V_{th} << 1$. Alternatively, it can also be expressed as a power series in terms of the current variable, *i.e.*,

$$V = \sum_{n=1}^{N} a_n I^n = a_1 I + a_2\, I^2 + a_3\, I^3 + a_4\, I^4 + \cdots \qquad \text{(A18-1)}$$

where a_n (n =1, 2, 3, , ,) are constant parameters. For a perfect square law device $a_n = 0$ for $n \geq 3$. The remaining terms for $n \geq 3$ are due to the higher order nonlinear characteristics of the detector.

Using (A18-1), the time-average voltage measured by the radiometer is

$$\langle V \rangle = a_1 \langle I \rangle + a_2 \langle I^2 \rangle + a_3 \langle I^3 \rangle + a_4 \langle I^4 \rangle + \cdots \qquad \text{(A18-2)}$$

In order to evaluate the different terms, ensemble averages are used to represent time-averages; *i.e.*,

$$\langle V \rangle = \int_{-\infty}^{\infty} V\, f(V)\, dV \quad , \quad \langle I^n \rangle = \int_{-\infty}^{\infty} I^n\, g(I)\, dI \ . \qquad \text{(A18-3)}$$

where $f(V)$ and $g(I)$ are the distribution functions obtained from histogram analysis of the detector measurements. Since thermal radiation is completely random, I(t) is a random variable where g(I) is represented by the normalized gaussian distribution $(1/\sigma\sqrt{2\pi})\,\mathrm{Exp}(-I^2/2\sigma^2)$ where σ is the standard deviation. It then follows that $\langle I^n \rangle = 0$ for odd n and $\langle I^n \rangle = \sigma^n (n-1)!!$ for even values of n. Furthermore, for n = 2, $\langle I^2 \rangle = \sigma^2$ so that for even values of n we obtain the important result $\langle I^n \rangle = \langle I^2 \rangle^{n/2} (n-1)!!$. Therefore, in summary,

$$\langle I^n \rangle = \begin{cases} 0 & n = \text{odd} \\ 1, 3, 15, \cdots\ (n-1)!! \langle I^2 \rangle^{n/2} & n = \text{even} \end{cases} \qquad \text{(A18-4)}$$

Substituting (A18-4) into (A18-2), only the terms containing even powers in current are non-zero. Keeping only these first two terms, the output voltage is given by

$$\langle V \rangle = \left[a_2 + 3a_4 \langle I^2 \rangle \right] \cdot \langle I^2 \rangle \qquad \text{(A18-5)}$$

which is a quadratic equation in terms of the mean-squared current.

From Nyquist's theorem, the mean-squared current is

$$\langle I^2 \rangle = KGB \left[T_R + T_b \right] \qquad \text{(A18-6)}$$

where K is Boltzman's constant, G is the amplifier gain and B is the bandwidth of the radiometers *IF* amplifier. The equation also contains the radiometric brightness temperature of the scene as viewed by the antenna T_b and the instrument temperature T_R. For a Dicke radiometer, this temperature is nearly also the reference temperature.

Combining (A18-5) and (A18-6) we obtain the result,

$$V \equiv \langle V \rangle = b_o + b_1\, T_b\, [1 + \mu\, T_b] \qquad \text{(A18-7)}$$

where V is the output radiometer voltage in terms of brightness temperature. The b_o, b_1 and μ parameters are

$$b_o = [a_2 + 3a_4\ KBT_R]\ KBG\ T_R\ , \tag{A18-8a}$$
$$b_I = [a_2 + 6a_4\ KBT_R]\ KBG\ , \tag{A18-8b}$$
$$\mu = \frac{3a_4}{a_2} KBG\ . \tag{A18-8c}$$

Equation (A18-7) expresses the output radiometer voltage in terms of the brightness temperature viewed by the antenna[9]. For perfect square-law detectors $a_4 = 0$ so that the voltage becomes $V = a_2\ KBG\ (T_R + T_b)$. However, the μ parameter characterizing the detector nonlinearity produces a quadratic brightness temperature response.

To calibrate the brightness temperature based on equation (A18-7), the parameters b_o, b_I and μ must be determined using radiometer measurements at three different temperatures. However, in actual practice most satellite and ground-based radiometers only use two calibration targets, viewing a warm load and cold space. In the case of the satellite instruments described in Chapter 11 the radiometer is calibrated by viewing each of the two targets at the beginning and end of every earth observation, *i.e.,* for each scan line. If V_C, T_C corresponds to the voltage and temperature of cold space and V_W, T_W the values for the warm load then from (A18-7) we obtain

$$V_C = b_o + b_I\ T_C\ [1 + \mu\ T_C] \tag{A18-9}$$
$$V_W = b_o + b_I\ T_W\ [1 + \mu\ T_W] \tag{A18-10}$$

Calibration of the radiometer is done frequently enough so the gain of the radiometer (*i.e.*, b_o, b_I and μ) does not change during the interval between the warm target and cold space observations. In this analysis the parameters b_o and b_I in (A18-7) are determined as a function of V_C, V_W, T_C, T_W and μ using the two calibration equations (A18-9) and (A18-10). The two parameters b_o and b_I are then substituted into (A18-7), resulting in a quadratic equation for the scene brightness temperature in terms of the scene voltage V and calibration voltages, *viz.*,

$$T_b + \mu T_b^{\ 2} = T_C(1 + \mu\ T_C) + [T_W(1 + \mu\ T_w) - T_C(1 + \mu\ T_C)]\left(\frac{V - V_C}{V_W - V_C}\right) \tag{A18-11}$$

Rather than solve the quadratic equation for T_b, an accurate solution is obtained by noting that the detectors nonlinearity is very small. The μT_b^2 term is then accurately approximated using T_b for $\mu = 0$, *i.e.*, $T_b \cong T_C + [T_W - T_C](V - V_C)/(V_W - V_C)$. Upon substitution, the brightness temperature solution of (A18-11) can be written as

$$\boxed{T_b = (I + S\ V)\ +\ \mu\ S^2 (V - V_C)(V_w - V)} \tag{A18-12}$$

where the Intercept, I, and Slope, S, are those of the linear calibration procedure, *i.e.*,

$$S = \left(\frac{T_W - T_C}{V_W - V_C}\right)\ , \qquad I = T_C - S\ V_C\ . \tag{A18-13}$$

[9] Equation (A18-7) is written for a total power radiometer. For a Dicke radiometer the output voltage is $V = b'_o + b_I\ T_b\ [1 + \mu\ T_b]$ where $b'_o = -b_I\ T_R\ [1 + \mu\ T_R]$ so V has the same form for both instruments.

Equation (A18-12) includes the nonlinearity due an imperfect square law detector. Since μ is very small $T_b \cong I + S\,V$ so the difference between the nonlinear and linear calibration is

$$\Delta T_b = \mu\, S^2 (V - V_C)(V_w - V) \cong \mu\,(T_b - T_C)(T_w - T_b). \qquad \text{(A18-14)}$$

This difference quantity is a parabolic function of the linearized brightness temperature T_b. It is zero when $T_b = T_W$ and $T_b = T_C$ and has maximum value $(\mu/4)(T_W - T_C)^2$ when $T_b = (T_W + T_C)/2$. The equation is plotted in Figure A18-1 as a function of brightness temperature for Tw=280 K and Tc=3 K. Separate plots are shown for $\mu = 0.5\text{x}10^{-4}\ K^{-1}$ and $\mu = 1.5\text{x}10^{-4}\ K^{-1}$. As explained below, these μ parameters cover the range measured in the laboratory for different prelaunch MSU satellite radiometers. Its effect on the radiometer measurements is also discussed next.

1. Nonlinear Calibration Parameter

Satellite radiometers generally contain two calibration targets at different temperatures T_W and T_C to measure the slope and intercept parameters. However, to obtain the nonlinear parameter in (A18-12) one must measure a third target at a different temperature. As an alternate approach, the calibration parameters of the MSU's are obtained from prelaunch laboratory radiometer measurements of a variable temperature target. This procedure is discussed in a 1995 paper by Dr. Tsan Mo [47]. The paper lists the measured μ parameters in Table V to be about $10^{-4}\ K^{-1}$ for all channels. Additional laboratory data is shown in Figure A10-2 for the MSU radiometers on TIROS-N and NOAA satellites. These results were published in 2001 by Mo *et al.*, (see citation in Figure) and show the μ parameter at 53.74 GHz for all MSU's plotted as a function of instrument temperature. Note, that the NOAA-12 MSU has the smallest nonlinearity of $10^{-4}\ K^{-1}$. Excluding NOAA-11 measurements, the μ parameters generally range between about $0.5\ \text{x}10^{-4}\ K^{-1}$ and $1.5\text{x}10^{-4}\ K^{-1}$.

Although laboratory measurements are important to provide initial calibration parameters, it is also important to evaluate the accuracy and adjust the radiometer measurements for possible calibration changes following the satellite launch. Such a post-launch procedure was developed by Dr. Konstantin Vinnikov using the small bias observed at 53.74 GHz between different MSU's at low and high latitudes. This latitudinal bias between overlapping satellite radiometers was shown to depend on the detector nonlinearity in addition to errors in the high emissivity calibration target and cold space views. An analysis as well as the procedure for obtaining the calibration corrections was developed [48]. Last but not least, the recalibrated series of MSU's was then analyzed to obtain the brightness temperature trend of 0.17 K/decade shown in Figure 64 for the 53.74 GHz channel. This work was performed by Dr. Konstantin Vinnikov [23], who also showed how well the results compared with surface temperature and climate models. A brief review of the calibration procedure and its analysis is discussed next.

The improved calibration of the 53.74 GHz channel for all MSU's was obtained by partitioning the data into two equal area latitude bands of 30^0 N to 30^0S and between 30^0 and 82^0 in both hemispheres. The analysis approximately accounted for the different observing times (*i.e.*, diurnal variations) seen by each MSU by averaging their ascending and descending orbit measurements which are 12 hours apart. The calibration parameters in (A18-12) were then determined by minimizing the difference between overlapping MSU measurements covering the same time period and latitude bands. For the first nine satellite instruments, Table 1 in the paper [23] shows the change in nonlinear calibration parameter (relative to prelaunch values) varies between $0.2\text{x}10^{-4}$ and $10^{-4}\ K^{-1}$ for the 53.74 GHz channel. Upon recalibration, the globally averaged T_b trend was then found to be the above mentioned 0.17 K/decade. Most significant was the fact that this same trend was found using surface temperature measurements. The paper also discusses the underestimated trend obtained by other researchers using an empirical calibration correction $\Delta T_b = c_0 + c_1(T_w - T_0)$ where T_0 is a fixed temperature rather than $\Delta T_b = \mu (T_b - T_C)(T_w - T_b)$.

To demonstrate the importance of the correction $\mu (T_b - T_C)(T_w - T_b)$, Figure A18-1 plots it as a function of T_b with μ as a parameter for T_W = 280 K and T_C = 3 K. The two curves are for μ = 0.5x10^{-4} K^{-1} and 1.5x10^{-4} K^{-1} which results in a corresponding maximum correction $(\mu/4)(T_W - T_C)^2$ of 0.96 K and 2.88 K when $T_b = (T_W + T_C)/2$ = 141.5 K. However, since the globally averaged brightness temperature at 53.74 GHz is about 250 K in Figure 64, the nonlinear correction is 1.25 K for μ = 1.5x10^{-4} K^{-1}. Although this correction may seem small, it is important to include when measuring small climatic temperature trends. Also, as discussed below, other studies have observed different nonlinear calibration effects. For example, to determine the calibration accuracy of temperature sounders the measurements are compared with those calculated using co-located radiosonde observations (RAOB's) of temperature. These RAOB's are routinely balloon launched to high altitude twice a day from weather stations around the globe to obtain meteorological data. As indicated by (64a), the calculated brightness temperatures are obtained by integrating the RAOB vertical temperature profiles over the weighting functions. However, while the RAOB and satellite results compare well when plotted against another, one observes a small bias and non-unity slope, particularly when using linear calibration. These measurements can be used to correct for calibration error as described below.

2. Nonlinear Calibration Effect

In summary, three different methods can be used to correct for calibration errors, the laboratory procedure, the inter-satellite procedure and the more traditional RAOB comparison method. An early example using RAOB comparisons is described in my paper [49] on severe storm measurements. The study used the first MSU which was launched aboard TIROS-N in 1978. For illustration, Figure A18-3 shows the brightness temperature measurements plotted against all RAOB calculated values having a time difference less than 3 hours between MSU observations. Note that the slope between measured and calculated brightness temperature is smallest for the 53.74 GHz channel and largest for the higher sounding channels at 54.96 and 57.95 GHz whose measurements are lower. This feature is consistent with that shown in Figure A18-1, where the 54.96 and 57.95 GHz measurements are affected more by detector nonlinearity than the 53.74 GHz channel. Such characteristics were also found in other case studies using later MSU's and originally attributed to errors in the peak weighting function height. However, laboratory measurements established that the oxygen absorption model used to calculate the weighting functions is accurate. Therefore, the discrepancy is now considered due to detector nonlinearity error. Interestingly, it was only in the late 1990's, prompted by MSU climate studies, that greater attention was placed on the nonlinear calibration issue and (A18-12) was first implemented.

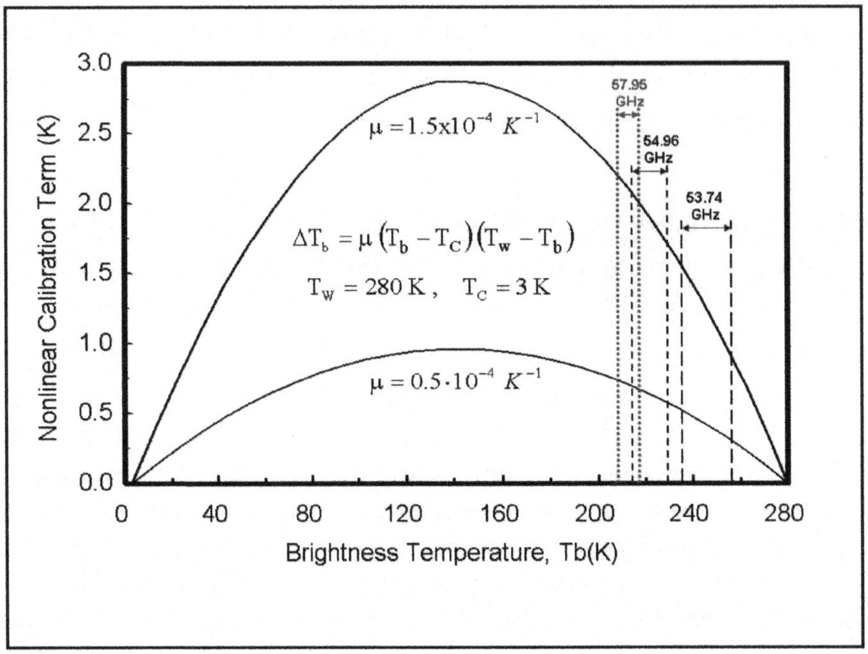

Figure A18-1. Nonlinear calibration term (A18-14) plotted as a function of T_b. Separate plots are shown for $\mu = 1.5\text{x}10^{-4}$ and $0.5\ \text{x}10^{-4}\ K^{-1}$. The measurement range for the three MSU channels is shown by the dashed vertical lines which are based on data shown in Figure A18-3.

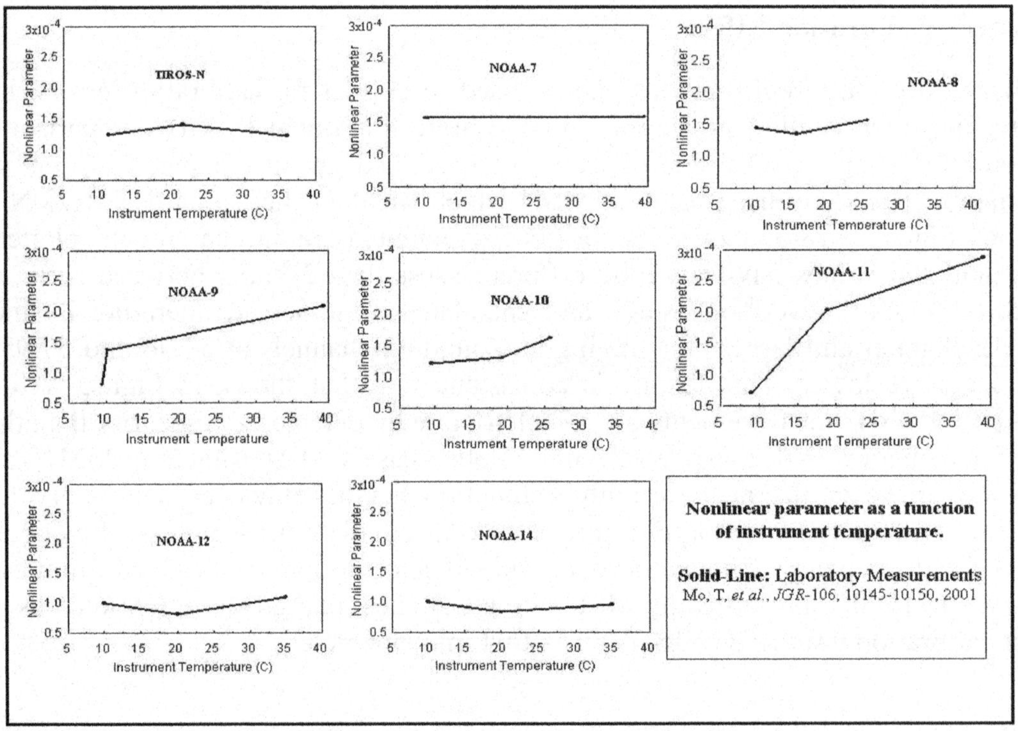

Figure A18-2. Laboratory measured nonlinear calibration parameter μ at 53.74 GHz for the MSU radiometers flown on TIROS-N and NOAA polar orbiting satellites. Except for the NOAA-11 instrument, the nonlinearity mainly varies from about $0.5\ \text{x}10^{-4}\ K^{-1}$ to $1.5\text{x}10^{-4}\ K^{-1}$.

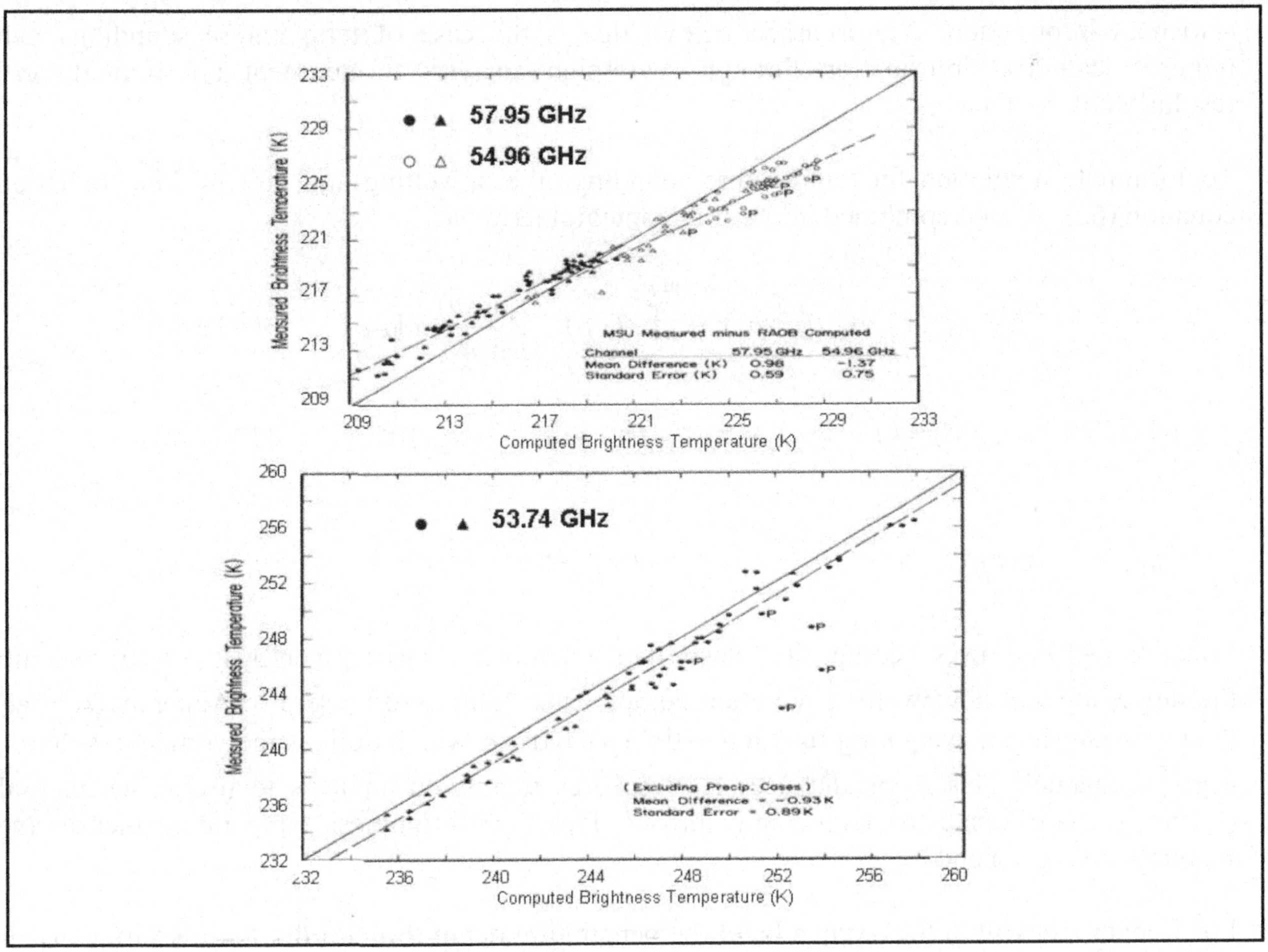

Figure A18-3. MSU brightness temperatures plotted against twice daily RAOB computed measurements for the lower atmospheric channel at 53.74 GHz (Bottom). Similar plots are shown for the 54.96 and 57.95 GHz channels (Top) which sense the upper troposphere and lower stratosphere, respectively. These scatter plots of 67 RAOB match-ups were obtained using twice daily observations (denoted by circles and triangles) for a severe storm over the central United States on April 11, 1979. The data shown here contain all RAOB match-ups over the United States including those around the storm. Also shown are the least squares straight line fit to the data. The analysis is described in the paper [49] referenced in the Text. It is notable that excluding the 5 match-ups containing precipitation (P), the MSU measurements for the lowest sounding channel at 53.74 GHz displays a barely noticeable positive slope when compared with computations. However, the other two MSU channels with higher peaking weighting functions (see Figure 63) display a significant negative slope between the measured and RAOB calculated values. These slopes are consistent with errors in detector nonlinearity since for the same μ parameter the 54.96 and 57.95 GHz channels are seen to produce larger corrections in Figure A18-1 than the 53.74 GHz channel.

A19. Atmospheric Temperature Retrievals

Least squares regression analysis was used in Section 4.1 to calibrate ground-based radiometers by accurately fitting the measurements to the calibration target temperature. In other applications, regression analysis is used to determine geophysical parameters from radiometer measurements. This is often referred to as an *inverse problem* which in general indirectly relates the measurements to geophysical parameters. However, before discussing this I must mention that an example of direct measurements is that of water vapor or *TPW* and cloud liquid water, *Q*. These measurements and its equations were developed both analytically and by regression analysis in Chapter 8. While these vertically integrated quantities have a strong physical connection to radiometer measurements, other parameters are less directly related and must be obtained by incorporating statistical information. As discussed below, this is the case of temperature soundings, which requires statistical information through regression analysis to augment the limited vertical resolution of radiometers.

To formulate a solution for temperature soundings the upwelling and downwelling radiation in equation (62) are first combined into a single integral term, *viz.*,

$$T_b(\nu,\theta) = \tau'_\nu(p_S,\theta)T(p_S) - \int_{-\infty}^{\ln p_S} T(p)\frac{d\tau'_\nu(p,\theta)}{d\ln p}\,d\ln p \qquad \text{(A19-1a)}$$

where $$\tau'_\nu(p,\theta) = \left\{1-(1-\varepsilon_S)\left[\hat{\tau}_\nu(P_S)/\hat{\tau}_\nu(p)\right]^{2Sec\theta}\right\}\hat{\tau}_\nu(p)^{Sec\theta} \qquad \text{(A19-1b)}$$

with $$\hat{\tau}_\nu(p) = e^{-\int_z^{\infty}\gamma_\nu(z')\,dz'} = e^{-\frac{1}{g}\int_0^{p}\frac{\gamma_\nu(p')}{\rho}dp'}\ . \qquad \text{(A19-1c)}$$

Equation (A19-1a) uses a composite transmittance function $\tau'_\nu(p,\theta)$ given by (A19-1b) to combine the upwelling and downwelling radiation components. More will be said in Appendix A20 about the Gaussian shaped weighting function $-d\tau'_\nu(p,\theta)/d\ln p$ which defines the vertical resolution of a given channel. This Appendix uses least squares regression analysis to invert the integral in (A19-1a) and estimate the temperature profile T(p) from brightness temperature measurements within the oxygen band.

For frequencies within the oxygen band the penetration depth through the Earth's surface is small so $T(p_S)$ in (A19-1a) is the skin temperature. Since the Gaussian-shaped weighting function varies slowly with pressure, the integral can be approximated by a limited number of terms. The brightness temperature is then given by

$$T_b(\nu_m,\theta) = \sum_{n=1}^{N} c_n(\nu_m,\theta)\,T(p_n) + \tilde{T}_b(\nu_m) \qquad m = 1,\cdots M \qquad \text{(A19-2)}$$

with $c_1(\nu,\theta) = \varepsilon_S\,\hat{\tau}_\nu(p_s)^{Sec\theta}$ where p_1 is the surface pressure p_S . All other coefficients $c_n(\nu_m,\theta)$ depend on the weighting functions and interpolation formula (*e.g.*, trapezoidal rule) used in the series representation. In addition to quadrature error, which reduces as the number of terms increase, the brightness temperature contains instrumental noise and biases due to calibration error. These errors are given by the term $\tilde{T}_b(\nu_m)$.

Other than temperature, which is the major parameter, the other variables in (A19-1) are the viewing angle θ and surface emissivity. As discussed in Chapter 10, many temperature sounders use cross-track scanners so that the angular scan produces an upward vertical shift of the weighing function distribution as shown in Figure 63 for MSU. As mentioned in Chapter 11, this effect can be reduced using limb corrections to normalize the measurements to nadir viewing, or by simply determining separate coefficients $c_n(\nu_m,\theta)$ for each angle. Therefore, in what follows, the angular dependence in (A19-2) is omitted. On the other hand, the emissivity mainly affects oxygen channels with frequencies less than about 53.7 GHz since then $\hat{\tau}_\nu(p_S) > 0.1$. For these lower sounding channels the emissivity must be estimated based on surface type or determined more accurately from measurements. The following describes an example of the procedure using MSU.

Since the MSU only contained channels within the oxygen band, its most transparent channel at 50.3 GHz was used together with the more opaque channel at 53.74 GHz to estimate emissivity.

The equation is

$$\varepsilon_S = a_0 + a_1\, T_b(50.30) - a_2\, T_b(53.74) \qquad \text{(A19-3)}$$

where the a_n coefficients are obtained from regression analysis using simulated measurements from radiative transfer calculations [49]. At nadir viewing a_0 =3.62, a_1 = 0.00952 and a2 = 0.02095. The algorithm is loosely based on the equation $\varepsilon_S = T_b / T_S$ where T_b is the lowest frequency MSU channel at 50.3 GHz and the surface temperature T_S is inferred from the 53.74 GHz channel. This inference is based on the fact that although the 53.74 GHz channel measures temperature above the surface (~700 mb) it is coupled to surface temperature through vertical convection. In fact, as mentioned in Chapter 11, this coupling produces the same climatic temperature trend as surface temperature measurements [23].

The 50 GHz emissivity obtained using (A19-3) is shown in Figure A19-1 on the left. As expected, the emissivity over land is about 0.95, decreasing to less than 0.9 due to surface wetness from the heavy rain. The rain area is shown by the radar map on the right. Lowest emissivities between 0.8 and 0.9 are seen along the Mississippi river, Great Lakes areas and coastal boundaries. All of the low emissivity regions are due to water but enlarged due to the large MSU field of view of 110 km at nadir. However, this spatial averaging of the 50 GHz emissivity is not a problem when used to correct the 53.74 GHz channel which has the same resolution. More accurate emissivity is obtained from sounders such as AMSU, which has higher spatial resolution and contains more transparent window channels. Such higher resolution and additional channels are important for measuring the surface features discussed in Chapter 12.

Returning back to the temperature retrieval problem, microwave sounders use multiple frequencies within the oxygen band to invert equation (A19-2) using simultaneous equations. To obtain a solution, the equations are expressed in matrix form containing K columns for each pair of coincident brightness temperature and temperature measurements. The T_b matrix has M rows for each frequency measurement while the T matrix contains N rows for each pressure level, *i.e.*,

$$\begin{vmatrix} T_{b1}(\nu_1) & \cdots & T_{bK}(\nu_1) \\ T_{b1}(\nu_2) & \cdots & T_{bK}(\nu_2) \\ \vdots & \cdots & \vdots \\ T_{b1}(\nu_M) & \cdots & T_{bK}(\nu_M) \end{vmatrix} = \begin{bmatrix} c_{1,1} & \cdots & c_{1,N} \\ c_{2,1} & \cdots & c_{2,N} \\ \vdots & \cdots & \vdots \\ c_{M,1} & \cdots & c_{M,N} \end{bmatrix} \cdot \begin{vmatrix} T_1(p_1) & \cdots & T_K(p_1) \\ T_1(p_2) & \cdots & T_K(p_2) \\ \vdots & \cdots & \vdots \\ T_1(p_N) & \cdots & T_K(p_N) \end{vmatrix} \qquad \text{(A19-4)}$$

The elements $T_{bK}(\nu_m)$ also contain the radiometer bias and its noise. Alternatively, one can include these errors as a separate column matrix of M rows with elements $\widetilde{T}_b(\nu_m)$. The corresponding temperature matrix on the right has elements $T_k(p_n)$. Connecting these is the matrix of coefficients $c_{m,n} = c_n(\nu_m)$ which has M rows with N columns. Due to the overlapping and smooth characteristic of the Gaussian-shaped weighting functions the inversion of (A19-3) to obtain T is ill-conditioned and particularly sensitive to the measurement noise in T_b. Also, the matrices are rectangular rather than square so a pseudoinverse solution is required to estimate T.

To obtain the inverse solution, (A19-4) is first written in matrix notation, *i.e.*,

$$B_{M,K} = C_{M,N} T_{N,K} \qquad \text{(A19-5)}$$

where $B_{M,K}$, $C_{M,N}$ and $T_{N,K}$ are the respective brightness temperature, coefficient, and temperature matrices. Using matrix multiplication, each element in $B_{M,K}$ has the form of (A19-2). Also, since (A19-5) is linear, its inverse is given by the linear equation, *i.e.*,

$$\hat{T}_{N,K} = A_{N,M}\, \hat{B}_{M,K}\ , \qquad \text{(A19-6a)}$$

$$\text{where } \hat{T}_{N,K} = T_{N,K} - \overline{T}_N \text{ and } \hat{B}_{M,K} = B_{M,K} - \overline{B}_M\ . \qquad \text{(A19-6b)}$$

The matrix $A_{N,M}$ is rectangular with elements referred to as regression coefficients. Also, $\overline{T}_N$ and $\overline{B}_M$ are column matrices of the *a-priori mean* of temperature (at each pressure level) and brightness temperature (at each frequency), averaged over the K data samples. These quantities are related by $\overline{T}_N = A_{N,M}\overline{B}_M$. Also, $T_{N,K}$ is referred to as the dependent variable or predictand while $B_{M,K}$ is called the independent variable or predictor. Alternatively, using remote-sensing terminology $T_{N,K}$ is referred to as the **retrieved temperature** with $A_{N,M}$ the **retrieval coefficients**. Moreover, the procedure to obtain the retrieval coefficients is referred to as **regression analysis**.

Using matrix operation[10], the $A_{N,M}$ matrix is obtained by multiplying the right sides of (A19-5a) by the transpose, t, of $\hat{B}_{M,K}$ so

$$\hat{T}_{N,K}\hat{B}^t_{M,K} = A_{N,M}\hat{B}_{M,K}\hat{B}^t_{M,K}\,. \qquad \text{(A19-7)}$$

As such, $\hat{B}_{M,K}\hat{B}^t_{M,K}$ is a square matrix of dimensions MxM. Assuming this matrix is non-singular, it can be inverted so

$$A_{N,M} = \hat{T}_{N,K}\hat{B}^t_{M,K}[\hat{B}_{M,K}\hat{B}^t_{M,K}]^{-1}\ . \qquad \text{(A19-8)}$$

To avoid $[\hat{B}_{M,K}\hat{B}^t_{M,K}]^{-1}$ being singular, the K pair of brightness temperatures must be independent by using a wide distribution of atmospheric soundings from the tropics, mid-latitudes and polar regions. Also, the $\hat{T}_{N,K}$ matrix elements in (A19-6a) are linear combinations of brightness temperatures at different frequencies, *i.e.*, $A_{N,1}\hat{T}_{bK}(\nu_1)+\cdots+A_{N,M}\,\hat{T}_{bK}(\nu_M)$. This is the same form as equation (66) in Chapter 11. The regression coefficients $A_{N,M}$ in (A19-8) depend on the square covariance matrix $\hat{B}_{M,K}\hat{B}^t_{M,K}$, with each element being the product of brightness temperatures at

---

[10] The inverse of (A19-5) written as (A19-6a) is called a pseudoinverse since it only minimizes the least square error or Euclidian distance between $\hat{T}_{N,K}$ and $A_{N,M}\hat{B}_{M,K}$. Using this criteria, $A_{N,M}$ is obtained by minimizing the function $F(A) = \mathrm{tr}[(\hat{T} - A\,\hat{B})(\hat{T} - A\,\hat{B})^t]$ with the subscripts removed. Upon differentiating $F(A)$ with respect to A and setting it to zero we obtain $2\hat{T}\hat{B}^t - 2A\hat{B}\hat{B}^t = 0$ so $A = \hat{T}\hat{B}^t[\hat{B}\hat{B}^t]^{-1}$, which is also given by (A19-8).

M different frequencies. Also, the rectangular cross-covariance matrix $\hat{T}_{N,K}\hat{B}^t_{M,K}$ .of dimensions NxM contains the correlation or product of atmospheric temperature and brightness temperatures.

The $A_{N,M}$ coefficients can be obtained using actual measurements of co-located temperatures and brightness temperatures, or from simulations obtained using radiation transfer calculations. When using simulations, the T_b measurements in B_{MK} are calculated using an historical collection of global RAOBS of temperature and water vapor soundings. These same temperatures T are contained in $T_{N,K}$ to obtain the covariance matrix. In contrast to this physically-based approach, the empirical method uses co-located global match-ups of actual T_b and coincident RAOB temperatures to obtain the B_{MK} and $T_{N,K}$ matrices in (A19-8). In summary, both techniques employ regression analysis to generate the relationship (A19-6a) or equation (66), although each has their advantage and disadvantage. For simulations to be accurate they must include instrumental noise in addition to variations due to clouds and emissivity by modeling these parameters. In comparison, the empirical method contains match-up errors due to differences in observing time and spatial averaging of satellite measurements compared to localized RAOB temperatures. However, except for rapidly changing meteorological events such as severe storms the space-time averaging is generally small.

As an example of the simulation approach, Figure A19-2 shows results for the MSU temperature sounder. The right-most Figure displays the weighting functions for its three sounding channels at 53.74, 54.96 and 57.95 GHz. It also shows the weighting functions at additional frequencies of 53.4, 54.35 and 55.50 GHz which are similar to those included in more advanced radiometers such as AMSU and SSMI/S. These additional channels were used in an early study [60] to simulate the improved temperature accuracy when using a hypothetical five-channel instrument at 53.4, 54.35, 54.96, 55.50 and 57.95 GHz. The plot on the left-side of the Figure shows the standard error of temperature retrievals for the 3-channel MSU compared with the 5-channel instrument. Both results are based on the above-mentioned regression analysis which uses radiative transfer to calculate brightness temperatures using an historical sample of 400 global RAOB temperature profiles together with simulated cloud and surface emissivity variations. The data is then used to derive the retrieval coefficients $A_{N,M}$ for both sounders and retrieve temperature profiles. These simulated retrievals are then compared against the same data used to obtain the error distribution in Figure A19-2. Therefore, there is no bias or mean difference so the standard error is the same as the RMS error. Incidentally, the improved accuracy shown when adding two more channels to MSU is very similar to that found from actual measurements when comparing the MSU with SSM/T soundings in Figure 68 on page 99.

1. Retrieval Accuracy

The following, more fully analyzes the effect of additional channels on temperature accuracy. To begin the analysis, the temperature retrieval is written as

$$T'(p) = \overline{T}(p) + \sum_{m=1}^{M} A_m(p)\left[T_b(\nu_m) - \overline{T}_b(\nu_m)\right] \qquad \text{(A19-9)}$$

where $T'(p)$ is the temperature at pressure p obtained from the linear combination of brightness temperatures at frequencies ν_m. This temperature is designated as $T'(p)$ to distinguish it from the true temperature $T(p)$. As with (A19-6a), $\bar{T}(p)$ and $\bar{T}_b(\nu_m)$ are the sample-averaged temperature and corresponding brightness temperature used to obtain the coefficients, written as $A_m(p)$. Substituting (A19-1a) into (A19-9) to eliminate brightness temperatures we obtain

$$T'(p) - \bar{T}(p) = \Delta T(p_S)\,\Im(p, p_S) \;-\; \int_{-\infty}^{\ln Ps} \Delta T(p')\frac{d\,\Im(p,p')}{d\ln p'}\,d\ln p' \qquad \text{(A19-10a)}$$

$$\text{where}\quad \Delta T(p) = T(p) - \bar{T}(p) \quad\text{and}\quad \Im(p,p') = \sum_{m=1}^{M} A_m(p)\,\tau'_{\nu m}(p') \qquad \text{(A19-10b)}$$

The left side of (A19-10a) is the difference between the estimated and *a-priori* mean profile, while the right side contains the difference between the true and mean temperature profile, $\Delta T(p)$. The difference on the left is zero when $\Delta T(p) = 0$ or when the Kernel $-d\,\Im(p,p')/d\ln p'$ approaches a delta function. Also, when the Kernel is unimodal and centered near $p' = p$, the integral in (A19-10a) represents the average temperature near pressure p. This occurs above or below the tropopause since the temperature then generally varies with a nearly constant lapse rate. Conversely, the largest errors appear near the tropopause where the temperature profile has large curvature. This explains why the errors in Figure A19-2 are minimal near 500 mb and 70 mb for both sounders and largest near 200 mb. Also, for tropical atmospheres the brightness temperature from single channels generally follow its smoother temperature profile, while the same weighting functions result in measurements less capable of resolving the more abrupt changes in polar and mid-latitude regions without introducing statistical information. This point will be expanded upon in a case study performed below. However, we begin this study by first examining the effect of statistical correlation on retrieval accuracy.

The statistical correlation between temperature and brightness temperature in the cross-correlation matrix of (A19-8) produce regression coefficients that alter the Kernel from unimodal to bimodal. This is seen in Figure A19-3 by displaying the Kernels at 850, 500, 250 and 150-mb using the $A_m(p)$ coefficients obtained for the 3- and 5-channel sounder. Observe that while the 500-mb function is unimodal and significantly reduced in width by additional channels, little improvement occurs for the 850- and 250-mb levels. On the other hand the 150-mb Kernel is bimodal with a sharper positive and negative mode for the five-channel sounder due to strong inter-level correlation between stratospheric and tropospheric temperature. Also, the 70-mb Kernel (not shown) is mainly defined by the 57.95 GHz weighting function for both the 3- and 5-channel sounder. These different Kernels explain why the errors are much less at 500- and 150 mb for the 5- versus 3-channel sounder, while the errors are nearly the same at 850- and 250 mb..

2. Case Study of a Frontal System

To examine the measurements and retrievals in detail a case study [50] was performed for a cold front that formed over the Central United States on April 5, 1979. This study was performed the first year after MSU was launched on the TIROS-N satellite and was used to demonstrate the capability and limitations of NOAA's first operational microwave sounder.

Figure A19-4 displays a map on the top-left of the MSU scan coverage (o) and radiosonde locations (Δ) over the area of interest. It also shows a north-south transit passing through the radiosonde (RAOB) locations that will be used to study the difference between the actual MSU measurements at 1000 GMT and the RAOB's at 1200 GMT. The other maps in the Figure display the contours of the limb-corrected MSU brightness temperatures for its channels at 53.74, 54.96

and 57.95 GHz. Also indicated on each map is the difference between the measured and calculated brightness temperatures from RAOB temperature soundings. Figure A19-5 shows the RAOB soundings (solid line) and MSU temperature retrievals (dotted line) along the transit shown in Figure A19-4 on the top-left.

What is most interesting is how well the satellite retrieved temperatures follow the RAOB soundings even though the MSU only contains three sounding channels. This is particularly true for the southern-most or tropical sounding at Chihuahua Mexico (RAOB #225) and northern-most or polar sounding at Mosonee Ontario (RAOB #836). Conversely, the largest difference occurs in the transition region between the tropical and polar air mass where the MSU retrieval fails to capture the temperature inversion near the surface and abrupt changes in lapse rate at Green Bay Wisconsin (RAOB #645). For additional comparisons, the Figure shows the soundings retrieved using simulated brightness temperatures for the MSU and the 5-channel sounder discussed previously (see Figure A19-2). The simulated temperatures are indicated at the standard pressure levels by the symbols Δ for the 3-channel MSU and o for the 5-channel sounder .As expected, the simulated and actual MSU temperature retrievals are almost the same. Also, as expected, the 5-channel temperature retrievals follow the RAOB profiles much better than MSU, particularly for the central RAOB's in Peoria Illinois (# 532) and Topeka Kansas (# 456).

These comparisons between the 3- and 5-channel temperature soundings is shown better in Figure A19-6 by plotting the RMS temperature difference (Left) and mean difference (Right) relative to the seven RAOB measurements. Note that both the RMS and difference is much less for the 5-channel microwave sounder at all pressure levels, with the largest improvement around 500 mb and 150 mb. This is the same result found in the independent analysis of retrieval errors shown in Figure A19-1 for the two sounders as well as in Figure 68 between the MSU and SSM/T microwave temperature sounders

3. Summary

These satellite derived temperature retrievals from microwave radiometers were used to provide input data to numerical forecast models by organizations such as the National Centers for Enviromental Prediction[11] of NOAA and the European Center of Medium-Range Weather Forecast (ECMWF). However, soon after the launch of AMSU the ECMWF felt that the satellite measurements are best applied directly into numerical forecast models using the raw brightness temperature measurements rather than using the temperature retrievals. This different approach was based on the fact that the radiometers provide their greatest accuracy when being used directly through their vertically weighted temperatures (*i.e.*, brightness temperatures) rather than indirectly using the retrieved temperatures at discrete pressure levels.

Therefore, to best utilize satellite temperature measurements the data assimilation component of these forecast models began using a variational approach to constrain or minimize the difference between satellite brightness temperature measurements and similar vertical weighted measurements acquired from conventianal sources such as radiosondes, ships, buoys, aircraft, etc.,. Furthermore, these satellite measurements are also found to provide the greatest impact on forecasting and analysis when used over the remote data sparse oceans as well as the Southern Hemisphere which suffers from limited surrounding *in-situe* observations. However, in these data sparse areas, satellite retrieved temperature soundings are important since conventional observations are unavailable for input to the forecast models.

---

[11] Originally the National Meteorological Center (NMC), now called NCEP.

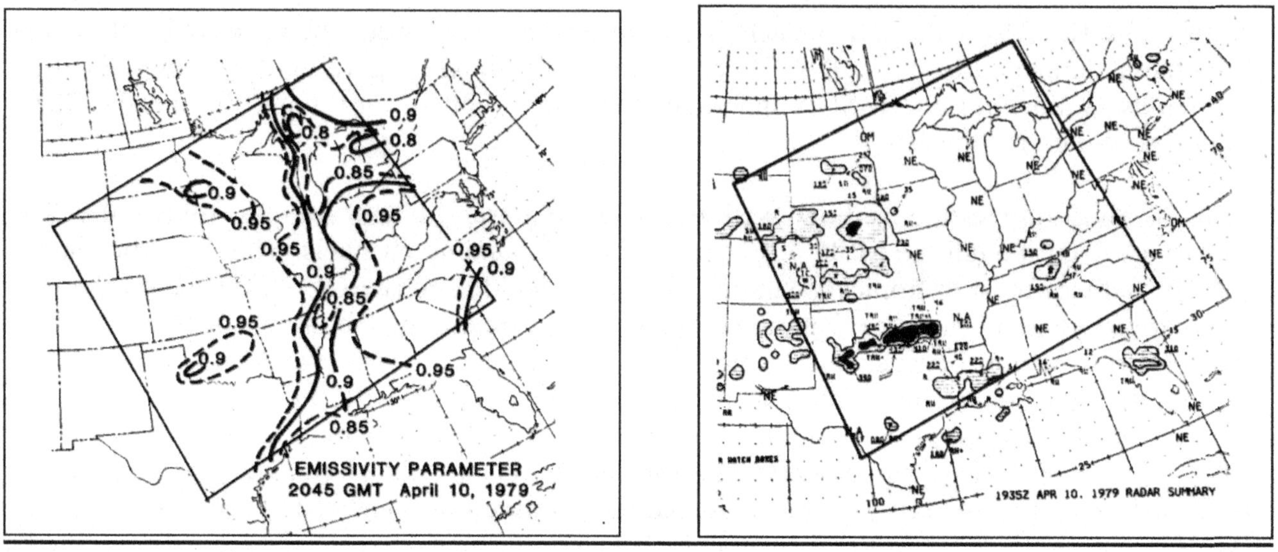

Figure A19-1. The left-most figure shows a contour map of the MSU emissivity at 50 GHz over the Eastern United States on April 10, 1979 at 2045 GMT. The emissivities are derived using the 50.3 and 53.74 GHz measurements in accordance with equation (A19-3). Besides the Mississippi river and Great lakes, emissivities less than 0.95 result from surface wetness due to rainfall. This is particularly evident over north Texas and Oklahoma as seen by the radar measurements on the right, which was obtained at 1935 GMT.

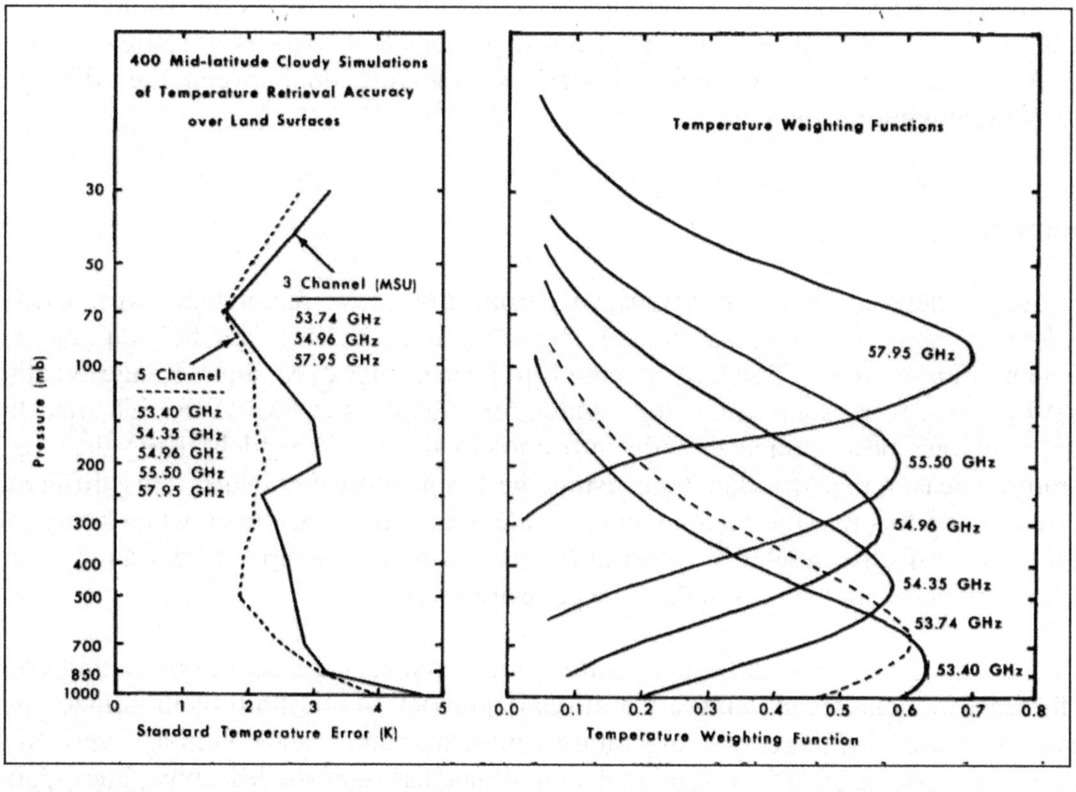

Figure A19-2. Weighting functions (Right) for the MSU channels (53.74, 54.96, 57.95 GHz) and a 5-channel sounder (53.40, 54.35, 54.96, 55.50, 57.95 GHz). Also shown on the Left are the simulated standard errors of temperature obtained using a 3-channel MSU instrument (solid line) and 5-channel sounder (dotted line).

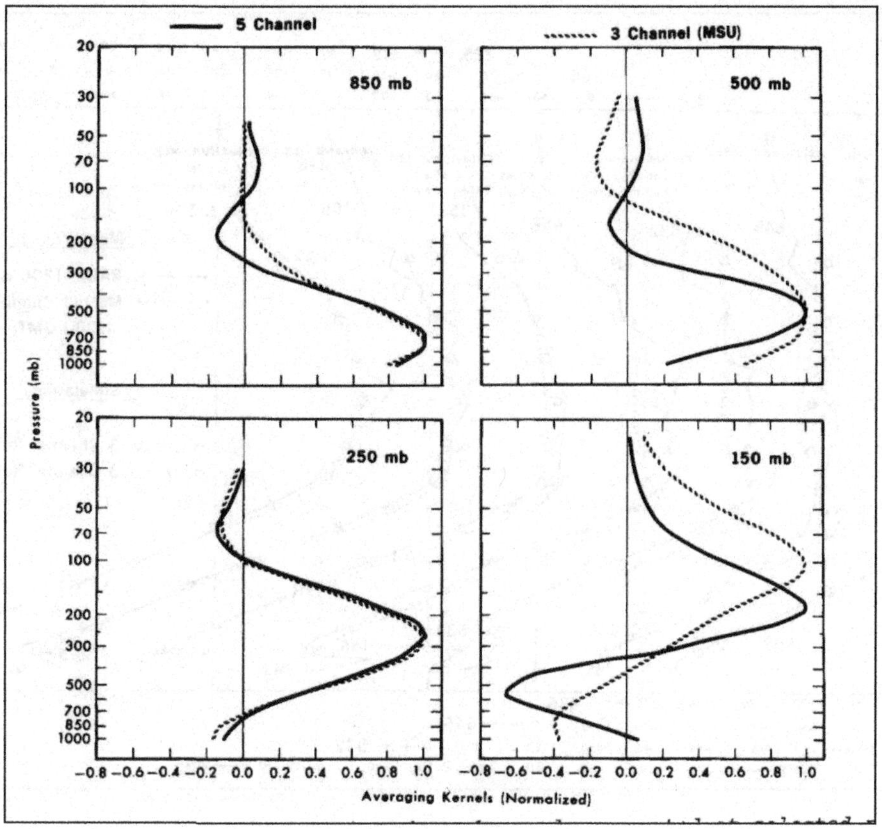

Figure A19-3. Kernel functions at different pressure levels for MSU (dotted) and a five-channel sounder (solid) whose frequencies are shown in Figure A19-2.

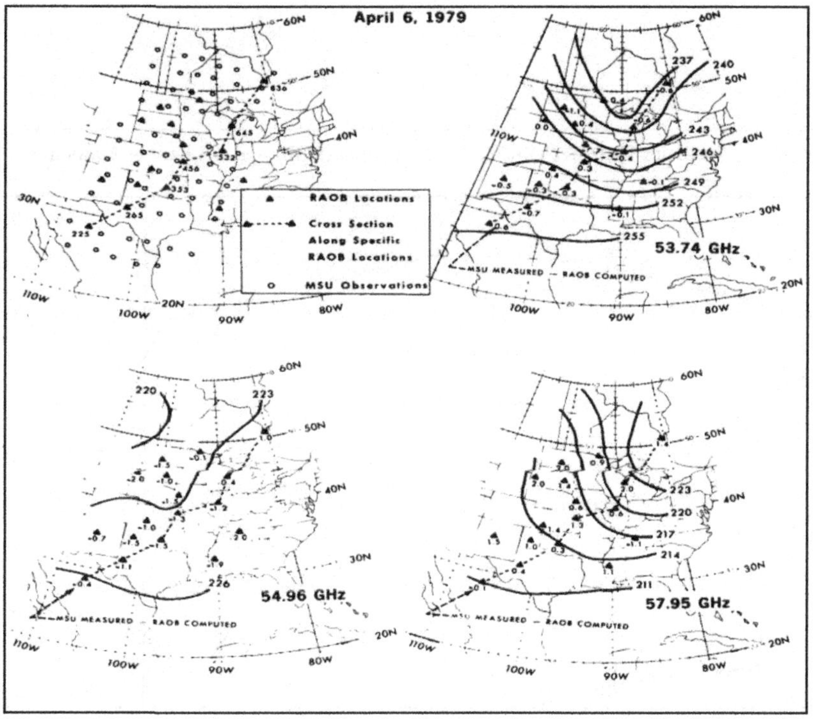

Figure A19-4. Radiosonde and satellite (MSU) observations of a cold front over the Central United States on April 5, 1979. The upper left corner shows the MSU observations at 1000 GMT (○) and radiosonde locations at 1200 GMT (Δ). The other figures show contour maps of the MSU brightness temperatures at 53.74, 54.96 and 57.95 GHz as well as brightness temperature differences relative to radiosonde – computed values (K).

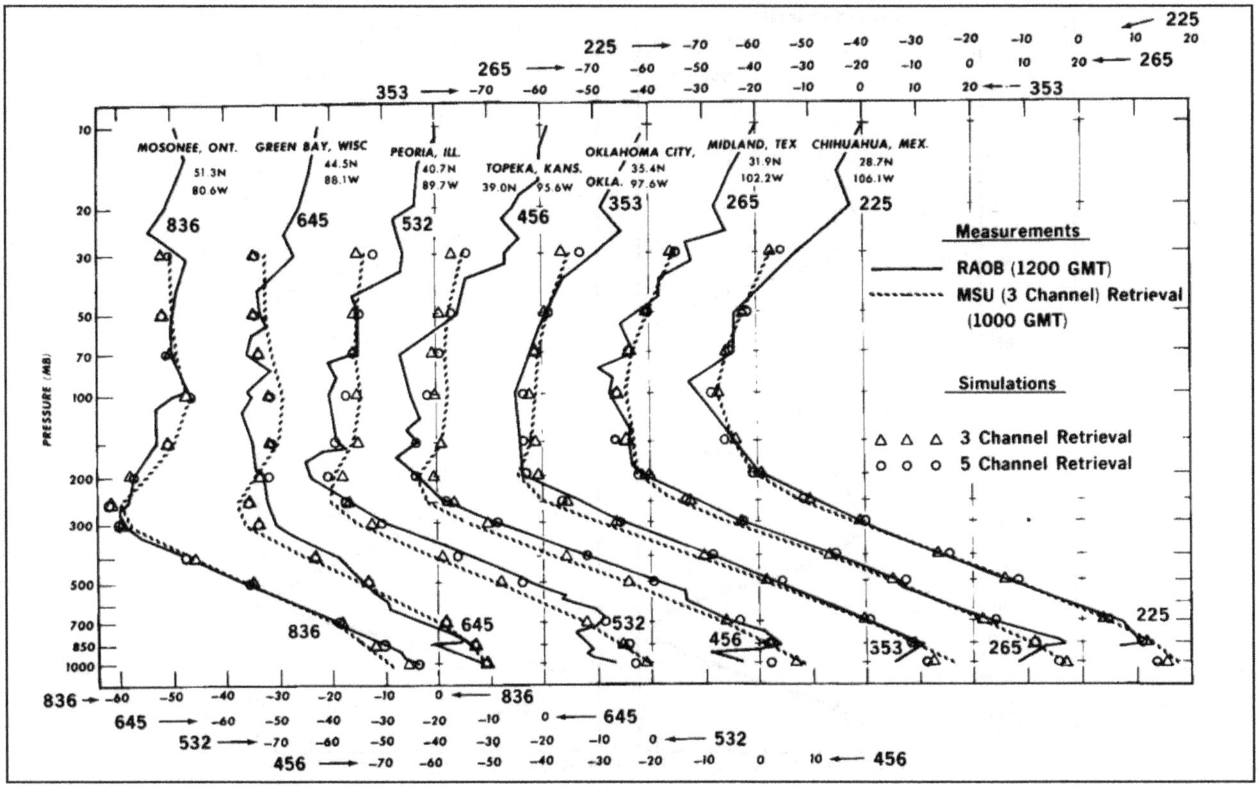

Figure A19-5. Comparisons between soundings from radiosondes (solid) and those retrieved from MSU measurements (dotted) for the cross-section shown in Figure A19-4 (upper left corner). Radiosonde locations are referred to by station number. Also shown are the temperatures retrieved using simulated brightness temperatures for the three MSU channels (Δ) and a five channel sounder ($\circ$) whose weighting functions are shown in Figure (A19-2). Note that the temperature scale is above and below the plots.

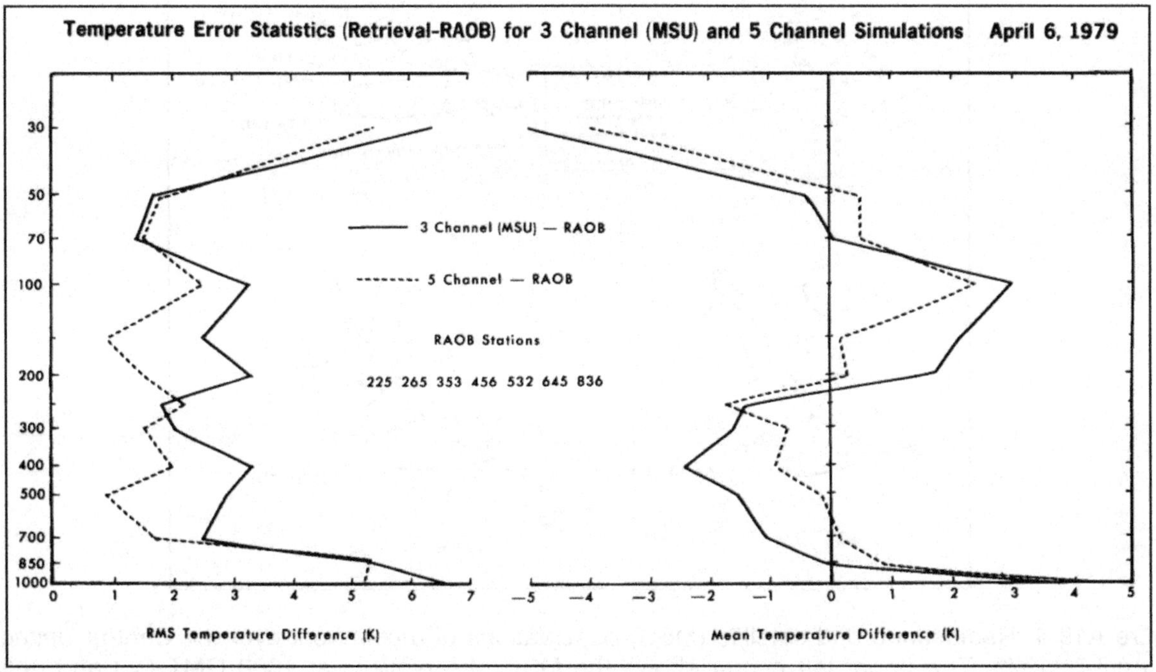

Figure A19-6. Vertical distribution of root-mean-square error (Left) and mean difference (Right) relative to radiosondes for the 3-channels of MSU (solid) and 5-channel sounder (dashed). The results use simulated brightness temperatures to produce the soundings shown in Figure (A19-5).

A20. Temperature Weighting Function

This appendix develops the temperature weighting functions resulting from oxygen absorption. It was first mentioned in Chapter 11 and approximated by equations (65a) and (65b), whose derivation is now given. However, we begin the discussion by first referring to the water vapor absorption mentioned in Section 8.7. Its absorption is defined by a single line centered at 22.235 GHz with its line shape approximated by equation (52b). In comparison to water vapor, molecular oxygen contains a complex of 45 lines centered at 60 GHz with a total spread of ±10 GHz. The line separation ranges from 500 MHz for the weaker lines to 120 MHz between the most intense lines near 60 GHz. Therefore, the absorption coefficient for oxygen contains a series of contributions $\gamma_{i,j}$,

$$\gamma_v(p) = \sum_{i,j} \gamma_{i,j} \qquad \text{(A20-2)}$$

where the absorption due to each line depends on the frequency, molecular density, line strength, as well as other factors. From quantum theory, the absorption coefficient due to a single spectral line is

$$\gamma_{i,j} = \frac{8\pi^3 \nu N}{3hc} f(\nu_{i,j}, \nu) \left|\mu_{i,j}\right|^2 N_i \qquad \text{(A20-3)}$$

where h is Plank's constant, c is the speed of light, N is the molecular density. Equation (A20-3) also contains the matrix elements $\mu_{i,j}$ which define the dipole strength of the oxygen molecule. It represents the transition probability connecting two stationary states of energy E_i and E_j. The frequency and separation between spectral lines $\nu_{i,j}$ is given by the Bohr condition $h\nu_{i,j} = E_i - E_j$. Also, from Boltzmann's theory the population of energy levels N_i for each line is proportional to $\mathrm{Exp}[-E_i/KT]$, having statistical weighting factors accounting for degeneracy. This relationship assumes thermodynamic equilibrium and holds well into the mesosphere for microwave transitions.

As in the case of water vapor, the resonant character of the oxygen absorption spectrum is described by a line shape factor $f(\nu_{i,j}, \nu)$. In all but the very highest regions of the atmosphere (≥ 70 km), it is collisions between molecules that determines $f(\nu_{i,j}, \nu)$. This form of line shape then depends on the response of a molecule following a collision which results in the pressure broadening of the spectral lines. At microwave frequencies the line shape factor can be expressed as

$$f(\nu_{i,j}, \nu) = \frac{\nu}{\pi \nu_{i,j}} \left[\frac{\delta\nu_{i,j}}{(\nu - \nu_{i,j})^2 + \delta\nu_{i,j}^{\,2}} + \frac{\delta\nu_{i,j}}{(\nu + \nu_{i,j})^2 + \delta\nu_{i,j}^{\,2}} \right] \qquad \text{(A20-4)}$$

where $\delta\nu_{i,j}$ is the linewidth due to collisions. Similar to water vapor, the linewidth is proportional to the density or pressure of the atmosphere ($\delta\nu_{i,j} = \Gamma_{ij} P$ where $\Gamma_{ij} \approx 2.5$ MHz / mbar). However, in addition to collisional processes, the detailed behavior of the shape factor can also depend on interference effects between spectral lines as developed by Dr. Phil. Rosenkranz [51]. He also provides a complete analysis on the microwave absorption by atmospheric gasses in Chapter 2 of the book edited by Michael Janssen [7].

The absorption coefficient γ_{ij} is obtained using (A20-3) and (A20-4), with the matrix elements $\mu_{i,j}$ linewidths $\delta\nu_{i,j}$ and line centers $\nu_{i,j}$ determined using laboratory measurements and theoretical calculations. For frequencies near the line centers, the major contribution results from the 1st (resonant) term in the brackets of (A20-4). Conversely, for frequencies outside the oxygen and water vapor regions (*i.e.*, window channels), the absorption is characterized by the 2nd (non-resonant) term. However, an even larger simplification occurs when the absorption is dominated by one spectral line. This results in the same Lorentz line shape used for water vapor absorption and was approximated by equation (52b). If we further consider that the channel response is dominated by one line, the absorption coefficient (A20-3) is expressed by a single term so

$$\gamma_\nu = k_{ij}\, N \frac{\nu^2\, \delta\nu_{ij}}{(\nu-\nu_{ij})^2 + \delta\nu_{ij}^{\;2}} \quad \text{with} \quad k_{ij} = \frac{8\pi^3}{3hc}\, \frac{|\mu_{i,j}|^2}{\pi\nu_{i,j}} \qquad \text{(A20-5)}$$

where $k_{i,j}$, $\delta\nu_{i,j}$ and $\nu_{i,j}$ is the strength, width and line centers. While this is a good approximation at low pressures, the strong collisional or pressure broadening of individual lines begins to overlap at high pressures so one must include effects of additional lines for a more accurate representation. Nevertheless, we shall use this Lorentz line shape to obtain the weighting function and examine its characteristics. Furthermore, at frequencies far from the line center (*i.e.*, $|\nu - \nu_{i,j}| > \delta\nu_{i,j}$) the absorption coefficient can be further approximated by neglecting $\delta\nu_{i,j}^{\;2}$ in the denominator of (A20-5a) compared to the term $(\nu-\nu_{ij})^2$ so

$$\gamma_\nu \approx k_{ij}\, N \frac{\nu^2\, \delta\nu_{i,j}}{(\nu-\nu_{i,j})^2}\,. \qquad \text{(A20-6)}$$

For a uniformly mixed gas such as oxygen, the ratio of gas density to that of dry air N/ρ is constant so upon substituting (A20-6) into (A19-1c) we obtain

$$\hat{\tau}_\nu(p)^{\mathrm{Sec}\theta} = e^{-\left(\frac{p}{P_\nu\sqrt{\mathrm{Cos}\theta}}\right)^2} \quad \text{where} \quad P_\nu = \left(\frac{\nu-\nu_{i,j}}{\nu}\right)\sqrt{\frac{2g}{k_{i,j}\Gamma_{i,j}}\left(\frac{\rho}{N}\right)}\,. \qquad \text{(A20-7)}$$

Finally, the weighting function given by $-d\hat{\tau}_\nu(p)^{\mathrm{Sec}\theta}/d\ln p$ then becomes,

$$-\frac{d\hat{\tau}_\nu(p)^{\mathrm{Sec}\theta}}{d\ln p} = 2\,X_\nu\, e^{-X_\nu} \quad \text{with} \quad X_\nu = \left(\frac{p}{P_\nu\sqrt{\mathrm{Cos}\theta}}\right)^2 \qquad \text{(A20-8)}$$

This weighting function approximates the result obtained using the exact absorption model [51]. However, to obtain a better fit with the exact model the transmittance (A20-7) is modified as $\tau_\nu = \mathrm{Exp} -[p/P_\nu\sqrt{\mathrm{Cos}\theta}\,]^{\mathrm{n}}$ where values of n different than 2 is used to account for the added absorption from overlapping lines at higher pressures. Figure 65 shows the weighting functions are approximately Gaussian-shaped with a maximum of $2e^{-1}$ at pressure $P_\nu\sqrt{\mathrm{Cos}\theta}$, and its full width at half maximum is also $P_\nu\sqrt{\mathrm{Cos}\theta}$. Although not displayed in Figure 65, the AMSU weighting functions peak higher in the atmosphere, or at lower pressures, as the scan angle increases relative to the nadir viewing position. This limb effect is shown for the MSU weighting functions in Figure 63 and contained in (A20-8), where the pressure at peak height is $P_\nu(\theta) = P_\nu(0^{\circ})\sqrt{\mathrm{Cos}\theta}$. Also, from (A20-8), the peak height is proportional to the frequency displacement from the line center $\nu_{i,j}$ and inversely proportional to the line strength $\kappa_{i,j}$. Therefore, the weighting function is highest for frequencies near the most intense lines, which occurs near the center of the oxygen band at 60 GHz.

Prior to defining the AMSU frequencies listed in Table 7, extensive analysis using simulations was performed to arrive at the optimal frequencies and bandwidths necessary to obtain the highest temperature accuracy. The resulting weighting functions are shown in Figure 65 with the simulated brightness temperatures shown in Figure 66 (see expanded spectrum). Note that channels 10 through 14 have the highest peaking weighting functions with very narrow bandwidths. They therefore respond to the higher or warmer stratospheric temperatures seen by the profiles in Figure 61. Conversely, the lower peaking channels 3 through 9 have broader bandwidths and respond to tropospheric temperatures, which can be lower or colder than in the stratosphere. Also note that most of the channels are between two adjacent oxygen lines so that large bandwidths can be used to reduce instruments thermal noise without appreciably smoothing the spectral response or broadening the weighting functions. Of necessity, only the highest sounding channels reside along the wings of the most absorbing lines with narrower bandwidths.

A21. Atmospheric Wind Retrievals

Although sounders are primarily used to derive temperature profiles, they can also be used to estimate wind profiles under steady state conditions, and in regions where frictional forces (*e.g.*, at the surface) are negligible. Under these conditions a balance exists between the Coriolis force and pressure gradient. The relationship is called the thermal wind equation and can be expressed as,

$$\frac{\partial \mathbf{V}_{\mathrm{g}}(\mathbf{r},\mathrm{p})}{\partial \ln p} = -\frac{R}{f}\boldsymbol{k}\,\mathrm{x}\,\nabla \mathrm{T}(\mathbf{r},p) \qquad \text{(A21 -1)}$$

where $\mathbf{V}_{\mathrm{g}}(\mathbf{r},\mathrm{p})$ is the geostrophic wind, ∇ T is the temperature gradient, f is the Coriolis parameter, R is the gas constant, and $\boldsymbol{k}$ is the unit vector in the vertical direction. Multiplying (A21-1) by the weighting function of (A19-1a) and integrating through the atmosphere results in

$$\int_{-\infty}^{\ln p_s} \frac{\partial \mathbf{V}_{\mathrm{g}}(\mathbf{r},\mathrm{p})}{\partial \ln p}\,\frac{\mathrm{d}\tau'_{\mathrm{v}}(p)}{\mathrm{d}\ln p}\,d\ln p = -\frac{R}{f}\boldsymbol{k}\,\mathrm{x}\,\nabla \mathrm{T}_{\mathrm{b}}(\mathbf{r};\nu) \qquad \text{(A21 -2)}$$

so neglecting the temperature gradient at the surface, the left side is proportional to the horizontal brightness temperature gradient shown on the right. Integrating the left side by parts we obtain,

$$-\frac{R}{f}\boldsymbol{k}\,\mathrm{x}\,\nabla \mathrm{T}_{\mathrm{b}}(\mathbf{r};\nu) = \mathbf{V}_{\mathrm{g}}(\mathbf{r},p_{\mathrm{s}})\left.\frac{\mathrm{d}\tau'_{\mathrm{v}}(p)}{\mathrm{d}\ln p}\right]_{p_s} - \int_{-\infty}^{\ln p_s} \mathbf{V}_{\mathrm{g}}(\mathbf{r},\mathrm{p})\frac{\mathrm{d}^2\,\tau'_{\mathrm{v}}(p)}{\mathrm{d}\ln p^2}\,\mathrm{d}\ln p\,. \qquad \text{(A21-3)}$$

Equation (A21-3) has the same form as (A20-1a) with the horizontal brightness temperature gradient related to wind profiles through a wind weighting function $-\mathrm{d}^2\tau'_{\mathrm{v}}(p,\theta)/\mathrm{d}\ln p^2$ which is the derivative of the temperature weighting function. As such it is a doublet having positive and negative modes of equal area with a crossover where the temperature weighting function peaks. Analogous to temperature retrievals using (A19-9), the wind profile can be expressed as a linear combination of brightness temperature gradients, *i.e.*,

$$\mathbf{V}_g'(p) = \overline{\mathbf{V}_g}(p) + \sum_{m=1}^{M} D_m(p)\, \boldsymbol{k} \times \left[\nabla T_b(\mathbf{r}; \nu_m) - \nabla \overline{T}_b(\mathbf{r}; \nu_m)\right] \qquad \text{(A21-4)}$$

where $\overline{\mathbf{V}_g}(p)$ and $\nabla\overline{T}_b$ are the *a-priori* mean values of the wind and brightness temperature gradient with $D_m(p)$ the retrieval coefficients. These coefficients are related to the temperature coefficients $A_m(p)$ in (A19-9) using equation (A21-1), *viz.*,

$$D_m(p) = -\frac{R}{f} \int_{-\infty}^{\ln p} A_m(p')\, d\ln p' \qquad \text{(A21 -5)}$$

However, unlike temperature, the accuracy of wind profiles depends on the quality of the brightness temperature gradient. Fortunately, clouds do not introduce appreciable perturbations in the microwave brightness temperatures so the gradient is of high quality. It is therefore possible to construct accurate wind fields independent of cloud cover. As one of the first examples, Figure A21-1 shows the 500 mb wind fields over the North Atlantic and Western Europe on August 24, 1975 derived using SCAMS measurements (Bottom). Both the wind direction and magnitude is indicated by the wind barbs. These results are comparable with that obtained by NCEP (formally NMC) whose wind field (Top) was mainly based on radiosondes with tracking (called rawinsonds or RAWIN), whose data was applied to numerical models for analysis. Also, as mentioned on page 93, the British later used the thermal gradient information provided by MSU to obtain winds during the Falkland war in 1982 when conventional observations were not available because of the war effort. I mention this example only because it was not publicized and probably one of the least recognized applications of microwave sounder measurements.

In addition to obtaining winds in the mid-latitudes, gradient measurements can also be used to derive geostrophic winds for tropical storms. In this case, equation (A21-3) is modified to include the centrifugal acceleration of air parcels in addition to the pressure gradient and Coriolis forces. The resulting equation for channels opaque to the surface then become

$$R\frac{dT_b(r;\nu)}{dr} = \int_{-\infty}^{\ln p_s} \left[\frac{V_\theta^{\,2}}{r} + f\, V_\theta\right] \frac{d^2 \tau_v'(p)}{d\ln p^2}\, d\ln p \qquad \text{(A21-6)}$$

where the brightness temperature gradient is in the radial direction and the wind velocity V_θ is the tangential component. This equation can be approximated as

$$R\frac{dT_b(r;\nu)}{dr} = \frac{<V_\theta>^2}{r} + f <V_\theta> \qquad \text{(A21-7a)}$$

$$\text{where} \quad <V_\theta> = \int_{-\infty}^{\ln p_s} V_\theta \frac{d^2 \tau_v'(p)}{d\ln p^2}\, d\ln p \qquad \text{(A21-7b)}$$

Equation (A21-7a) assumes that the mean squared tangential velocity equals the squared-mean velocity. This approximation is found to be nearly the case for the wind profiles in Figure A21-2.

Also, as with the thermal wind equation this model assumes steady-state conditions or a well developed meteorological system. To be more accurate, the radial distance contained in the centrifugal acceleration component should be replaced by the radius of trajectory curvature. Also, the tangential winds should exclude the component resulting from storm's translational motion. For this reason the velocity corresponds to the tangential component relative to the storm's motion.

For illustration, Figure A21-2 (Left) shows the temperature and wind weighting functions for the SCAMS 55.45 GHz channel at nadir viewing. Note that the temperature weighting function peaks near 200 mb while the bi-modal wind weighting function peaks at about 400 and 100 mb with a crossover at 200 mb where the temperature weighting function peaks. However, while the brightness temperature responds to temperature near 200 mb, the radial derivative of brightness temperature pertains to tangential winds around 400 mb where the lower mode of the wind weighting function.peaks. To understand this better, the right-most Figure shows a representative cross-section of tangential wind profiles for tropical cyclones. The winds are shown to diminish rapidly with height so the major contribution from the integral in (A21-7b) occurs from the lower mode. As such, *the brightness temperature gradient corresponds to winds around* 400 *mb while the brightness temperature itself correlates best with temperatures around* 200 *mb*. These different relationships between brightness temperature and its gradient are shown next in greater detail from a case study of Hurricane David performed using MSU data [53].

Following the launch of MSU aboard TIROS-N on October 13, 1978 numerous studies were conducted to evaluate its performance. Foremost, was the major study performed by the National Weather Service of NOAA to evaluate the impact of MSU temperature soundings on short-time numerical weather prediction. Other studies were done by individual researchers to evaluate other MSU applications. Having already observed the use of the SCAMS to observe the warm core temperature structure of Typhoon June, an extensive study was performed for Hurricane David. This hurricane was one of the most intense of the 1979 season and was observed by the MSU as well as by radiososndes, which measured both its vertical temperature and wind profiles.

Figure A21-3 (Right) shows the hurricane trajectory as well as the radiosonde locations (triangle) and MSU observationns (asteriks). The right side of the Figure shows the brightness temperature contours of its 54.96 GHz channel at observation times of 0900 GMT and 2000 GMT on September 3rd and 4th 1979. For comparison, the contour maps also show the calculated brightness temperatures using RAOB measurements. It also shows the 700 mb wind vectors by tracking the RAOB's (called rawinsonds or RAWINS). As a major part of this study, the brightness temperatures and its gradient measurement were compared against those computed at different distances from the hurricane. To begin this comparison, Figure A21-4 on the left displays the temperature weighting functions next to the change in the RAOB temperture profiles at different distances from the hurricane. Similarly, the right-most Figure shows the wind weighting function for the 54.96 GHz channel next to the wind profiles obtained at the same distances from the hurricane by RAWINS. From these plots, the brightness temperature gradient defined by the wind weighting function, corresponds to winds near 700 mb while the brightness temperature defined by the temperature weighting function, corresponds to temperatures near 300 mb.

For a more quantititave measure, Figure A21-5 on the left side shows the radial derivative dT_b/dr of MSU's 54.96 GHz brightness temperatures at the indicated dates and times. These gradients are determined from the measurements in Figure A21-3 (Right). Also shown are match-ups with the radial derivative computed from equation (A21-7a) using the rawinsonde data. For reference, the straight dashed line is the nadir satellite track. The right-most figure compares these measured radial derivatives with the rawinsonde computed values, with the solid line representing perfect correlation. Distances (nmi) between the measurements and hurricane center are inserted next to each data symbol. Note that the largest differences between the measured and computed values occur close to the hurricane center where the MSU has insufficient spatial resolution to accurately measure the radial derivative.

This last set of results from the study determines the optimal pressure level corresponding to the 54.96 GHz brightness temperature and its radial derivative. In order to elliminate any errors due to spatial averaging (see Figure A21-5), the analysis use the computed MSU measurements.rather than actual measurements. The top set of graphs in Figure A21-6 plots the 54.96 GHz brightness temperatures computed from the 21 RAOB temperature soundings against the RAOB temperatures at 700- , 500- , 300- and 200 mb. The dates and times of these match-ups is the same as in Figure A20-5 and indicated on the plots. Together with each scatter plot is shown the best fit line and correlation coefficient. From this data the best correlation ($r = 0.87$) occurs for the 300 mb temperature. Also, at the same pressure levels, the bottom graphs show the radial acceleration in the integral of equation (A21-6) plotted against the radial derivitive of brightness temperatures determined also using the equation. In this case the best correlation ($r = 0.98$) occurs at 500 mb. It also shows nearly the same high correlation ($r = 0.94$) at 700 mb.

This early study was one of the first in-depth analysis showing the use of microwave sounders for hurricane measurements. However, as mentioned in Chapter 11, later studies by Dr. Stan Kidder [24] using more advanced microwave sounders provided much better observations of hurricanes (see Figure 67). In addition to accurately measuring the temperature structure, the higher resolution available from AMSU provided accurate measurements of the central pressure of hurricanes and therefore its intensity and maximum winds. However, besides the use of microwave temperature sounders, microwave imagers were used to measure the sea surface winds of storms.

One of the earliest applications of satellite radiometers for hurricanes was done using microwave imagers. As shown in Table 7, beginning with the SMMR and followed by the SSM/I, microwave imagers were used by the Navy as well as other organizations to measure sea surface winds. The measurement results from the increased emissivity due to surface roughness at low speeds and foam generation at high speeds. This application appears to be originally investigated by Williams in 1969 [54] who artificially generated foam in a water-tank, and measured using radiometers at frequencies from 9.4 to 34 GHz. Around the same time, radiometer measurements of sea surface winds were measured from aircraft using a 19.4 GHz radiometer by Nordberg *et. al.*, [55]. One of the most interesting radiometers developed for this application was the Stepped Frequency Microwave Radiometer or SFMR. This C-Band Dicke radiometer has 6-frequencies between 4.6 and 7.2 GHz to measure ocean wind speed and precipitation for tropical storms and hurricanes [56]. It has been flown aboard hurricane-penetrating reconnassance aircraft for over a decade by the Atlantic Oceanographic and Meteorologic Laboratory (AOML) divison of NOAA. It has been used to measure wind speeds from about 10 to 60 m/sec or from 20 to 110 knots. Most importantly, it is used operationally by forecasters to make decisions on system classifiications, for example when to declare that a tropical storm has intensified to hurricane status.

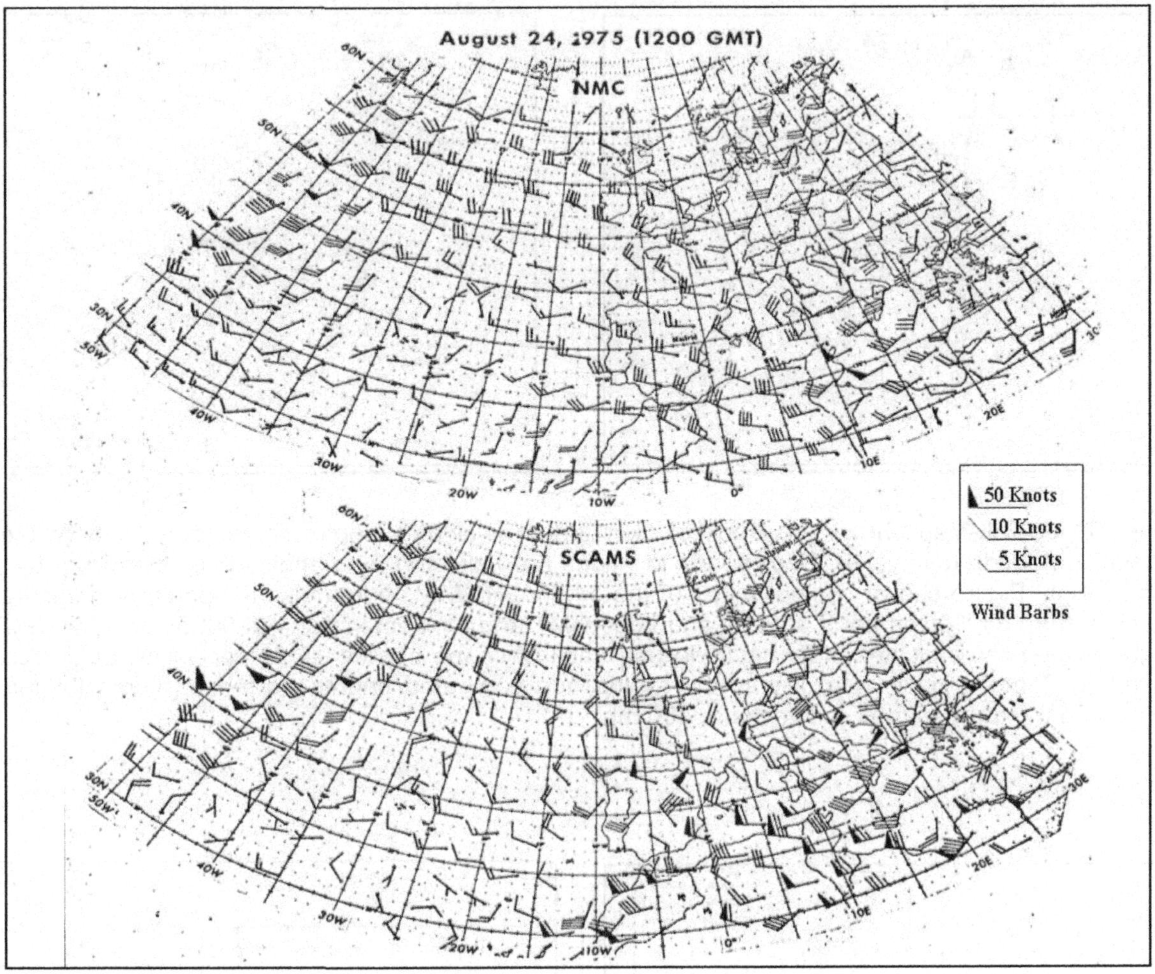

Figure A21-1. SCAMS derived wind field at 500 mb (Bottom) over the Atlantic and Western Europe compared with that from the National Meteorological Center (NMC)[11] of NOAA (Top) on August 24, 1975.

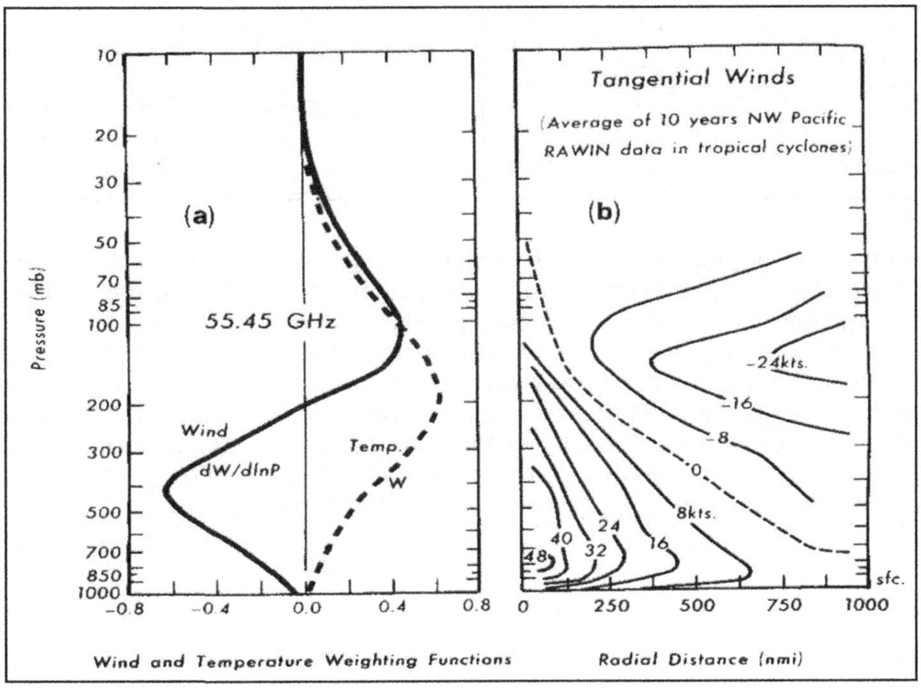

Figure A21-2. (a) SCAMS 55.45 GHz wind and temperature weighting functions. (b) Average tangential wind cross-sections for typical tropical cyclones after Frank [52].

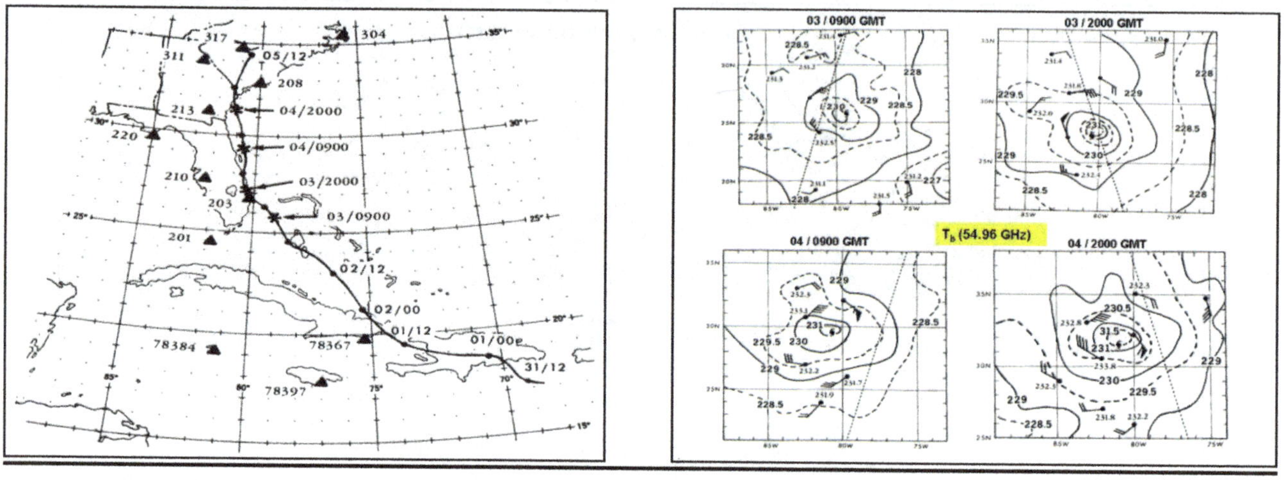

Figure A21-3. Left-Map shows the track of Hurricane David at synoptic times (dots) and at the MSU observation periods (asterisks). Radiosonde locations are indicated by triangles and referred to by station number. Right-Figure shows contour maps of the 54.96 GHz MSU brightness temperature measurements (limb corrected) for September 3 at 0900 and 2000 GMT and for September 4 at 0900 and 2000 GMT. The straight dashed line indicates MSU nadir viewing track while the hurricanes center position is indicated by its symbol. Contour maps also display the radiosonde computed brightness temperatures for this MSU channel and the rawinsonde wind vector at 700 mb.

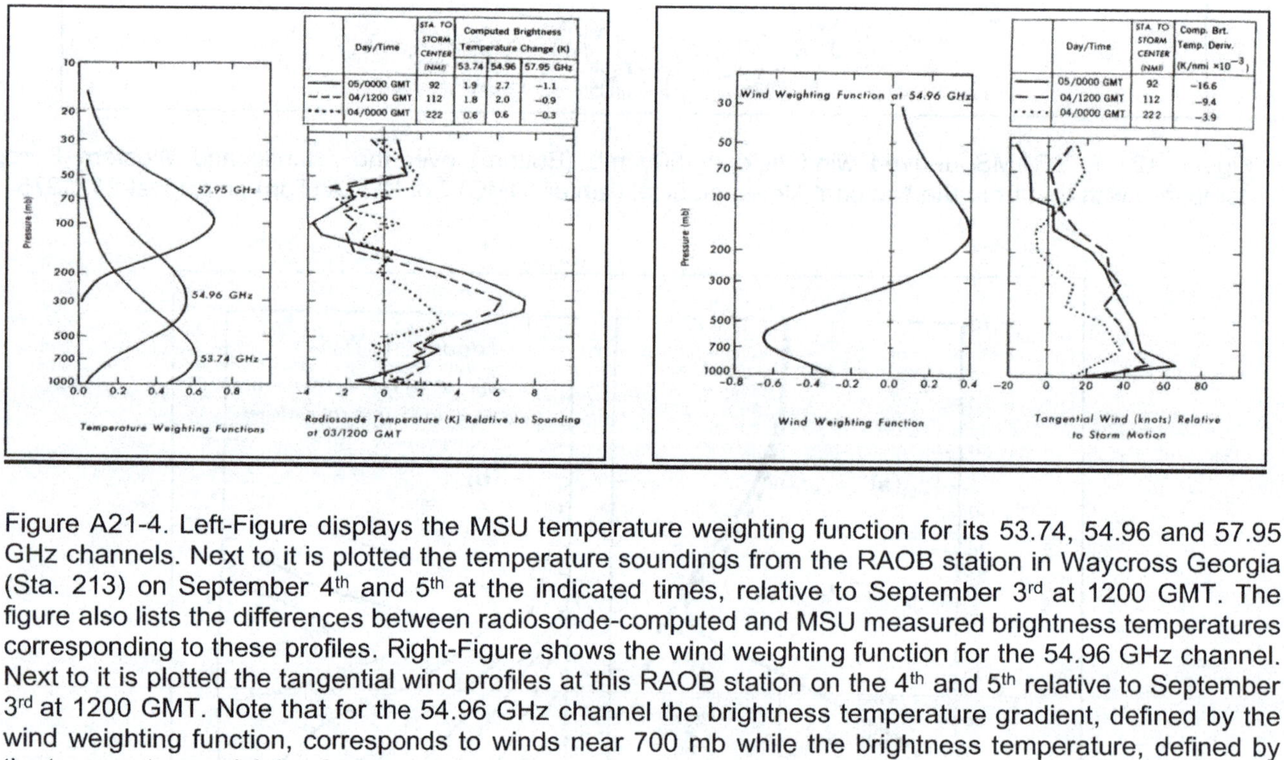

Figure A21-4. Left-Figure displays the MSU temperature weighting function for its 53.74, 54.96 and 57.95 GHz channels. Next to it is plotted the temperature soundings from the RAOB station in Waycross Georgia (Sta. 213) on September 4th and 5th at the indicated times, relative to September 3rd at 1200 GMT. The figure also lists the differences between radiosonde-computed and MSU measured brightness temperatures corresponding to these profiles. Right-Figure shows the wind weighting function for the 54.96 GHz channel. Next to it is plotted the tangential wind profiles at this RAOB station on the 4th and 5th relative to September 3rd at 1200 GMT. Note that for the 54.96 GHz channel the brightness temperature gradient, defined by the wind weighting function, corresponds to winds near 700 mb while the brightness temperature, defined by the temperature weighting function, corresponds to temperatures near 300 mb.

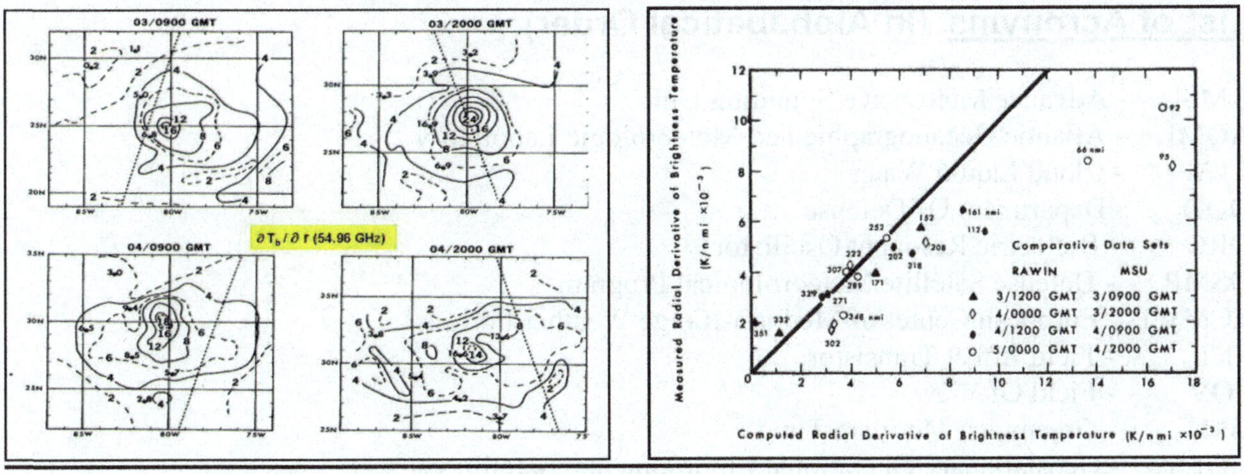

Figure A21-5. Left-Figure shows the radial derivative (K/nmi x10^{-3}) of MSU's 54.96 GHz brightness temperatures at indicated times. Contour maps are based on measurements in Figure A21-3 (Right). Dashed lines are the nadir satellite track. Also shown are computed radial derivatives based on equation (A21-7a) using rawinsonde winds at RAOB locations. Right-Figure shows the 21 computed values plotted against MSU measurements with the solid line representing perfect correlation. Distances (nmi) between RAOB's and hurricane center are inserted next to each data point. Largest differences between measured and computed values occur close to the hurricane center due to insufficient spatial resolution by MSU.

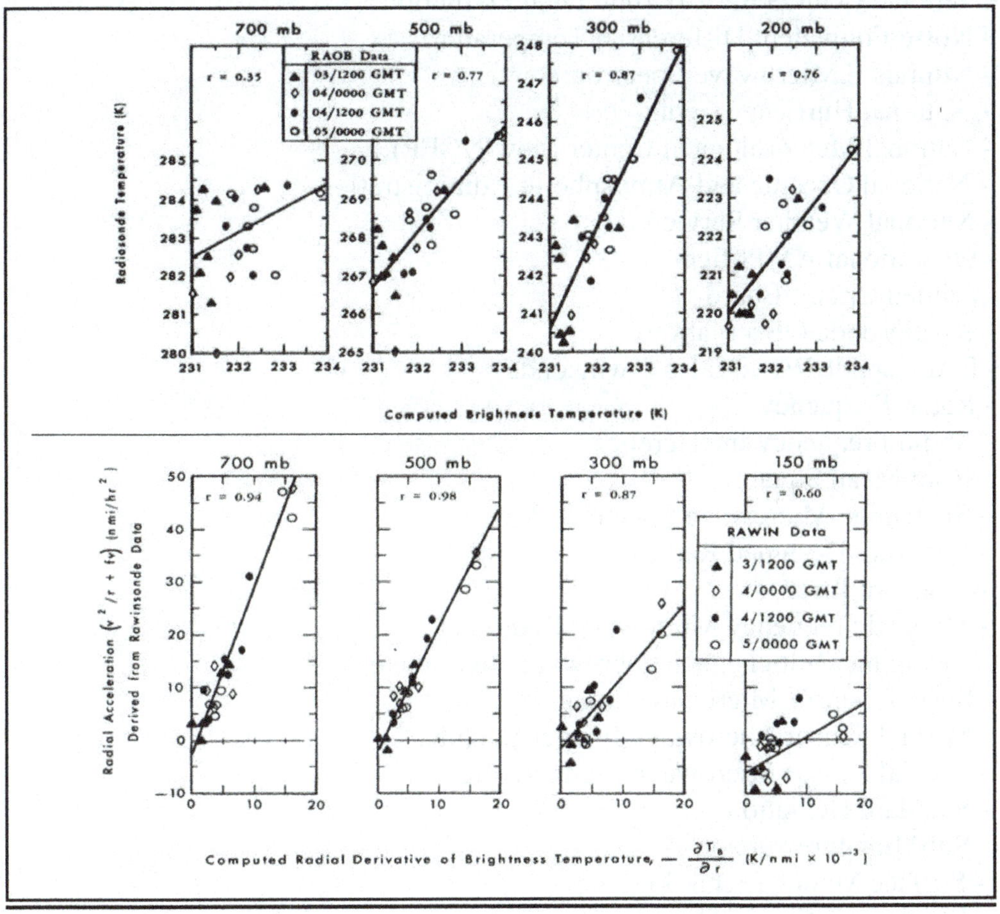

Figure A21-6. Top set of graphs show RAOB temperatures at different pressure levels plotted against the computed brightness temperatures for the 54.96 GHz channel. Bottom set of graphs show the rawinsonde derived radial acceleration at different pressure levels, plotted against the computed radial derivative using equation (A21-6). Note that the brightness temperature correlates best (r = 0.87) with temperature near 300 mb while the gradient correlates best (r $\geq$ 0.94) with winds at 500 mb and 700 mb.

List of Acronyms (In Alphabetical Order)

AMSU - Advance Microwave Sounding Unit
AOML - Atlantic Oceanographic and Meteorologic Laboratory
CLW - Cloud Liquid Water
DOD - Department Of Defense
DRO - Dielectric Resonant Oscillator
DSMP - Defense Satellite Meteorological Program
ECMWF - European Center of Medium-Range Weather Forecast
FET - Field Effect Transistor
FOV - Field Of View
GMT - Greenwich Meridian Time
GOES - Geostationary Operational Environmental Satellites
IF - Intermediate Frequency
JPL - Jet Propulsion Laboratory
LNB - Low Noise Block
LO - Local Oscillator
MMIC - Monolithic Microwave Integrated Circuit
MOSFET - Metal Oxide Semiconductor Field Effect Transistor
MSU - Microwave Sounding Unit
NASA - National Aeronautic and Space Administration
NCEP - National Centers for Environmental Prediction
NEΔT - Noise Equivalent Differential Temperature
NEMS - Nimbus E Microwave Spectrometer
NHC - National Hurricane Center
NMC - National Meteorological Center (now NCEP)
NOAA - National Oceanic and Atmospheric Administration
NWS - National Weather Service
OPAMP - OPerational AMPlifier
PCB - Printed Circuit Board
RAOB - RAdiosonde OBservation
RAWIN - RAdiosonde WINd (also rawinsonde)
RF - Radio Frequency
RFI - Radio Frequency Interference
RMS - Root Mean Square
SCAMS - SCAnning Microwave Spectrometer
SDR - Software Designed Radio
SE - Standard Error
SFMR - Stepped Frequency Microwave Radiometer
SMMR - Scanning Multichannel Microwave Radiometer
SSMI - Special Sensor Microwave Imager
SSMIS - Special Sensor Microwave Imager Sounder
SSMT - Special Sensor Microwave Temperature
STD - Standard Deviation
SMA - SubMiniature version-A
SMT - Surface Mount Technology
SPST - Single Pole Single Throw
TOVS - T IROS Operational Vertical Sounder
TPW - Total Precipitable Water
VSWR - Voltage Standing Wave Ratio